The Shortwave Guide

VOLUME 2

Editor: Sean Gilbert
Contributing Editors: Mauno Ritola, Bernd Trutenau
Design: Richard Boxall Design Associates
Database engineering: Ionian Business Software
Publisher: Nicholas Hardyman

With grateful thanks to all our contributors, and
particular thanks to Anker Petersen of DSWCI, Bob
Padula of EDXP, the British DX Club, and the Asian
Broadcasting Institute

ISBN: 0-9535864-5-6 paper

WRTH Publications Limited
P.O. Box 290, Oxford OX2 7FT, UK
E-mail: wrthdx@aol.com
Web: www.wrth.com

Printed and bound in Great Britain by
Bath Press Limited

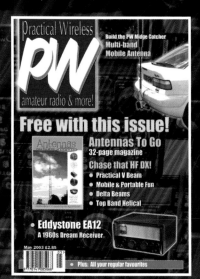

Contents

Introduction to Shortwave Radio

by **Bernd Trutenau**

Why listen to shortwave radio?

Shortwave radio is an exciting way of listening to the world. There is a thrill to hearing someone talking to you from the other side of the world. You can find out about foreign countries and their peoples; listen to great music; hear the news from someone else's point of view; and discover what is going on on the other side of the world. Perhaps you would like to know more about a country's culture or politics. Or, when you are travelling, keep in touch with what is happening back home.

Shortwave radio brings the whole world into your living room. And you do not always need to understand foreign languages. International stations broadcast specifically for people abroad, in the languages of their target audience, and music is, as has often been said, a universal language, whether it is chamber music from Vienna, carnival from Rio, film music from Mumbai, Chinese opera from Beijing, or the latest sound in world music from Africa.

Listening to shortwave radio can also be a fascinating hobby, as the thousands of enthusiasts worldwide will tell you. Shortwave hobbyists are known as 'Short Wave Listeners' or 'SWLs'. The more serious listener who is specifically trying to hear broadcasts from weak low power/domestic stations often in far away places is known as a 'DXer' (DX being the radio abbreviation for 'long distance'). You can get to know SWLs and DXers from all over the world who share your passion. You can exchange news, successes and listening tips with DXers around the globe. There are now many DX/SWL groups and clubs on the internet. You can also report back to the radio stations that you have tuned in to, to let them know you picked up their signal. You send in reception reports on cards known as QSL cards (QSL, as now used, is the radio code for 'confirmation of reception') or by e-mail or letter, and the station often confirms your reception with its own QSL card. For many enthusiasts, collecting these colourful QSL cards is a major part of their hobby. This is a great hobby which makes you part of an international network.

How to listen to shortwave radio

Before starting it is useful to understand the frequencies that are used for shortwave broadcasts. The shortwave radio (SW) broadcast bands start above the mediumwave (MW) band. They cover the frequency range between about 1700 and 30000 kHz, but the broadcasting stations you will want to listen to are found between 2300 and 26000

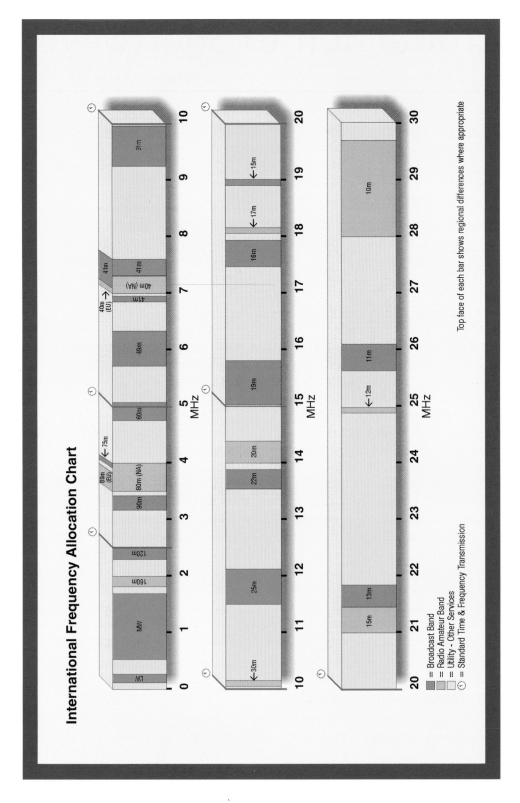

International Frequency Allocation Chart

kHz in sections of the band that have been set aside for their use, and most stations keep their transmissions within this range. These SW broadcast bands are shown in green on the chart opposite. As you will see on the chart, the rest of the band is used for other kinds of communication, such as amateur radio (radio hams), and the Utility services such as coast guards and commercial airliners. You can also pick up these signals, if you have the right equipment.

Frequencies are generally given in kilohertz (kHz), although sometimes they appear in megahertz (MHz). 1 MHz is 1000 kHz, so a frequency of 7500 kHz is 7.5 MHz. Each SW broadcast band also has a wavelength name given in metres. The frequencies with their corresponding wavelength are:

Frequency (kHz)	Wavelength (metres)
2300-2500	120
3200-3400	90
3900-4000	75★
4750-5060	60
5730-6295	49
7100-7600	41†
9400-9900	31
11650-12100	25
13570-13870	22
15030-15600	19
17480-17900	16
18900-19020	15
21450-21850	13
25670-26100	11

★part of radio amateur band in North America

†from 7300 in North America

To listen to SW radio you will need a receiver and an antenna. Receivers vary in size, complexity, quality and price. A small, inexpensive portable receiver can provide perfectly good results, while a serious enthusiast might want a larger, more expensive communication receiver. For the best listening results it is important to use a proper antenna. Portable receivers usually have a telescopic antenna that will provide sufficient reception, but you can also connect separate antennas to some, which will usually improve reception. For catching small and distant home service or domestic broadcasts, it is a good idea to use an outdoor, or a separate indoor, antenna. A typical outdoor antenna is a longwire or, more correctly, random wire antenna. This consists of an isolated wire several metres (yards) long. At the height the average listener can install a random wire, they tend be omni-directional so orientation is not important.

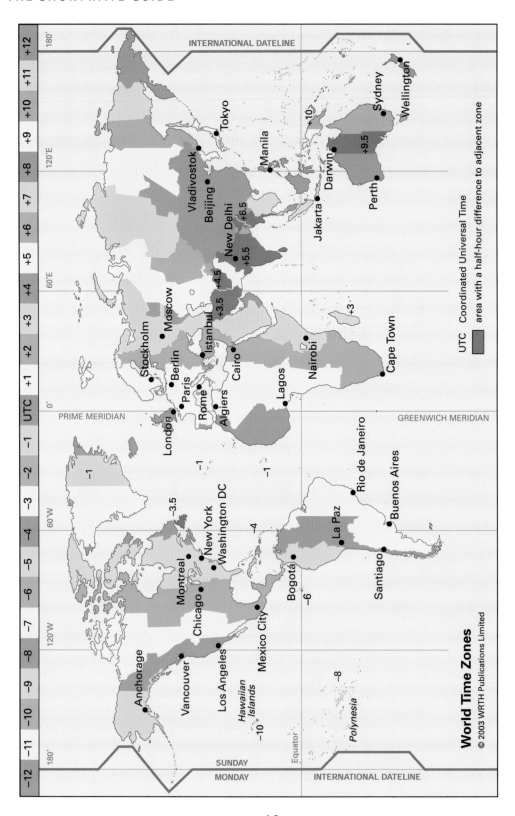

World Time Zones

© 2003 WRTH Publications Limited

This also means that the wire does not have to be perfectly straight. Best results will, however, be obtained with a wire antenna when it is as high as possible, and clear of objects such as trees and buildings. Wire antennas do start to become directional when the length is over a wavelength, with the effect becoming more noticeable as the length is increased to multiples of a wavelength. A random wire antenna becomes known as a longwire antenna when it is over a wavelength long at the lowest frequency it is to be used on. There are also active shortwave antennas available for use inside a building and when there is no space for a random wire antenna. These handy devices can be turned to the required direction.

What frequencies should you tune to in order to hear broadcasts? You can tune in to a station by directly entering a listed frequency into your receiver. But you may also like to scan the band in search of frequencies yourself. New stations are appearing on the bands all the time, and existing stations start using new frequencies. If you want to tune to a specific station, you need to choose the right frequency. There are several ways of finding out about these. An up-to-date frequency guide such as this book is the ideal reference. Another useful book is our sister publication, *World Radio TV Handbook* (*WRTH*), which is published in December each year and also gives information about MW and FM broadcasts. Other sources of information include DX chat groups, DX clubs and their websites, broadcasters' websites and printed schedules, other publications in book and CD form, and monthly magazines such as *Short Wave Magazine*, *Monitoring Times* and *Popular Communications*. Foreign service programmes are often transmitted on more than one frequency at the same time; you can experiment to find out which frequency reaches you best.

As well as finding the right frequency, you also need to know when the broadcast is being transmitted. To make this easier, all international broadcasters around the world give the times of their broadcasts in Coordinated Universal Time (abbreviated as UTC). This is the same as Greenwich Mean Time (GMT). Times are given using the 24-hour clock, 4am, for example, is given as 0400 hours, whereas 4pm is 1600. In order to know what time you would be able to hear a broadcast that is transmitted at 1600 UTC, you need to find out which time zone you are in (this will tell you how many hours your local time is ahead, or behind, UTC). Europe is one or two hours ahead of UTC whereas Argentina is three hours behind UTC. So in Europe you will be listening to the 1600 UTC broadcast at your local time which will be 5 or 6pm. An Argentine listener would hear the same broadcast at 1pm local time. The broadcaster will probably be in yet another time zone, and the 1600 UTC broadcast may actually be being transmitted at 3am the following morning (local time at the broadcaster)!

The time zones around the world are shown on the map opposite. An exception to the convention of using UTC is domestic broadcasts. As these are not intended for listeners abroad, times are given in the local time of each country. You can find out what the UTC time is by using the list on page 186 in the reference section which shows winter and summer times + or - UTC. Most serious listeners and DXers have a clock set to UTC by their receiver. You can also check the time in UTC by tuning in to one of the Standard Time and Frequency Transmissions. These are shown as black bands in   this guide and the first group of them is at 2500 kHz. All the times shown in this guide are UTC.

It is a good idea to keep a log of the stations you manage to tune in to. To identify an unknown station, wait for the identification which may be aired at the top of the hour, or during a programme. This may be an announcement such as 'This is the general overseas service of All India Radio', or a signature tune such as the Arabic melody on string instruments which identifies the Qatar Broadcasting Service. Make a note of the station's name and the date and time you heard the station on a particular frequency. It is also useful to record the strength and clarity of the signal. Working from your log you can find out the best times for listening to certain frequencies.

What to listen to on shortwave radio

Shortwave radio offers a wide range of interesting programming. There are two main kinds of broadcasting stations on the band: domestic stations and foreign stations.

Domestic stations broadcast programmes for listeners in their own country, in a variety of their native languages. Their transmitters are typically located in states with large, densely populated areas or where the terrain makes broadcasting difficult by other means, such as FM or MW. In Europe, the number of domestic stations on SW has decreased considerably in recent decades, as they have been replaced by transmissions on other broadcast bands. In other parts of the world, however, such as Asia and South America, domestic SW is still a widespread form of radio broadcasting.

Most domestic radio stations use the lower SW bands. Shortwave stations from the countries between the Tropic of Cancer and Tropic of Capricorn can be heard on the 120, 90 and 60 metre bands. These are known as the Tropical Bands, and many fascinating programmes can be picked up from here. As they are intended for internal listeners they are in local languages only, and they are rarely broadcast 24 hours a day. Typically they will shut down at midnight local time, and start again at dawn.

Since they are intended for a local audience, transmitters carrying domestic services have often moderate or low power. This makes them hard to receive abroad, which creates an extra challenge for the shortwave listener. To catch an exotic station from thousands of miles or kilometres away is an exciting moment for the shortwave DXer.

Foreign service stations broadcast programmes for listeners in other countries. The main aim of these broadcasts is to present the country to the outside world. These stations will often broadcast in several languages, to specific target audiences abroad. For example, broadcasts targeted at South America will be in Spanish; those to Europe may be in English, German and French. The programmes may be short news bulletins, or longer talk or music programmes. They are usually broadcast at the same time every day, or on specific days of the week. They may be repeated at various times of the day, so that listeners can tune in at a time that suits them. The transmitters carrying these services are typically very powerful, so reception quality can be very good. You might start by listening to the big international broadcasters such as Deutche Welle, Voice of America or All India Radio, and then move on to some of the smaller broadcasters. Reception is possible even if you are not in the target area. Shortwave signals know no boundaries and in the right conditions they can travel around the whole world.

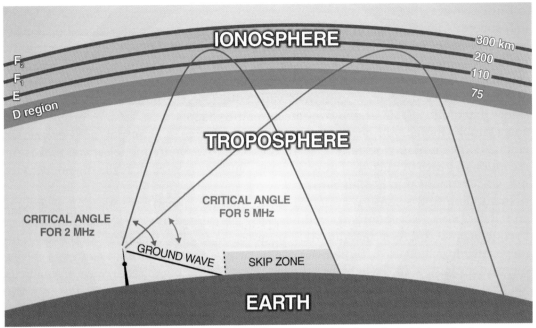

Diagram of radio propagation and the ionosphere. The critical angle is higher for lower frequencies than for higher frequencies.

In order to get the best reception results, and to minimise interference between transmitters, most foreign service stations meet to co-ordinate their frequencies at a special international radio conference (the HFCC), held twice a year. Most stations also change their frequencies in spring and autumn, in order to adjust to the different propagation conditions during the winter and summer months.

An introduction to propagation

An important factor in learning to be a successful SW listener is understanding how radio waves travel, or propagate, and the factors which influence this.

Radio waves travel in two ways at the same time: as groundwaves and as skywaves. Groundwaves travel along the ground all the way from the transmitter to the receiver. The maximum distance they can do this is about 100 miles or 160 kilometres. This is dependant on frequency, with lower frequencies travelling farthest. Skywaves, on the other hand, can travel extremely long distances using the properties of the ionosphere, enabling us, with the right equipment, to receive shortwave broadcasts from all over the world.

Between the groundwave and skywave of a transmission, there is a skip zone (see the diagram above) where the signal is virtually inaudible. In order to provide local listeners with optimal reception, domestic service broadcasters usually use omni-directional antennas, designed for a large groundwave zone and a moderate skywave zone. For long-distance, foreign-service transmissions, directional antennas are used that radiate only a short groundwave and focus more on skywave propagation.

With skywaves, the radio waves travel upwards and are reflected back to earth by the ionosphere, a layer of gas in the upper atmosphere. The angle at which the signal is beamed governs where it will come back to earth, making it possible for broadcasters to send a signal to a particular part of the world. The ionosphere itself comprises several unstable layers, whose ability to redirect shortwave signals is affected by activity on the surface of the Sun. This activity varies over a cycle of about 11 years, and the same timetable affects shortwave reception. Sunspots and solar storms have a considerable impact on the ionosphere. A high level of sunspot activity usually enhances the audibility of stations at the higher shortwave frequencies, and makes the reception of lower frequencies more difficult. During a period of high sunspot activity, many SW stations move to higher frequencies.

In addition to the sunspot cycle, other solar activity affects reception on a seasonal and daily basis. Reception possibilities are determined by the Maximum Usable Frequency (MUF). As a rule, the MUF is lowest during the dark hours of the day. This means that low shortwave frequencies can be received much better at night and during the winter months. In daylight and during the summer, higher shortwave frequencies are favoured. Reception can also be subject to heavy disturbance on specific days. Sun storms, for example, can be very severe and can make reception impossible in large parts of the shortwave band. Sometimes the entire band can be blacked out for many hours as a result of sudden ionospheric disturbances (SIDs), or the so-called Moegel-Dellinger effect.

Another factor is the different ways in which a signal is reflected before it reaches the receiver. Often shortwave signals travel around the globe via the 'long path'. For example, a signal is beamed from Europe towards Asia in a south-easterly direction, but a signal from Asia to Europe will often be received in Europe from the north-west, having travelled the long way round the globe.

Yet another factor affecting how well the signal is received is the power output of the transmitter. Many transmitters carrying foreign service broadcasts use a typical power of 100-500 kW, some even 1000 kW. Combined with a high-gain directional antenna, this can often produce almost perfect reception in the target area. Low or medium power transmitters have a smaller coverage area, and more distant reception depends on favourable propagation conditions.

How to Use this Guide

Welcome to volume 2 of *The Shortwave Guide*. This guide has been carefully designed to give the maximum information in a clear and easily read format. Here we explain how to use the main directory section:

The first column shows the frequency of the broadcasts in kHz.

The second column shows the name of the station making the broadcast or the broadcaster responsible for the broadcast. You can tell international from domestic broadcasts at a glance, as stations making domestic broadcasts are shown in italic type.

The column headed TXC shows the ITU code of the country from which the broadcast was transmitted. This may be different from the location of the radio station from which the broadcast originates.

The fourth column shows the power of the transmitter in kW.

The colour bars on the directory pages show the duration of each broadcast in UTC on the 24-hour clock. The colour of the bar shows the language of the broadcast. Twelve different languages are identified by different coloured bars, with the colour and language shown at the bottom of the page. Other languages, or combinations of languages, are shown above a buff-coloured bar. The black bars are the Standard Time & Frequency Transmissions (often called 'time signals'). Sometimes you will see a bar on the left that has no details above it. This is a broadcast that begins on the previous day and if you look on the same line to the right you will see the details of the broadcast

The target area or country at which the broadcast is aimed is given above the bar; AUS for Australia or SAf for southern Africa, for example. A full list of these abbreviations, the country codes and the target area codes, is given on page 224 in the reference section at the end of the book.

Also above the bar is an indication of the days on which the broadcast is made. Following a standard convention, Sunday = day 1, Monday = day 2, Tuesday = day 3, Wednesday = day 4, Thursday = day 5, Friday = day 6, and Saturday = day 7. If a broadcast is transmitted every day of the week then no numbers appear above the bar. If, however, the broadcast only happens on some days of the week then numbers indicating those days are shown above the bar. For example, a broadcast made from Sunday to Friday inclusive will be shown as 1-6, a broadcast made on Sunday, Wednesday and Friday will be shown as 1,4,6; and a broadcast on Sunday to Wednesday and Friday will be shown as 1-4,6.

You can use these pages to identify a broadcast you have heard on a specific frequency, or you can scan the colour bars to find broadcasts in your chosen language at a particular UTC time.

We hope you enjoy using this guide.

Abbreviations

The following abbreviations are used in this guide. We hope you will find this list useful when identifying broadcasters in the frequency lists and checking station details in the directory.

Admin.	=	Administration	Ext.	=	External
AFRTS	=	Armed Forces Radio &	F.M.	=	Frequency manager
		Television Service	Fax	=	Facsimile
AIR	=	All India Radio	FEBC	=	Far East Broadcasting
Ann:	=	Announcement			Company
Asst.	=	Assistant	FM	=	Frequency Modulation
Ave.	=	Avenue	Fr.	=	Father
AWR	=	Adventist World Radio	Freq(s)	=	Frequency(ies)
BBC	=	British Broadcasting	FS	=	Foreign Service
		Corporation	G.M.	=	General Manager
BBG	=	Broadcasting Board	GEN	=	General
		Governors	Gov.	=	Government
BC	=	Broadcasting Corporation	HS	=	Home Service
Bldg.	=	Building	I.T.	=	Information Technology
Broadc.	=	Broadcasting	Inc.	=	Incorporated
BS	=	Broadcasting Service	Inf.	=	information
BSKSA	=	Broadcasting Service of	Int.	=	International
		the Kingdom of Saudi	IRC	=	International Reply
		Arabia			Coupon
C.E.	=	Chief Executive	IS	=	Interval Signal
Clan.	=	Clandestine	Jamh.	=	Jamahiriya
CNR	=	China National Radio	KBS	=	Korean Broadcasting
Comm.	=	Commercial			System
Comms	=	Communications	KCBS	=	Korean Central
Co-ord.	=	Co-ordinator			Broadcasting Station
Corp.	=	Corporation	LP:	=	Leading Personnel
D.G.	=	Director General	Ltd.	=	Limited
D.Prgr(s)	=	Daily Programme(s)	M.D.	=	Managing Director
Dif.	=	Diffusion/Dif(f)usora	Mgr.	=	Manager
Dir.	=	Director	MW	=	Mediumwave
DMR	=	Dniester Moldavian	Nat./Nac.	=	National/Nacional
		Republic	NBC	=	National Broadcasting
DRM	=	Digital Radio Mondiale			Corporation
Em.	=	Emis(s)ora	NHK	=	Nippon Hoso Kyokai
Eng.	=	English, engineer(ing)	No.	=	Number
Exec.	=	Executive	NRK	=	Norsk Rikskringkasting

NSB	=	Nihon Shortwave Broadcasting
NTSS	=	Northern Territory Shortwave Service
Occ.	=	Occaisional
Ops.	=	Operations
ORTB	=	Office de Radiodiffusion et Télévision du Benin
P.D.	=	Programme Director
P.O.	=	Post Office
P.R.	=	Public Relations
PBS	=	People's Broadcasting Station
Pl.	=	Planned
Prgr.	=	Programme
Prgrs.	=	Programmes
Prod.	=	Producer
Pub.	=	Public
R(S)PDT	=	Radio (Siaran) Pemerintah Daerah Tingkat Dua
R.	=	Radio/Rádio/ Radiophonikos
RAE	=	Radiodifusión Argentina al Exterior
RAI	=	Radiotelevisione Italiana
Rdif.	=	Radiodiffusion/ Radiodif(f)usora
RDP	=	Radiodifusão Portuguesa
Re.	=	Reply
Rec. acc.	=	Recordings accepted
REE	=	Radio Exterior de España
Reg.	=	Region, Regional
Rel.	=	Relay
Rep.	=	Republic
Rev.	=	Reverend
RFPI	=	Radio for Peace International
Rlg.	=	Religious
RNW	=	Radio Nederland Wereldomroep
RNZ	=	Radio New Zealand
Rp.	=	Return Postage
Rpt.	=	Report

RRI	=	Radio Republik Indonesia
RTE	=	Radio Telefís Éireann
RTM	=	Radio Television Malaysia
RTNC	=	Radio-Télévision Nationale Congolaise
RVI	=	Radio Vlaanderen International
S/off	=	Sign off
S/on	=	Sign on
S.M.	=	Station Manager
SABC	=	South African Broadcasting Corporation
SADR	=	Saharan Arab Democratic Republic
SAE	=	Self Addressed Envelope
Sce	=	Service
SINPO	=	Signal Interference Noise Propagation Overall
SLBC	=	Sri Lanka Broadcasting Corporation
SLBS	=	Sierra Leone Broadcasting Service
SSB	=	Single Side Band
Stn.	=	Station
SW	=	Shortwave
T.D.	=	Technical Director
Tech.	=	Technical
Tel:	=	Telephone
Tr.	=	Transmitter
TV	=	Television
V.O.	=	Voice of
V.P.	=	Vice President
V:	=	Verification
WS	=	World Service
YLE	=	Yleisradio Oy

✉	=	Postal address
☎	=	Telephone
🖹	=	Fax

International &
Domestic Broadcasts

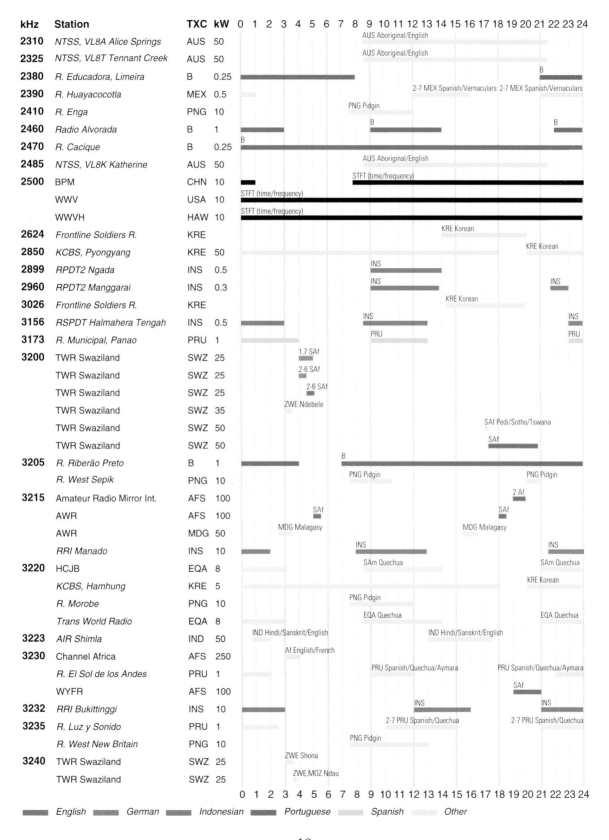

kHz	Station	TXC	kW	Schedule (0–24 h)
2310	NTSS, VL8A Alice Springs	AUS	50	AUS Aboriginal/English
2325	NTSS, VL8T Tennant Creek	AUS	50	AUS Aboriginal/English
2380	R. Educadora, Limeira	B	0.25	B
2390	R. Huayacocotla	MEX	0.5	2-7 MEX Spanish/Vernaculars ... 2-7 MEX Spanish/Vernaculars
2410	R. Enga	PNG	10	PNG Pidgin
2460	Radio Alvorada	B	1	B ... B
2470	R. Cacique	B	0.25	B
2485	NTSS, VL8K Katherine	AUS	50	AUS Aboriginal/English
2500	BPM	CHN	10	STFT (time/frequency)
	WWV	USA	10	STFT (time/frequency)
	WWVH	HAW	10	STFT (time/frequency)
2624	Frontline Soldiers R.	KRE		KRE Korean
2850	KCBS, Pyongyang	KRE	50	KRE Korean
2899	RPDT2 Ngada	INS	0.5	INS
2960	RPDT2 Manggarai	INS	0.3	INS ... INS
3026	Frontline Soldiers R.	KRE		KRE Korean
3156	RSPDT Halmahera Tengah	INS	0.5	INS ... INS
3173	R. Municipal, Panao	PRU	1	PRU ... PRU
3200	TWR Swaziland	SWZ	25	1,7 SAf
	TWR Swaziland	SWZ	25	2-6 SAf
	TWR Swaziland	SWZ	25	2-6 SAf
	TWR Swaziland	SWZ	35	ZWE Ndebele
	TWR Swaziland	SWZ	50	SAf Pedi/Sotho/Tswana
	TWR Swaziland	SWZ	50	SAf
3205	R. Riberão Preto	B	1	B
	R. West Sepik	PNG	10	PNG Pidgin ... PNG Pidgin
3215	Amateur Radio Mirror Int.	AFS	100	2 Af
	AWR	AFS	100	SAf ... SAf
	AWR	MDG	50	MDG Malagasy ... MDG Malagasy
	RRI Manado	INS	10	INS ... INS
3220	HCJB	EQA	8	SAm Quechua ... SAm Quechua
	KCBS, Hamhung	KRE	5	KRE Korean
	R. Morobe	PNG	10	PNG Pidgin
	Trans World Radio	EQA	8	EQA Quechua ... EQA Quechua
3223	AIR Shimla	IND	50	IND Hindi/Sanskrit/English ... IND Hindi/Sanskrit/English
3230	Channel Africa	AFS	250	Af English/French
	R. El Sol de los Andes	PRU	1	PRU Spanish/Quechua/Aymara ... PRU Spanish/Quechua/Aymara
	WYFR	AFS	100	SAf
3232	RRI Bukittinggi	INS	10	INS ... INS
3235	R. Luz y Sonido	PRU	1	2-7 PRU Spanish/Quechua ... 2-7 PRU Spanish/Quechua
	R. West New Britain	PNG	10	PNG Pidgin
3240	TWR Swaziland	SWZ	25	ZWE Shona
	TWR Swaziland	SWZ	25	ZWE,MOZ Ndau

English   German   Indonesian   Portuguese   Spanish   Other

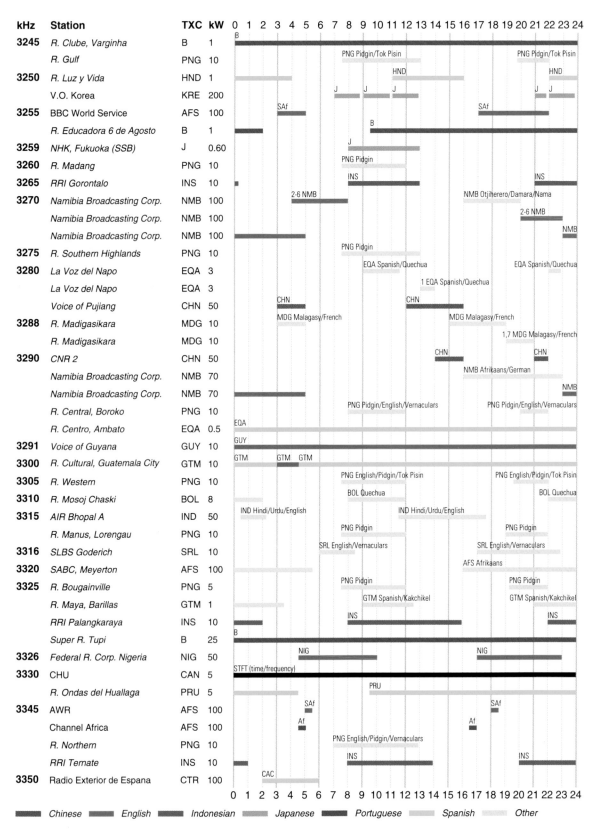

kHz	Station	TXC	kW
3245	R. Clube, Varginha	B	1
	R. Gulf	PNG	10
3250	R. Luz y Vida	HND	1
	V.O. Korea	KRE	200
3255	BBC World Service	AFS	100
	R. Educadora 6 de Agosto	B	1
3259	NHK, Fukuoka (SSB)	J	0.60
3260	R. Madang	PNG	10
3265	RRI Gorontalo	INS	10
3270	Namibia Broadcasting Corp.	NMB	100
	Namibia Broadcasting Corp.	NMB	100
	Namibia Broadcasting Corp.	NMB	100
3275	R. Southern Highlands	PNG	10
3280	La Voz del Napo	EQA	3
	La Voz del Napo	EQA	3
	Voice of Pujiang	CHN	50
3288	R. Madigasikara	MDG	10
	R. Madigasikara	MDG	10
3290	CNR 2	CHN	50
	Namibia Broadcasting Corp.	NMB	70
	Namibia Broadcasting Corp.	NMB	70
	R. Central, Boroko	PNG	10
	R. Centro, Ambato	EQA	0.5
3291	Voice of Guyana	GUY	10
3300	R. Cultural, Guatemala City	GTM	10
3305	R. Western	PNG	10
3310	R. Mosoj Chaski	BOL	8
3315	AIR Bhopal A	IND	50
	R. Manus, Lorengau	PNG	10
3316	SLBS Goderich	SRL	10
3320	SABC, Meyerton	AFS	100
3325	R. Bougainville	PNG	5
	R. Maya, Barillas	GTM	1
	RRI Palangkaraya	INS	10
	Super R. Tupi	B	25
3326	Federal R. Corp. Nigeria	NIG	50
3330	CHU	CAN	5
	R. Ondas del Huallaga	PRU	5
3345	AWR	AFS	100
	Channel Africa	AFS	100
	R. Northern	PNG	10
	RRI Ternate	INS	10
3350	Radio Exterior de Espana	CTR	100

Chinese   English   Indonesian   Japanese   Portuguese   Spanish   Other

20

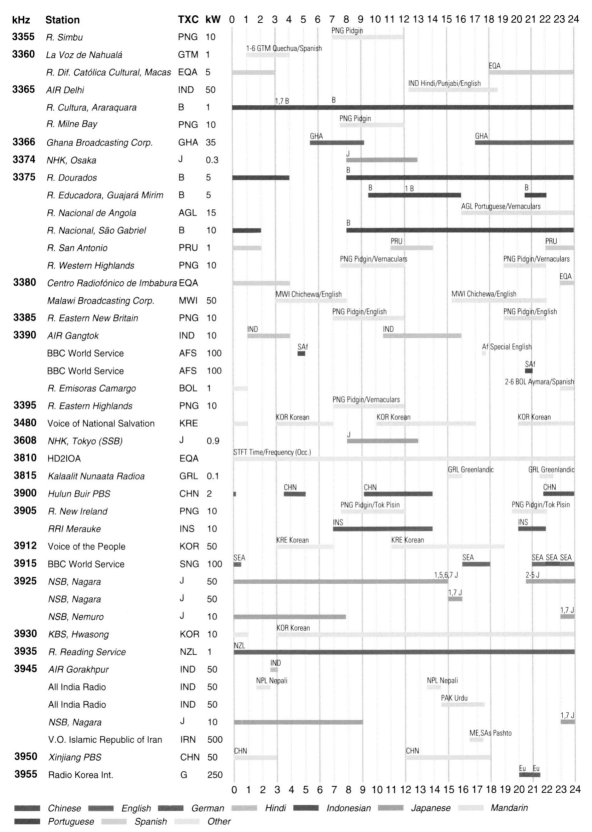

kHz	Station	TXC	kW	0 1 2 3 4 5 6 7 8 9 10 11 12 13 14 15 16 17 18 19 20 21 22 23 24
3355	R. Simbu	PNG	10	PNG Pidgin
3360	La Voz de Nahualá	GTM	1	1-6 GTM Quechua/Spanish
	R. Dif. Católica Cultural, Macas	EQA	5	EQA
3365	AIR Delhi	IND	50	IND Hindi/Punjabi/English
	R. Cultura, Araraquara	B	1	1,7 B · B
	R. Milne Bay	PNG	10	PNG Pidgin
3366	Ghana Broadcasting Corp.	GHA	35	GHA · GHA
3374	NHK, Osaka	J	0.3	J
3375	R. Dourados	B	5	B
	R. Educadora, Guajará Mirim	B	5	B · 1 B · B
	R. Nacional de Angola	AGL	15	AGL Portuguese/Vernaculars
	R. Nacional, São Gabriel	B	10	B
	R. San Antonio	PRU	1	PRU · PRU
	R. Western Highlands	PNG	10	PNG Pidgin/Vernaculars · PNG Pidgin/Vernaculars
3380	Centro Radiofónico de Imbabura	EQA		EQA
	Malawi Broadcasting Corp.	MWI	50	MWI Chichewa/English · MWI Chichewa/English
3385	R. Eastern New Britain	PNG	10	PNG Pidgin/English · PNG Pidgin/English
3390	AIR Gangtok	IND	10	IND · IND
	BBC World Service	AFS	100	SAf · Af Special English
	BBC World Service	AFS	100	SAf
	R. Emisoras Camargo	BOL	1	2-6 BOL Aymara/Spanish
3395	R. Eastern Highlands	PNG	10	PNG Pidgin/Vernaculars
3480	Voice of National Salvation	KRE		KOR Korean · KOR Korean · KOR Korean
3608	NHK, Tokyo (SSB)	J	0.9	J
3810	HD2IOA	EQA		STFT Time/Frequency (Occ.)
3815	Kalaalit Nunaata Radioa	GRL	0.1	GRL Greenlandic · GRL Greenlandic
3900	Hulun Buir PBS	CHN	2	CHN · CHN · CHN
3905	R. New Ireland	PNG	10	PNG Pidgin/Tok Pisin · PNG Pidgin/Tok Pisin
	RRI Merauke	INS	10	INS · INS
3912	Voice of the People	KOR	50	KRE Korean · KRE Korean
3915	BBC World Service	SNG	100	SEA · SEA · SEA SEA SEA
3925	NSB, Nagara	J	50	1,5,6,7 J · 2-5 J
	NSB, Nagara	J	50	1,7 J
	NSB, Nemuro	J	10	1,7 J
3930	KBS, Hwasong	KOR	10	KOR Korean
3935	R. Reading Service	NZL	1	NZL
3945	AIR Gorakhpur	IND	50	IND
	All India Radio	IND	50	NPL Nepali · NPL Nepali
	All India Radio	IND	50	PAK Urdu
	NSB, Nagara	J	10	1,7 J
	V.O. Islamic Republic of Iran	IRN	500	ME,SAs Pashto
3950	Xinjiang PBS	CHN	50	CHN · CHN
3955	Radio Korea Int.	G	250	Eu Eu

0 1 2 3 4 5 6 7 8 9 10 11 12 13 14 15 16 17 18 19 20 21 22 23 24

▬ Chinese	▬ English	▬ German	▬ Hindi	▬ Indonesian	▬ Japanese	▬ Mandarin
▬ Portuguese	▬ Spanish	▬ Other				

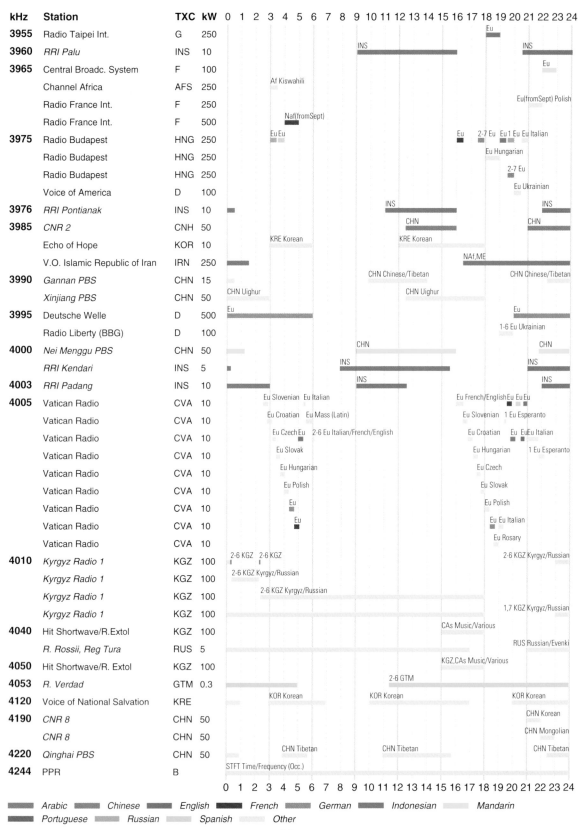

kHz	Station	TXC	kW
3955	Radio Taipei Int.	G	250
3960	*RRI Palu*	INS	10
3965	Central Broadc. System	F	100
	Channel Africa	AFS	250
	Radio France Int.	F	250
	Radio France Int.	F	500
3975	Radio Budapest	HNG	250
	Radio Budapest	HNG	250
	Radio Budapest	HNG	250
	Voice of America	D	100
3976	*RRI Pontianak*	INS	10
3985	*CNR 2*	CNH	50
	Echo of Hope	KOR	10
	V.O. Islamic Republic of Iran	IRN	250
3990	*Gannan PBS*	CHN	15
	Xinjiang PBS	CHN	50
3995	Deutsche Welle	D	500
	Radio Liberty (BBG)	D	100
4000	*Nei Menggu PBS*	CHN	50
	RRI Kendari	INS	5
4003	*RRI Padang*	INS	10
4005	Vatican Radio	CVA	10
	Vatican Radio	CVA	10
	Vatican Radio	CVA	10
	Vatican Radio	CVA	10
	Vatican Radio	CVA	10
	Vatican Radio	CVA	10
	Vatican Radio	CVA	10
	Vatican Radio	CVA	10
	Vatican Radio	CVA	10
4010	*Kyrgyz Radio 1*	KGZ	100
	Kyrgyz Radio 1	KGZ	100
	Kyrgyz Radio 1	KGZ	100
	Kyrgyz Radio 1	KGZ	100
4040	Hit Shortwave/R.Extol	KGZ	100
	R. Rossii, Reg Tura	RUS	5
4050	Hit Shortwave/R. Extol	KGZ	100
4053	*R. Verdad*	GTM	0.3
4120	Voice of National Salvation	KRE	
4190	*CNR 8*	CHN	50
	CNR 8	CHN	50
4220	*Qinghai PBS*	CHN	50
4244	PPR	B	

Legend: Arabic, Chinese, English, French, German, Indonesian, Mandarin, Portuguese, Russian, Spanish, Other

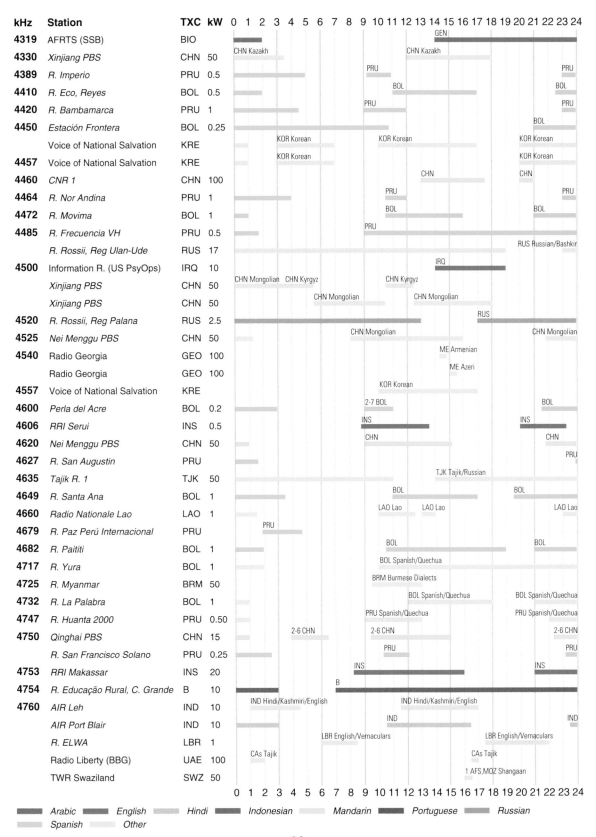

kHz	Station	TXC	kW
4319	AFRTS (SSB)	BIO	
4330	Xinjiang PBS	CHN	50
4389	R. Imperio	PRU	0.5
4410	R. Eco, Reyes	BOL	0.5
4420	R. Bambamarca	PRU	1
4450	Estación Frontera	BOL	0.25
	Voice of National Salvation	KRE	
4457	Voice of National Salvation	KRE	
4460	CNR 1	CHN	100
4464	R. Nor Andina	PRU	1
4472	R. Movima	BOL	1
4485	R. Frecuencia VH	PRU	0.5
	R. Rossii, Reg Ulan-Ude	RUS	17
4500	Information R. (US PsyOps)	IRQ	10
	Xinjiang PBS	CHN	50
	Xinjiang PBS	CHN	50
4520	R. Rossii, Reg Palana	RUS	2.5
4525	Nei Menggu PBS	CHN	50
4540	Radio Georgia	GEO	100
	Radio Georgia	GEO	100
4557	Voice of National Salvation	KRE	
4600	Perla del Acre	BOL	0.2
4606	RRI Serui	INS	0.5
4620	Nei Menggu PBS	CHN	50
4627	R. San Augustin	PRU	
4635	Tajik R. 1	TJK	50
4649	R. Santa Ana	BOL	1
4660	Radio Nationale Lao	LAO	1
4679	R. Paz Perú Internacional	PRU	
4682	R. Paititi	BOL	1
4717	R. Yura	BOL	1
4725	R. Myanmar	BRM	50
4732	R. La Palabra	BOL	1
4747	R. Huanta 2000	PRU	0.50
4750	Qinghai PBS	CHN	15
	R. San Francisco Solano	PRU	0.25
4753	RRI Makassar	INS	20
4754	R. Educação Rural, C. Grande	B	10
4760	AIR Leh	IND	10
	AIR Port Blair	IND	10
	R. ELWA	LBR	1
	Radio Liberty (BBG)	UAE	100
	TWR Swaziland	SWZ	50

Legend: Arabic · English · Hindi · Indonesian · Mandarin · Portuguese · Russian · Spanish · Other

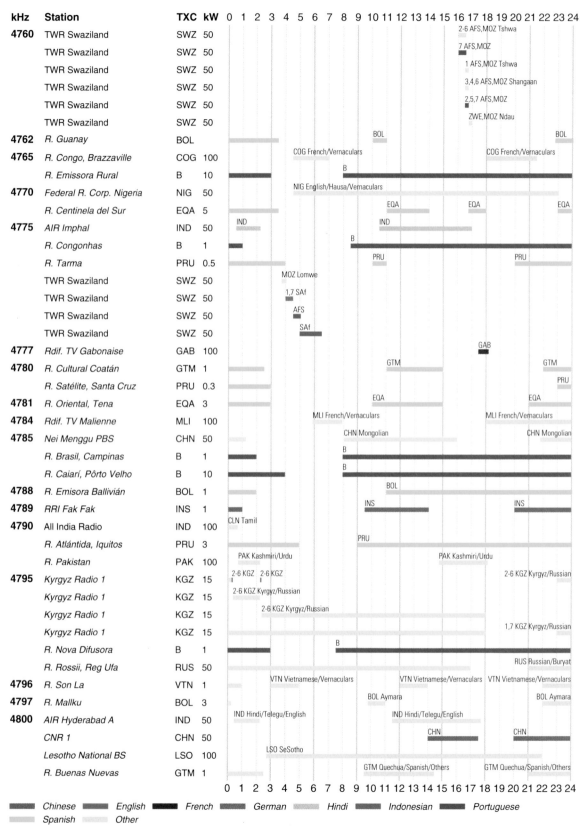

kHz	Station	TXC	kW	Schedule (0–24)
4760	TWR Swaziland	SWZ	50	2-6 AFS,MOZ Tshwa
	TWR Swaziland	SWZ	50	7 AFS,MOZ
	TWR Swaziland	SWZ	50	1 AFS,MOZ Tshwa
	TWR Swaziland	SWZ	50	3,4,6 AFS,MOZ Shangaan
	TWR Swaziland	SWZ	50	2,5,7 AFS,MOZ
	TWR Swaziland	SWZ	50	ZWE,MOZ Ndau
4762	R. Guanay	BOL		BOL
4765	R. Congo, Brazzaville	COG	100	COG French/Vernaculars
	R. Emissora Rural	B	10	B
4770	Federal R. Corp. Nigeria	NIG	50	NIG English/Hausa/Vernaculars
	R. Centinela del Sur	EQA	5	EQA
4775	AIR Imphal	IND	50	IND
	R. Congonhas	B	1	B
	R. Tarma	PRU	0.5	PRU
	TWR Swaziland	SWZ	50	MOZ Lomwe
	TWR Swaziland	SWZ	50	1,7 SAf
	TWR Swaziland	SWZ	50	AFS
	TWR Swaziland	SWZ	50	SAf
4777	Rdif. TV Gabonaise	GAB	100	GAB
4780	R. Cultural Coatán	GTM	1	GTM
	R. Satélite, Santa Cruz	PRU	0.3	PRU
4781	R. Oriental, Tena	EQA	3	EQA
4784	Rdif. TV Malienne	MLI	100	MLI French/Vernaculars
4785	Nei Menggu PBS	CHN	50	CHN Mongolian
	R. Brasil, Campinas	B	1	B
	R. Caiarí, Pôrto Velho	B	10	B
4788	R. Emisora Ballivián	BOL	1	BOL
4789	RRI Fak Fak	INS	1	INS
4790	All India Radio	IND	100	CLN Tamil
	R. Atlántida, Iquitos	PRU	3	PRU
	R. Pakistan	PAK	100	PAK Kashmiri/Urdu
4795	Kyrgyz Radio 1	KGZ	15	2-6 KGZ / 2-6 KGZ / 2-6 KGZ Kyrgyz/Russian
	Kyrgyz Radio 1	KGZ	15	2-6 KGZ Kyrgyz/Russian
	Kyrgyz Radio 1	KGZ	15	2-6 KGZ Kyrgyz/Russian
	Kyrgyz Radio 1	KGZ	15	1,7 KGZ Kyrgyz/Russian
	R. Nova Difusora	B	1	B
	R. Rossii, Reg Ufa	RUS	50	RUS Russian/Buryat
4796	R. Son La	VTN	1	VTN Vietnamese/Vernaculars
4797	R. Mallku	BOL	3	BOL Aymara
4800	AIR Hyderabad A	IND	50	IND Hindi/Telegu/English
	CNR 1	CHN	50	CHN
	Lesotho National BS	LSO	100	LSO SeSotho
	R. Buenas Nuevas	GTM	1	GTM Quechua/Spanish/Others

Chinese   English   French   German   Hindi   Indonesian   Portuguese
Spanish   Other

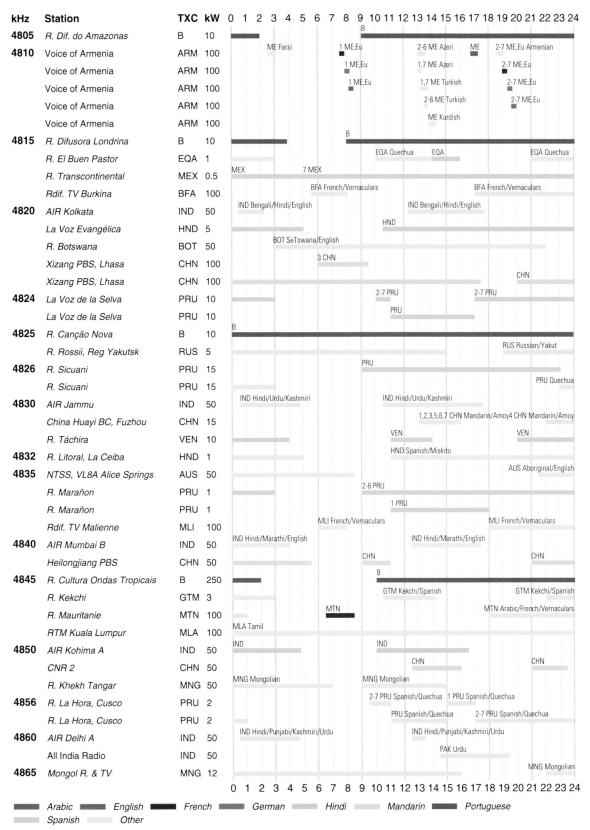

kHz	Station	TXC	kW	Schedule notes
4805	R. Dif. do Amazonas	B	10	B
4810	Voice of Armenia	ARM	100	ME Farsi; 1 ME,Eu; 2-6 ME Azeri; ME; 2-7 ME,Eu Armenian
	Voice of Armenia	ARM	100	1 ME,Eu; 1,7 ME Azeri; 2-7 ME,Eu
	Voice of Armenia	ARM	100	1 ME,Eu; 1,7 ME Turkish; 2-7 ME,Eu
	Voice of Armenia	ARM	100	2-6 ME Turkish; 2-7 ME,Eu
	Voice of Armenia	ARM	100	ME Kurdish
4815	R. Difusora Londrina	B	10	B
	R. El Buen Pastor	EQA	1	EQA Quechua; EQA; EQA Quechua
	R. Transcontinental	MEX	0.5	MEX; 7 MEX
	Rdif. TV Burkina	BFA	100	BFA French/Vernaculars; BFA French/Vernaculars
4820	AIR Kolkata	IND	50	IND Bengali/Hindi/English; IND Bengali/Hindi/English
	La Voz Evangélica	HND	5	HND
	R. Botswana	BOT	50	BOT SeTswana/English
	Xizang PBS, Lhasa	CHN	100	3 CHN
	Xizang PBS, Lhasa	CHN	100	CHN
4824	La Voz de la Selva	PRU	10	2-7 PRU; 2-7 PRU
	La Voz de la Selva	PRU	10	PRU
4825	R. Canção Nova	B	10	B
	R. Rossii, Reg Yakutsk	RUS	5	RUS Russian/Yakut
4826	R. Sicuani	PRU	15	PRU
	R. Sicuani	PRU	15	PRU Quechua
4830	AIR Jammu	IND	50	IND Hindi/Urdu/Kashmiri; IND Hindi/Urdu/Kashmiri
	China Huayi BC, Fuzhou	CHN	15	1,2,3,5,6,7 CHN Mandarin/Amoy4 CHN Mandarin/Amoy
	R. Táchira	VEN	10	VEN; VEN
4832	R. Litoral, La Ceiba	HND	1	HND Spanish/Miskito
4835	NTSS, VL8A Alice Springs	AUS	50	AUS Aboriginal/English
	R. Marañon	PRU	1	2-6 PRU
	R. Marañon	PRU	1	1 PRU
	Rdif. TV Malienne	MLI	100	MLI French/Vernaculars; MLI French/Vernaculars
4840	AIR Mumbai B	IND	50	IND Hindi/Marathi/English; IND Hindi/Marathi/English
	Heilongjiang PBS	CHN	50	CHN; CHN
4845	R. Cultura Ondas Tropicais	B	250	B
	R. Kekchi	GTM	3	GTM Kekchi/Spanish; GTM Kekchi/Spanish
	R. Mauritanie	MTN	100	MTN; MTN Arabic/French/Vernaculars
	RTM Kuala Lumpur	MLA	100	MLA Tamil
4850	AIR Kohima A	IND	50	IND; IND
	CNR 2	CHN	50	CHN; CHN
	R. Khekh Tangar	MNG	50	MNG Mongolian; MNG Mongolian
4856	R. La Hora, Cusco	PRU	2	2-7 PRU Spanish/Quechua; 1 PRU Spanish/Quechua
	R. La Hora, Cusco	PRU	2	PRU Spanish/Quechua; 2-7 PRU Spanish/Quechua
4860	AIR Delhi A	IND	50	IND Hindi/Punjabi/Kashmiri/Urdu; IND Hindi/Punjabi/Kashmiri/Urdu
	All India Radio	IND	50	PAK Urdu
4865	Mongol R. & TV	MNG	12	MNG Mongolian

Legend: Arabic | English | French | German | Hindi | Mandarin | Portuguese | Spanish | Other

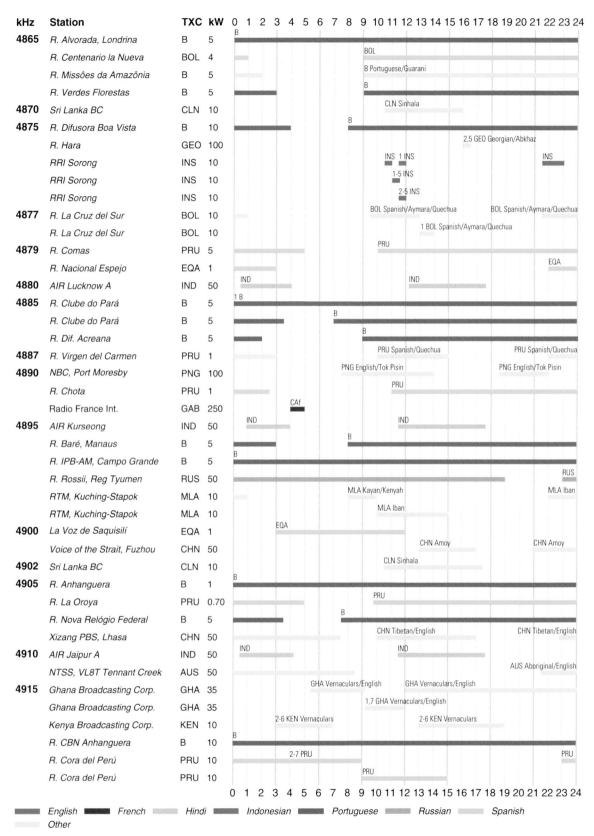

kHz	Station	TXC	kW
4865	R. Alvorada, Londrina	B	5
	R. Centenario la Nueva	BOL	4
	R. Missôes da Amazônia	B	5
	R. Verdes Florestas	B	5
4870	Sri Lanka BC	CLN	10
4875	R. Difusora Boa Vista	B	10
	R. Hara	GEO	100
	RRI Sorong	INS	10
	RRI Sorong	INS	10
	RRI Sorong	INS	10
4877	R. La Cruz del Sur	BOL	10
	R. La Cruz del Sur	BOL	10
4879	R. Comas	PRU	5
	R. Nacional Espejo	EQA	1
4880	AIR Lucknow A	IND	50
4885	R. Clube do Pará	B	5
	R. Clube do Pará	B	5
	R. Dif. Acreana	B	5
4887	R. Virgen del Carmen	PRU	1
4890	NBC, Port Moresby	PNG	100
	R. Chota	PRU	1
	Radio France Int.	GAB	250
4895	AIR Kurseong	IND	50
	R. Baré, Manaus	B	5
	R. IPB-AM, Campo Grande	B	5
	R. Rossii, Reg Tyumen	RUS	50
	RTM, Kuching-Stapok	MLA	10
	RTM, Kuching-Stapok	MLA	10
4900	La Voz de Saquisilí	EQA	1
	Voice of the Strait, Fuzhou	CHN	50
4902	Sri Lanka BC	CLN	10
4905	R. Anhanguera	B	1
	R. La Oroya	PRU	0.70
	R. Nova Relógio Federal	B	5
	Xizang PBS, Lhasa	CHN	50
4910	AIR Jaipur A	IND	50
	NTSS, VL8T Tennant Creek	AUS	50
4915	Ghana Broadcasting Corp.	GHA	35
	Ghana Broadcasting Corp.	GHA	35
	Kenya Broadcasting Corp.	KEN	10
	R. CBN Anhanguera	B	10
	R. Cora del Perú	PRU	10
	R. Cora del Perú	PRU	10

English   French   Hindi   Indonesian   Portuguese   Russian   Spanish   Other

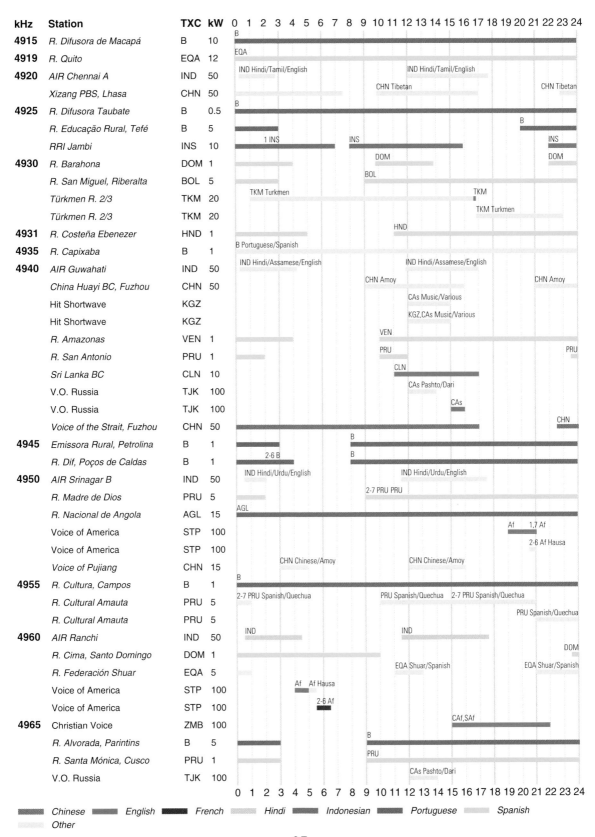

kHz	Station	TXC	kW
4915	R. Difusora de Macapá	B	10
4919	R. Quito	EQA	12
4920	AIR Chennai A	IND	50
	Xizang PBS, Lhasa	CHN	50
4925	R. Difusora Taubate	B	0.5
	R. Educação Rural, Tefé	B	5
	RRI Jambi	INS	10
4930	R. Barahona	DOM	1
	R. San Miguel, Riberalta	BOL	5
	Türkmen R. 2/3	TKM	20
	Türkmen R. 2/3	TKM	20
4931	R. Costeña Ebenezer	HND	1
4935	R. Capixaba	B	1
4940	AIR Guwahati	IND	50
	China Huayi BC, Fuzhou	CHN	50
	Hit Shortwave	KGZ	
	Hit Shortwave	KGZ	
	R. Amazonas	VEN	1
	R. San Antonio	PRU	1
	Sri Lanka BC	CLN	10
	V.O. Russia	TJK	100
	V.O. Russia	TJK	100
	Voice of the Strait, Fuzhou	CHN	50
4945	Emissora Rural, Petrolina	B	1
	R. Dif, Poços de Caldas	B	1
4950	AIR Srinagar B	IND	50
	R. Madre de Dios	PRU	5
	R. Nacional de Angola	AGL	15
	Voice of America	STP	100
	Voice of America	STP	100
	Voice of Pujiang	CHN	15
4955	R. Cultura, Campos	B	1
	R. Cultural Amauta	PRU	5
	R. Cultural Amauta	PRU	5
4960	AIR Ranchi	IND	50
	R. Cima, Santo Domingo	DOM	1
	R. Federación Shuar	EQA	5
	Voice of America	STP	100
	Voice of America	STP	100
4965	Christian Voice	ZMB	100
	R. Alvorada, Parintins	B	5
	R. Santa Mónica, Cusco	PRU	1
	V.O. Russia	TJK	100

Legend: Chinese — English — French — Hindi — Indonesian — Portuguese — Spanish — Other

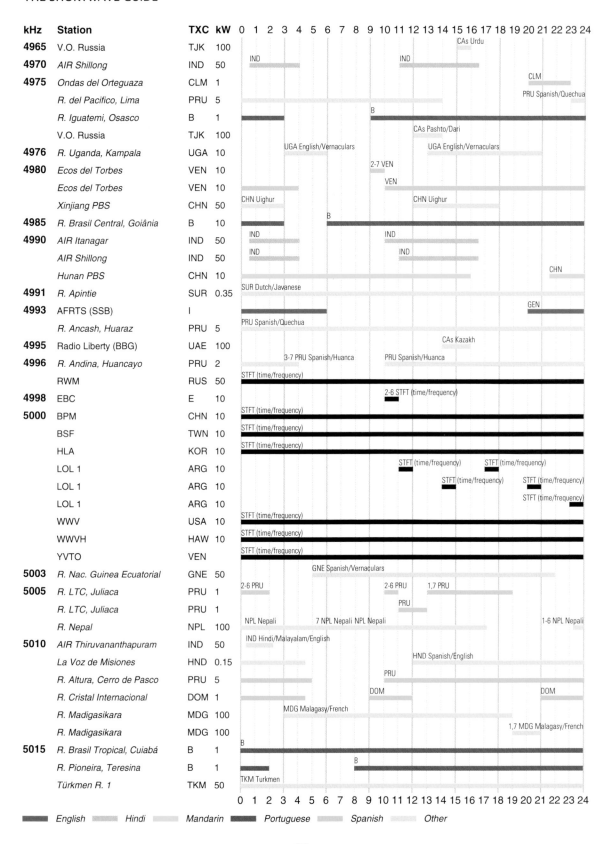

kHz	Station	TXC	kW	Schedule (0–24)
4965	V.O. Russia	TJK	100	CAs Urdu
4970	*AIR Shillong*	IND	50	IND / IND
4975	*Ondas del Orteguaza*	CLM	1	CLM
	R. del Pacifico, Lima	PRU	5	PRU Spanish/Quechua
	R. Iguatemi, Osasco	B	1	B
	V.O. Russia	TJK	100	CAs Pashto/Dari
4976	*R. Uganda, Kampala*	UGA	10	UGA English/Vernaculars / UGA English/Vernaculars
4980	*Ecos del Torbes*	VEN	10	2-7 VEN
	Ecos del Torbes	VEN	10	VEN
	Xinjiang PBS	CHN	50	CHN Uighur / CHN Uighur
4985	*R. Brasil Central, Goiânia*	B	10	B
4990	*AIR Itanagar*	IND	50	IND / IND
	AIR Shillong	IND	50	IND / IND
	Hunan PBS	CHN	10	CHN
4991	*R. Apintie*	SUR	0.35	SUR Dutch/Javanese
4993	AFRTS (SSB)	I		GEN
	R. Ancash, Huaraz	PRU	5	PRU Spanish/Quechua
4995	Radio Liberty (BBG)	UAE	100	CAs Kazakh
4996	*R. Andina, Huancayo*	PRU	2	3-7 PRU Spanish/Huanca / PRU Spanish/Huanca
	RWM	RUS	50	STFT (time/frequency)
4998	EBC	E	10	2-6 STFT (time/frequency)
5000	BPM	CHN	10	STFT (time/frequency)
	BSF	TWN	10	STFT (time/frequency)
	HLA	KOR	10	STFT (time/frequency)
	LOL 1	ARG	10	STFT (time/frequency) / STFT (time/frequency)
	LOL 1	ARG	10	STFT (time/frequency) / STFT (time/frequency)
	LOL 1	ARG	10	STFT (time/frequency)
	WWV	USA	10	STFT (time/frequency)
	WWVH	HAW	10	STFT (time/frequency)
	YVTO	VEN		STFT (time/frequency)
5003	*R. Nac. Guinea Ecuatorial*	GNE	50	GNE Spanish/Vernaculars
5005	*R. LTC, Juliaca*	PRU	1	2-6 PRU / 2-6 PRU / 1,7 PRU
	R. LTC, Juliaca	PRU	1	PRU
	R. Nepal	NPL	100	NPL Nepali / 7 NPL Nepali NPL Nepali / 1-6 NPL Nepali
5010	*AIR Thiruvananthapuram*	IND	50	IND Hindi/Malayalam/English
	La Voz de Misiones	HND	0.15	HND Spanish/English
	R. Altura, Cerro de Pasco	PRU	5	PRU
	R. Cristal Internacional	DOM	1	DOM / DOM
	R. Madigasikara	MDG	100	MDG Malagasy/French
	R. Madigasikara	MDG	100	1,7 MDG Malagasy/French
5015	*R. Brasil Tropical, Cuiabá*	B	1	B
	R. Pioneira, Teresina	B	1	B
	Türkmen R. 1	TKM	50	TKM Turkmen

Legend: ▇ English ░ Hindi ▒ Mandarin ▇ Portuguese ▒ Spanish ░ Other

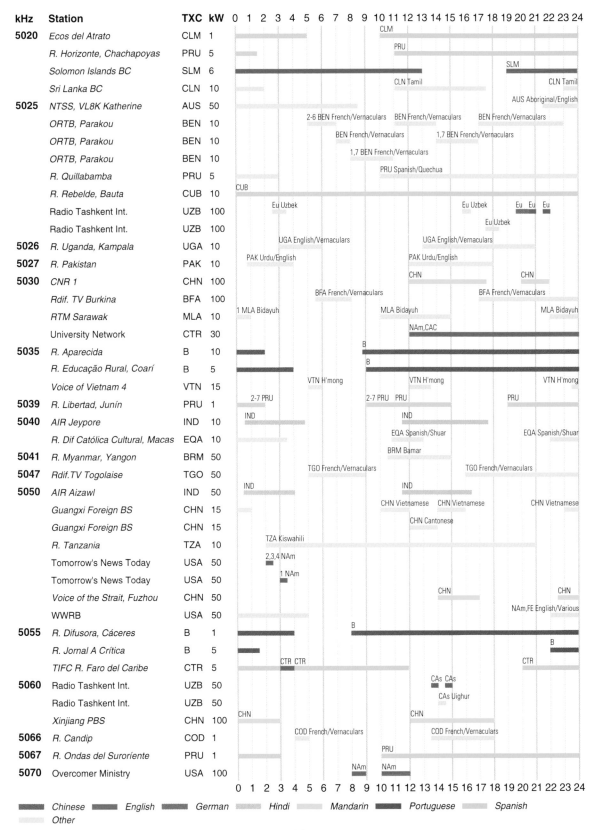

kHz	Station	TXC	kW	Timeline notes
5020	Ecos del Atrato	CLM	1	CLM
	R. Horizonte, Chachapoyas	PRU	5	PRU
	Solomon Islands BC	SLM	6	SLM
	Sri Lanka BC	CLN	10	CLN Tamil; CLN Tamil
5025	NTSS, VL8K Katherine	AUS	50	AUS Aboriginal/English
	ORTB, Parakou	BEN	10	2-6 BEN French/Vernaculars; BEN French/Vernaculars; BEN French/Vernaculars
	ORTB, Parakou	BEN	10	BEN French/Vernaculars; 1,7 BEN French/Vernaculars
	ORTB, Parakou	BEN	10	1,7 BEN French/Vernaculars
	R. Quillabamba	PRU	5	PRU Spanish/Quechua
	R. Rebelde, Bauta	CUB	10	CUB
	Radio Tashkent Int.	UZB	100	Eu Uzbek; Eu Uzbek; Eu Eu Eu
	Radio Tashkent Int.	UZB	100	Eu Uzbek
5026	R. Uganda, Kampala	UGA	10	UGA English/Vernaculars; UGA English/Vernaculars
5027	R. Pakistan	PAK	10	PAK Urdu/English; PAK Urdu/English
5030	CNR 1	CHN	100	CHN; CHN
	Rdif. TV Burkina	BFA	100	BFA French/Vernaculars; BFA French/Vernaculars
	RTM Sarawak	MLA	10	1 MLA Bidayuh; MLA Bidayuh; MLA Bidayuh
	University Network	CTR	30	NAm,CAC
5035	R. Aparecida	B	10	B
	R. Educação Rural, Coarí	B	5	B
	Voice of Vietnam 4	VTN	15	VTN H'mong; VTN H'mong; VTN H'mong
5039	R. Libertad, Junín	PRU	1	2-7 PRU; 2-7 PRU PRU; PRU
5040	AIR Jeypore	IND	10	IND; IND
	R. Dif Católica Cultural, Macas	EQA	10	EQA Spanish/Shuar; EQA Spanish/Shuar
5041	R. Myanmar, Yangon	BRM	50	BRM Bamar
5047	Rdif.TV Togolaise	TGO	50	TGO French/Vernaculars; TGO French/Vernaculars
5050	AIR Aizawl	IND	50	IND; IND
	Guangxi Foreign BS	CHN	15	CHN Vietnamese; CHN Vietnamese; CHN Vietnamese
	Guangxi Foreign BS	CHN	15	CHN Cantonese
	R. Tanzania	TZA	10	TZA Kiswahili
	Tomorrow's News Today	USA	50	2,3,4 NAm
	Tomorrow's News Today	USA	50	1 NAm
	Voice of the Strait, Fuzhou	CHN	50	CHN; CHN
	WWRB	USA	50	NAm,FE English/Various
5055	R. Difusora, Cáceres	B	1	B
	R. Jornal A Crítica	B	5	B
	TIFC R. Faro del Caribe	CTR	5	CTR CTR; CTR
5060	Radio Tashkent Int.	UZB	50	CAs CAs
	Radio Tashkent Int.	UZB	50	CAs Uighur
	Xinjiang PBS	CHN	100	CHN; CHN
5066	R. Candip	COD	1	COD French/Vernaculars; COD French/Vernaculars
5067	R. Ondas del Suroríente	PRU	1	PRU
5070	Overcomer Ministry	USA	100	NAm; NAm

Chinese　English　German　Hindi　Mandarin　Portuguese　Spanish　Other

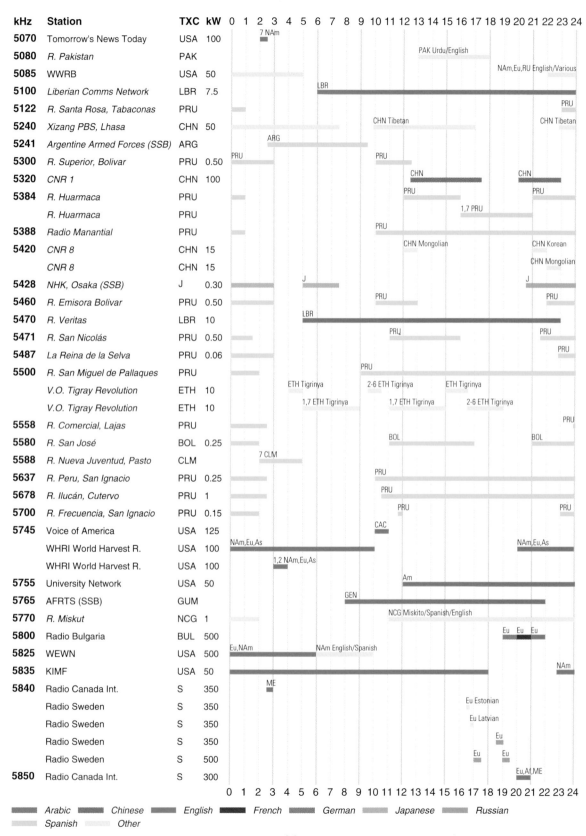

kHz	Station	TXC	kW
5070	Tomorrow's News Today	USA	100
5080	R. Pakistan	PAK	
5085	WWRB	USA	50
5100	Liberian Comms Network	LBR	7.5
5122	R. Santa Rosa, Tabaconas	PRU	
5240	Xizang PBS, Lhasa	CHN	50
5241	Argentine Armed Forces (SSB)	ARG	
5300	R. Superior, Bolivar	PRU	0.50
5320	CNR 1	CHN	100
5384	R. Huarmaca	PRU	
	R. Huarmaca	PRU	
5388	Radio Manantial	PRU	
5420	CNR 8	CHN	15
	CNR 8	CHN	15
5428	NHK, Osaka (SSB)	J	0.30
5460	R. Emisora Bolivar	PRU	0.50
5470	R. Veritas	LBR	10
5471	R. San Nicolás	PRU	0.50
5487	La Reina de la Selva	PRU	0.06
5500	R. San Miguel de Pallaques	PRU	
	V.O. Tigray Revolution	ETH	10
	V.O. Tigray Revolution	ETH	10
5558	R. Comercial, Lajas	PRU	
5580	R. San José	BOL	0.25
5588	R. Nueva Juventud, Pasto	CLM	
5637	R. Peru, San Ignacio	PRU	0.25
5678	R. Ilucán, Cutervo	PRU	1
5700	R. Frecuencia, San Ignacio	PRU	0.15
5745	Voice of America	USA	125
	WHRI World Harvest R.	USA	100
	WHRI World Harvest R.	USA	100
5755	University Network	USA	50
5765	AFRTS (SSB)	GUM	
5770	R. Miskut	NCG	1
5800	Radio Bulgaria	BUL	500
5825	WEWN	USA	500
5835	KIMF	USA	50
5840	Radio Canada Int.	S	350
	Radio Sweden	S	350
	Radio Sweden	S	350
	Radio Sweden	S	350
	Radio Sweden	S	500
5850	Radio Canada Int.	S	300

Legend: Arabic, Chinese, English, French, German, Japanese, Russian, Spanish, Other

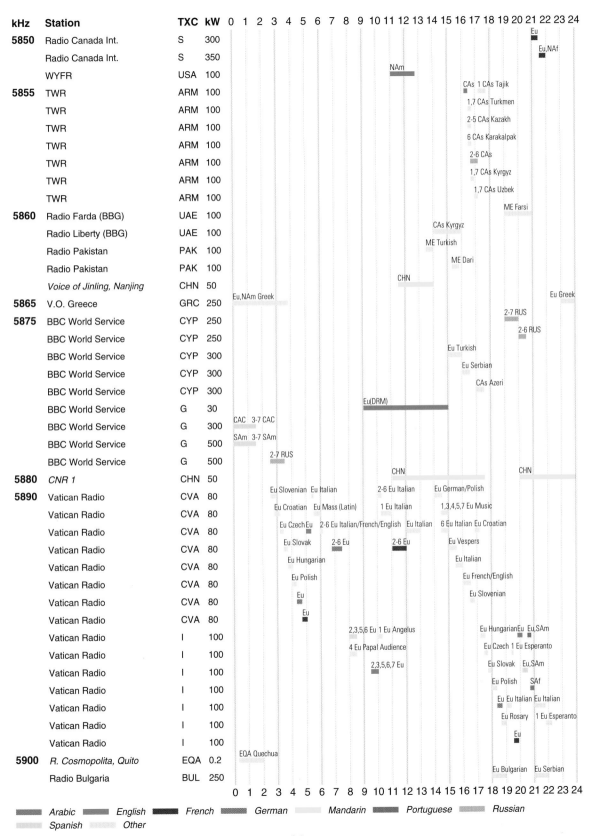

kHz	Station	TXC	kW
5850	Radio Canada Int.	S	300
	Radio Canada Int.	S	350
	WYFR	USA	100
5855	TWR	ARM	100
	TWR	ARM	100
	TWR	ARM	100
	TWR	ARM	100
	TWR	ARM	100
	TWR	ARM	100
	TWR	ARM	100
5860	Radio Farda (BBG)	UAE	100
	Radio Liberty (BBG)	UAE	100
	Radio Pakistan	PAK	100
	Radio Pakistan	PAK	100
	Voice of Jinling, Nanjing	CHN	50
5865	V.O. Greece	GRC	250
5875	BBC World Service	CYP	250
	BBC World Service	CYP	250
	BBC World Service	CYP	300
	BBC World Service	CYP	300
	BBC World Service	CYP	300
	BBC World Service	G	30
	BBC World Service	G	300
	BBC World Service	G	500
	BBC World Service	G	500
5880	*CNR 1*	CHN	50
5890	Vatican Radio	CVA	80
	Vatican Radio	CVA	80
	Vatican Radio	CVA	80
	Vatican Radio	CVA	80
	Vatican Radio	CVA	80
	Vatican Radio	CVA	80
	Vatican Radio	CVA	80
	Vatican Radio	CVA	80
	Vatican Radio	I	100
	Vatican Radio	I	100
	Vatican Radio	I	100
	Vatican Radio	I	100
	Vatican Radio	I	100
	Vatican Radio	I	100
	Vatican Radio	I	100
5900	*R. Cosmopolita, Quito*	EQA	0.2
	Radio Bulgaria	BUL	250

Legend: Arabic · English · French · German · Mandarin · Portuguese · Russian · Spanish · Other

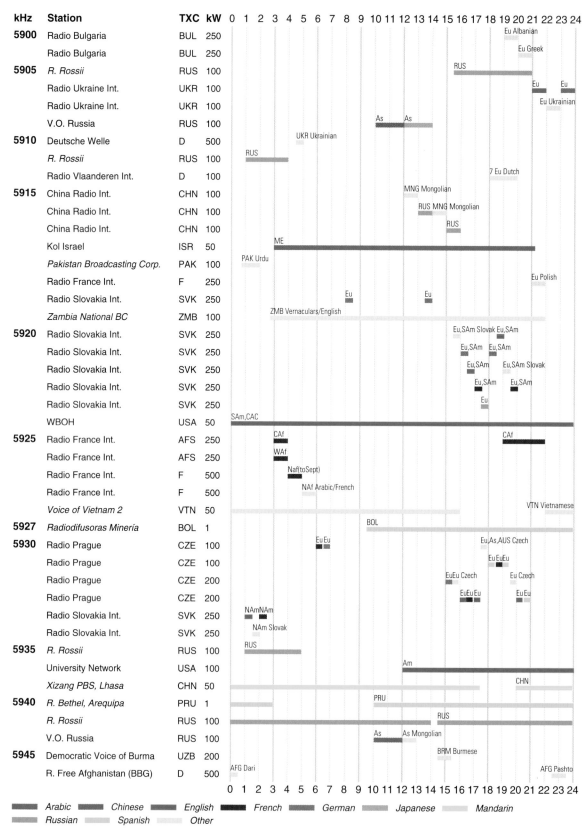

kHz	Station	TXC	kW
5900	Radio Bulgaria	BUL	250
	Radio Bulgaria	BUL	250
5905	*R. Rossii*	RUS	100
	Radio Ukraine Int.	UKR	100
	Radio Ukraine Int.	UKR	100
	V.O. Russia	RUS	100
5910	Deutsche Welle	D	500
	R. Rossii	RUS	100
	Radio Vlaanderen Int.	D	100
5915	China Radio Int.	CHN	100
	China Radio Int.	CHN	100
	China Radio Int.	CHN	100
	Kol Israel	ISR	50
	Pakistan Broadcasting Corp.	PAK	100
	Radio France Int.	F	250
	Radio Slovakia Int.	SVK	250
	Zambia National BC	ZMB	100
5920	Radio Slovakia Int.	SVK	250
	Radio Slovakia Int.	SVK	250
	Radio Slovakia Int.	SVK	250
	Radio Slovakia Int.	SVK	250
	Radio Slovakia Int.	SVK	250
	WBOH	USA	50
5925	Radio France Int.	AFS	250
	Radio France Int.	AFS	250
	Radio France Int.	F	500
	Radio France Int.	F	500
	Voice of Vietnam 2	VTN	50
5927	*Radiodifusoras Minería*	BOL	1
5930	Radio Prague	CZE	100
	Radio Prague	CZE	100
	Radio Prague	CZE	200
	Radio Prague	CZE	200
	Radio Slovakia Int.	SVK	250
	Radio Slovakia Int.	SVK	250
5935	*R. Rossii*	RUS	100
	University Network	USA	100
	Xizang PBS, Lhasa	CHN	50
5940	*R. Bethel, Arequipa*	PRU	1
	R. Rossii	RUS	100
	V.O. Russia	RUS	100
5945	Democratic Voice of Burma	UZB	200
	R. Free Afghanistan (BBG)	D	500

Legend: Arabic, Chinese, English, French, German, Japanese, Mandarin, Russian, Spanish, Other

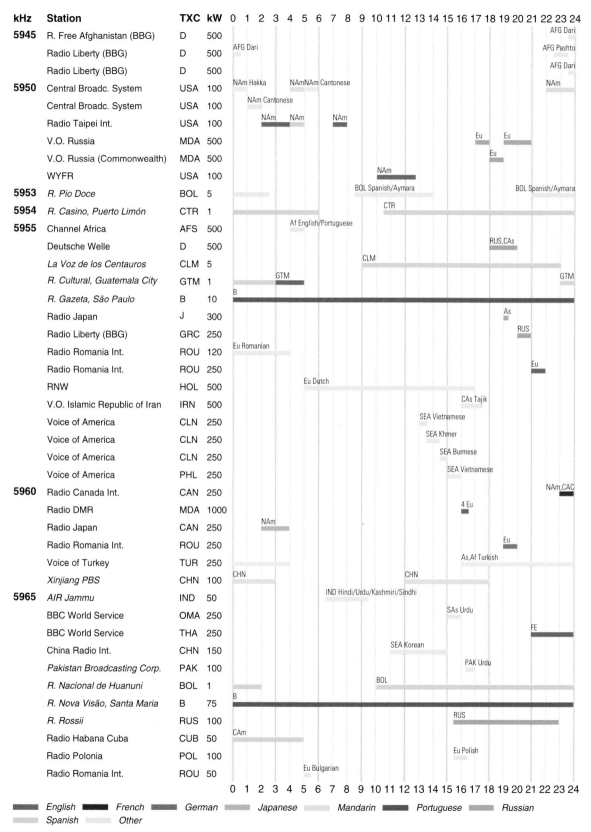

kHz	Station	TXC	kW	Time (UTC) / Notes
5945	R. Free Afghanistan (BBG)	D	500	AFG Dari
	Radio Liberty (BBG)	D	500	AFG Dari / AFG Pashto
	Radio Liberty (BBG)	D	500	AFG Dari
5950	Central Broadc. System	USA	100	NAm Hakka / NAmNAm Cantonese / NAm
	Central Broadc. System	USA	100	NAm Cantonese
	Radio Taipei Int.	USA	100	NAm / NAm / NAm
	V.O. Russia	MDA	500	Eu / Eu
	V.O. Russia (Commonwealth)	MDA	500	Eu
	WYFR	USA	100	NAm
5953	R. Pio Doce	BOL	5	BOL Spanish/Aymara / BOL Spanish/Aymara
5954	R. Casino, Puerto Limón	CTR	1	CTR
5955	Channel Africa	AFS	500	Af English/Portuguese
	Deutsche Welle	D	500	RUS,CAs
	La Voz de los Centauros	CLM	5	CLM
	R. Cultural, Guatemala City	GTM	1	GTM / GTM
	R. Gazeta, São Paulo	B	10	B
	Radio Japan	J	300	As
	Radio Liberty (BBG)	GRC	250	RUS
	Radio Romania Int.	ROU	120	Eu Romanian
	Radio Romania Int.	ROU	250	Eu
	RNW	HOL	500	Eu Dutch
	V.O. Islamic Republic of Iran	IRN	500	CAs Tajik
	Voice of America	CLN	250	SEA Vietnamese
	Voice of America	CLN	250	SEA Khmer
	Voice of America	CLN	250	SEA Burmese
	Voice of America	PHL	250	SEA Vietnamese
5960	Radio Canada Int.	CAN	250	NAm,CAC
	Radio DMR	MDA	1000	4 Eu
	Radio Japan	CAN	250	NAm
	Radio Romania Int.	ROU	250	Eu
	Voice of Turkey	TUR	250	As,Af Turkish
	Xinjiang PBS	CHN	100	CHN / CHN
5965	AIR Jammu	IND	50	IND Hindi/Urdu/Kashmiri/Sindhi
	BBC World Service	OMA	250	SAs Urdu
	BBC World Service	THA	250	FE
	China Radio Int.	CHN	150	SEA Korean
	Pakistan Broadcasting Corp.	PAK	100	PAK Urdu
	R. Nacional de Huanuni	BOL	1	BOL
	R. Nova Visão, Santa Maria	B	75	B
	R. Rossii	RUS	100	RUS
	Radio Habana Cuba	CUB	50	CAm
	Radio Polonia	POL	100	Eu Polish
	Radio Romania Int.	ROU	50	Eu Bulgarian

Legend: English | French | German | Japanese | Mandarin | Portuguese | Russian | Spanish | Other

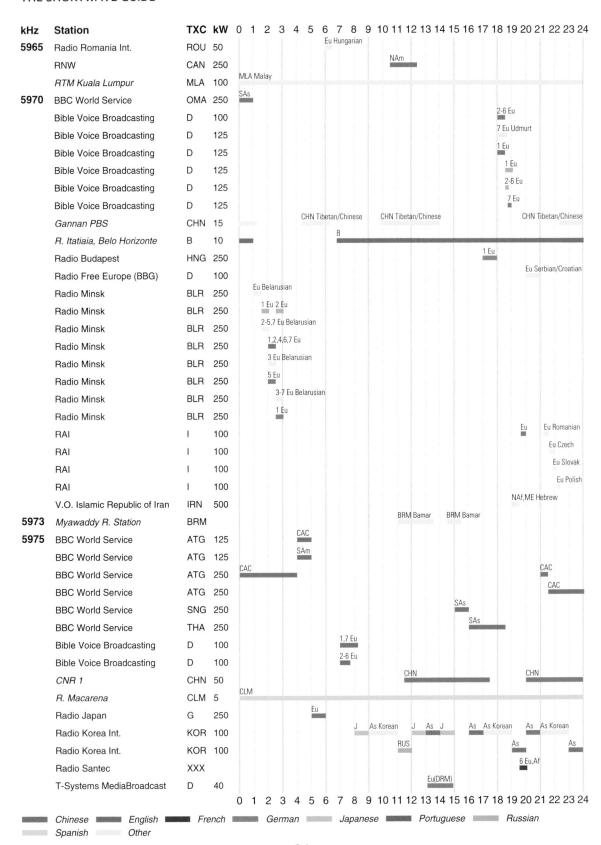

kHz	Station	TXC	kW	Notes (time bars)
5965	Radio Romania Int.	ROU	50	Eu Hungarian
	RNW	CAN	250	NAm
	RTM Kuala Lumpur	MLA	100	MLA Malay
5970	BBC World Service	OMA	250	SAs
	Bible Voice Broadcasting	D	100	2-6 Eu
	Bible Voice Broadcasting	D	125	7 Eu Udmurt
	Bible Voice Broadcasting	D	125	1 Eu
	Bible Voice Broadcasting	D	125	1 Eu
	Bible Voice Broadcasting	D	125	2-6 Eu
	Bible Voice Broadcasting	D	125	7 Eu
	Gannan PBS	CHN	15	CHN Tibetan/Chinese CHN Tibetan/Chinese CHN Tibetan/Chinese
	R. Itatiaia, Belo Horizonte	B	10	B
	Radio Budapest	HNG	250	1 Eu
	Radio Free Europe (BBG)	D	100	Eu Serbian/Croatian
	Radio Minsk	BLR	250	Eu Belarusian
	Radio Minsk	BLR	250	1 Eu 2 Eu
	Radio Minsk	BLR	250	2-5,7 Eu Belarusian
	Radio Minsk	BLR	250	1,2,4,6,7 Eu
	Radio Minsk	BLR	250	3 Eu Belarusian
	Radio Minsk	BLR	250	5 Eu
	Radio Minsk	BLR	250	3-7 Eu Belarusian
	Radio Minsk	BLR	250	1 Eu
	RAI	I	100	Eu Eu Romanian
	RAI	I	100	Eu Czech
	RAI	I	100	Eu Slovak
	RAI	I	100	Eu Polish
	V.O. Islamic Republic of Iran	IRN	500	NAf,ME Hebrew
5973	*Myawaddy R. Station*	BRM		BRM Bamar BRM Bamar
5975	BBC World Service	ATG	125	CAC
	BBC World Service	ATG	125	SAm
	BBC World Service	ATG	250	CAC CAC
	BBC World Service	ATG	250	CAC
	BBC World Service	SNG	250	SAs
	BBC World Service	THA	250	SAs
	Bible Voice Broadcasting	D	100	1,7 Eu
	Bible Voice Broadcasting	D	100	2-6 Eu
	CNR 1	CHN	50	CHN CHN
	R. Macarena	CLM	5	CLM
	Radio Japan	G	250	Eu
	Radio Korea Int.	KOR	100	J As Korean J As J As As Korean As As Korean
	Radio Korea Int.	KOR	100	RUS As As
	Radio Santec	XXX		6 Eu,Af
	T-Systems MediaBroadcast	D	40	Eu(DRM)

Legend: ▨ Chinese ▨ English ▨ French ▨ German ▨ Japanese ▨ Portuguese ▨ Russian ▨ Spanish ▨ Other

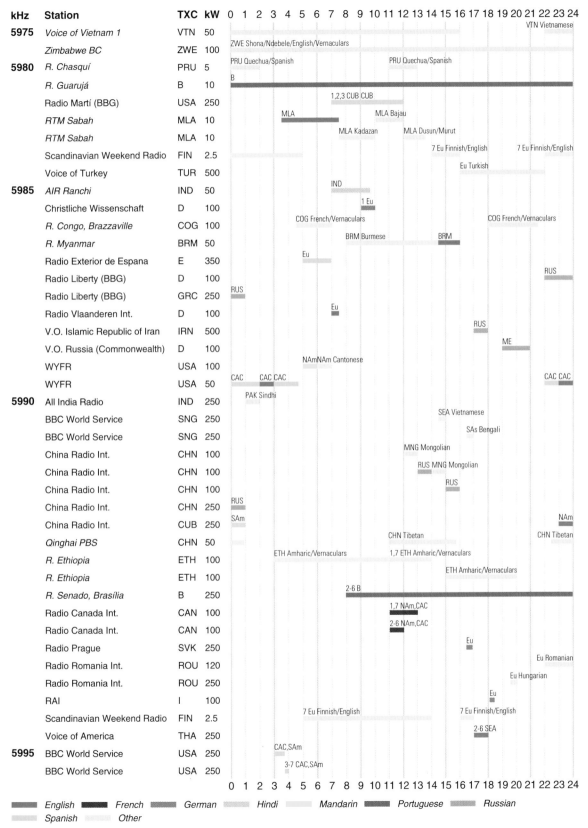

kHz	Station	TXC	kW	Broadcast times (0–24)
5975	*Voice of Vietnam 1*	VTN	50	VTN Vietnamese
	Zimbabwe BC	ZWE	100	ZWE Shona/Ndebele/English/Vernaculars
5980	*R. Chasquí*	PRU	5	PRU Quechua/Spanish · PRU Quechua/Spanish
	R. Guarujá	B	10	B
	Radio Martí (BBG)	USA	250	1,2,3 CUB CUB
	RTM Sabah	MLA	10	MLA · MLA Bajau
	RTM Sabah	MLA	10	MLA Kadazan · MLA Dusun/Murut
	Scandinavian Weekend Radio	FIN	2.5	7 Eu Finnish/English · 7 Eu Finnish/English
	Voice of Turkey	TUR	500	Eu Turkish
5985	*AIR Ranchi*	IND	50	IND
	Christliche Wissenschaft	D	100	1 Eu
	R. Congo, Brazzaville	COG	100	COG French/Vernaculars · COG French/Vernaculars
	R. Myanmar	BRM	50	BRM Burmese · BRM
	Radio Exterior de Espana	E	350	Eu
	Radio Liberty (BBG)	D	100	RUS
	Radio Liberty (BBG)	GRC	250	RUS
	Radio Vlaanderen Int.	D	100	Eu
	V.O. Islamic Republic of Iran	IRN	500	RUS
	V.O. Russia (Commonwealth)	D	100	ME
	WYFR	USA	100	NAmNAm Cantonese
	WYFR	USA	50	CAC · CAC CAC · CAC CAC
5990	All India Radio	IND	250	PAK Sindhi
	BBC World Service	SNG	250	SEA Vietnamese
	BBC World Service	SNG	250	SAs Bengali
	China Radio Int.	CHN	100	MNG Mongolian
	China Radio Int.	CHN	100	RUS MNG Mongolian
	China Radio Int.	CHN	100	RUS
	China Radio Int.	CHN	250	RUS
	China Radio Int.	CUB	250	SAm · NAm
	Qinghai PBS	CHN	50	CHN Tibetan · CHN Tibetan
	R. Ethiopia	ETH	100	ETH Amharic/Vernaculars · 1,7 ETH Amharic/Vernaculars
	R. Ethiopia	ETH	100	ETH Amharic/Vernaculars
	R. Senado, Brasília	B	250	2-6 B
	Radio Canada Int.	CAN	100	1,7 NAm,CAC
	Radio Canada Int.	CAN	100	2-6 NAm,CAC
	Radio Prague	SVK	250	Eu
	Radio Romania Int.	ROU	120	Eu Romanian
	Radio Romania Int.	ROU	250	Eu Hungarian
	RAI	I	100	Eu
	Scandinavian Weekend Radio	FIN	2.5	7 Eu Finnish/English · 7 Eu Finnish/English
	Voice of America	THA	250	2-6 SEA
5995	BBC World Service	USA	250	CAC,SAm
	BBC World Service	USA	250	3-7 CAC,SAm

English	French	German	Hindi	Mandarin	Portuguese	Russian	
Spanish	Other						

35

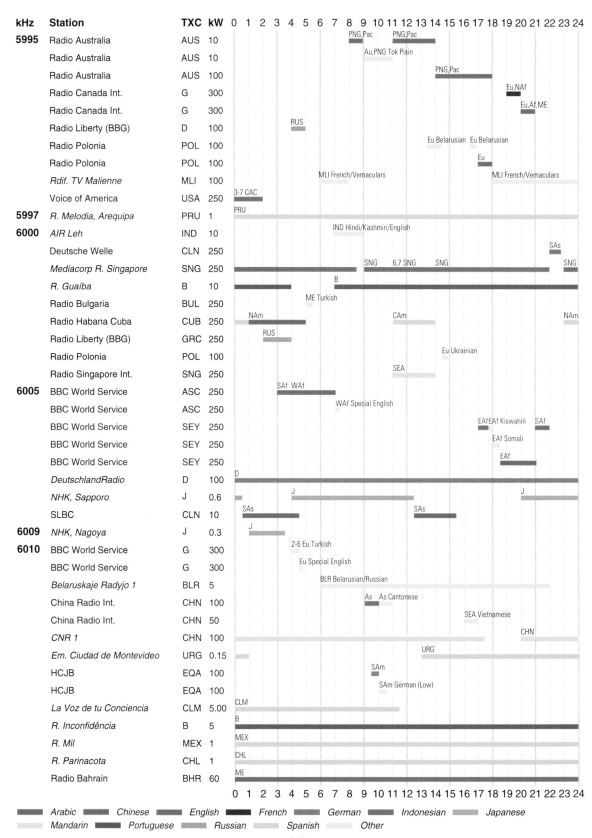

kHz	Station	TXC	kW	Schedule (0–24 h)
5995	Radio Australia	AUS	10	PNG,Pac (8–14)
	Radio Australia	AUS	10	Au,PNG Tok Pisin (9–11)
	Radio Australia	AUS	100	PNG,Pac (13–18)
	Radio Canada Int.	G	300	Eu,NAf (19–20)
	Radio Canada Int.	G	300	Eu,Af,ME (20–21)
	Radio Liberty (BBG)	D	100	RUS (4–5)
	Radio Polonia	POL	100	Eu Belarusian (13–15) Eu Belarusian (17–18)
	Radio Polonia	POL	100	Eu (18–19)
	Rdif. TV Malienne	MLI	100	MLI French/Vernaculars (6–9) MLI French/Vernaculars (18–24)
	Voice of America	USA	250	3-7 CAC (0–1)
5997	R. Melodia, Arequipa	PRU	1	PRU
6000	AIR Leh	IND	10	IND Hindi/Kashmiri/English (8–13)
	Deutsche Welle	CLN	250	SAs (22)
	Mediacorp R. Singapore	SNG	250	SNG (0–22) 6,7 SNG SNG SNG
	R. Guaíba	B	10	B
	Radio Bulgaria	BUL	250	ME Turkish (5)
	Radio Habana Cuba	CUB	250	NAm (1–5) CAm (11–14) NAm
	Radio Liberty (BBG)	GRC	250	RUS (3–5)
	Radio Polonia	POL	100	Eu Ukrainian (14)
	Radio Singapore Int.	SNG	250	SEA (11–14)
6005	BBC World Service	ASC	250	SAf WAf (4–8)
	BBC World Service	ASC	250	WAf Special English (7)
	BBC World Service	SEY	250	EAf EAf Kiswahili SAf (17–22)
	BBC World Service	SEY	250	EAf Somali (17)
	BBC World Service	SEY	250	EAf (18–21)
	DeutschlandRadio	D	100	D (0–24)
	NHK, Sapporo	J	0.6	J J J
	SLBC	CLN	10	SAs (0–5) SAs (11–14)
6009	NHK, Nagoya	J	0.3	J (0–4)
6010	BBC World Service	G	300	2-6 Eu Turkish
	BBC World Service	G	300	Eu Special English (4)
	Belaruskaje Radyjo 1	BLR	5	BLR Belarusian/Russian (6–24)
	China Radio Int.	CHN	100	As As Cantonese (9–11)
	China Radio Int.	CHN	50	SEA Vietnamese (16)
	CNR 1	CHN	100	CHN (0–17) CHN
	Em. Ciudad de Montevideo	URG	0.15	URG (13–24)
	HCJB	EQA	100	SAm (9)
	HCJB	EQA	100	SAm German (Low) (10)
	La Voz de tu Conciencia	CLM	5.00	CLM (0–11)
	R. Inconfidência	B	5	B (0–24)
	R. Mil	MEX	1	MEX (0–24)
	R. Parinacota	CHL	1	CHL (0–24)
	Radio Bahrain	BHR	60	ME (0–24)

Legend: Arabic · Chinese · English · French · German · Indonesian · Japanese · Mandarin · Portuguese · Russian · Spanish · Other

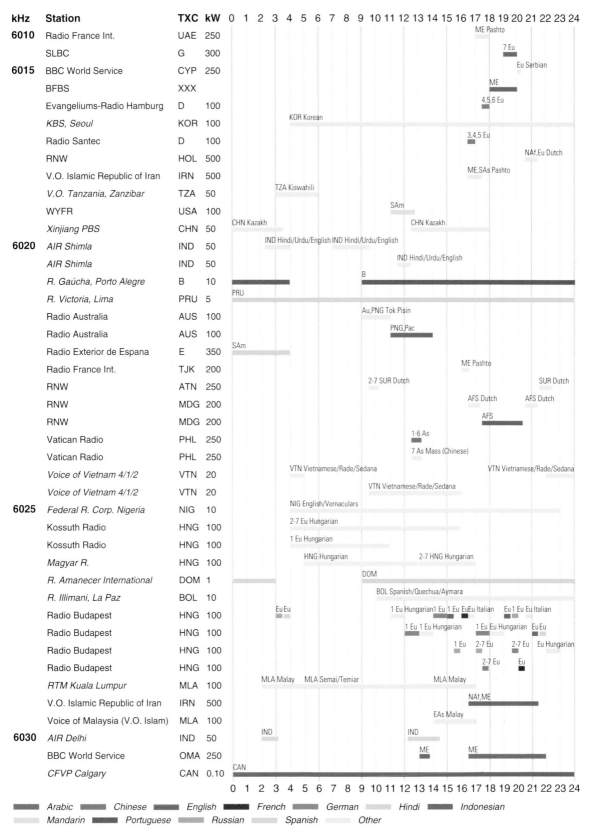

kHz	Station	TXC	kW
6010	Radio France Int.	UAE	250
	SLBC	G	300
6015	BBC World Service	CYP	250
	BFBS	XXX	
	Evangeliums-Radio Hamburg	D	100
	KBS, Seoul	KOR	100
	Radio Santec	D	100
	RNW	HOL	500
	V.O. Islamic Republic of Iran	IRN	500
	V.O. Tanzania, Zanzibar	TZA	50
	WYFR	USA	100
	Xinjiang PBS	CHN	50
6020	*AIR Shimla*	IND	50
	AIR Shimla	IND	50
	R. Gaúcha, Porto Alegre	B	10
	R. Victoria, Lima	PRU	5
	Radio Australia	AUS	100
	Radio Australia	AUS	100
	Radio Exterior de Espana	E	350
	Radio France Int.	TJK	200
	RNW	ATN	250
	RNW	MDG	200
	RNW	MDG	200
	Vatican Radio	PHL	250
	Vatican Radio	PHL	250
	Voice of Vietnam 4/1/2	VTN	20
	Voice of Vietnam 4/1/2	VTN	20
6025	*Federal R. Corp. Nigeria*	NIG	10
	Kossuth Radio	HNG	100
	Kossuth Radio	HNG	100
	Magyar R.	HNG	100
	R. Amanecer International	DOM	1
	R. Illimani, La Paz	BOL	10
	Radio Budapest	HNG	100
	Radio Budapest	HNG	100
	Radio Budapest	HNG	100
	Radio Budapest	HNG	100
	RTM Kuala Lumpur	MLA	100
	V.O. Islamic Republic of Iran	IRN	500
	Voice of Malaysia (V.O. Islam)	MLA	100
6030	*AIR Delhi*	IND	50
	BBC World Service	OMA	250
	CFVP Calgary	CAN	0.10

Arabic   Chinese   English   French   German   Hindi   Indonesian

Mandarin   Portuguese   Russian   Spanish   Other

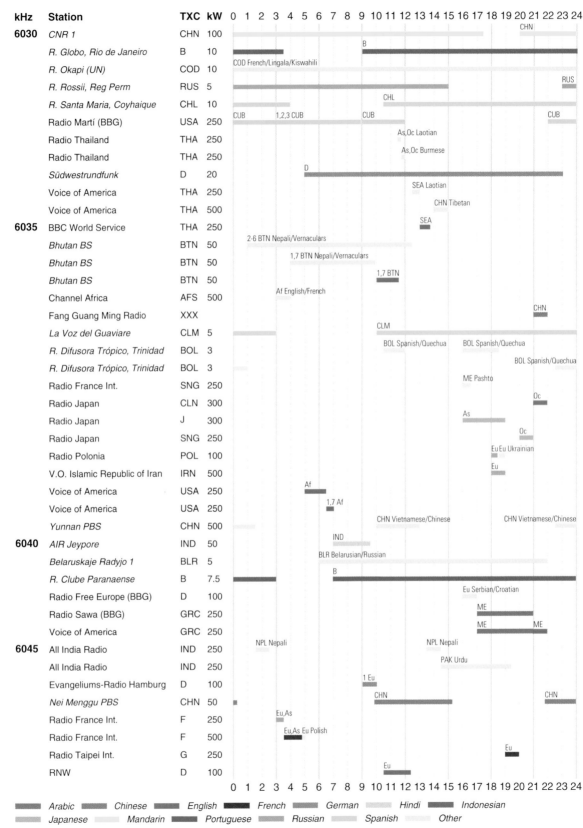

kHz	Station	TXC	kW
6030	CNR 1	CHN	100
	R. Globo, Rio de Janeiro	B	10
	R. Okapi (UN)	COD	10
	R. Rossii, Reg Perm	RUS	5
	R. Santa Maria, Coyhaique	CHL	10
	Radio Martí (BBG)	USA	250
	Radio Thailand	THA	250
	Radio Thailand	THA	250
	Südwestrundfunk	D	20
	Voice of America	THA	250
	Voice of America	THA	500
6035	BBC World Service	THA	250
	Bhutan BS	BTN	50
	Bhutan BS	BTN	50
	Bhutan BS	BTN	50
	Channel Africa	AFS	500
	Fang Guang Ming Radio	XXX	
	La Voz del Guaviare	CLM	5
	R. Difusora Trópico, Trinidad	BOL	3
	R. Difusora Trópico, Trinidad	BOL	3
	Radio France Int.	SNG	250
	Radio Japan	CLN	300
	Radio Japan	J	300
	Radio Japan	SNG	250
	Radio Polonia	POL	100
	V.O. Islamic Republic of Iran	IRN	500
	Voice of America	USA	250
	Voice of America	USA	250
	Yunnan PBS	CHN	500
6040	AIR Jeypore	IND	50
	Belaruskaje Radyjo 1	BLR	5
	R. Clube Paranaense	B	7.5
	Radio Free Europe (BBG)	D	100
	Radio Sawa (BBG)	GRC	250
	Voice of America	GRC	250
6045	All India Radio	IND	250
	All India Radio	IND	250
	Evangeliums-Radio Hamburg	D	100
	Nei Menggu PBS	CHN	50
	Radio France Int.	F	250
	Radio France Int.	F	500
	Radio Taipei Int.	G	250
	RNW	D	100

Labels on chart: CHN; B; COD French/Lingala/Kiswahili; RUS; CHL; CUB; 1,2,3 CUB; CUB; CUB; As,Oc Laotian; As,Oc Burmese; D; SEA Laotian; CHN Tibetan; SEA; 2-6 BTN Nepali/Vernaculars; 1,7 BTN Nepali/Vernaculars; 1,7 BTN; Af English/French; CHN; CLM; BOL Spanish/Quechua; BOL Spanish/Quechua; BOL Spanish/Quechua; ME Pashto; Oc; As; Oc; Eu Eu Ukrainian; Eu; Af; 1,7 Af; CHN Vietnamese/Chinese; CHN Vietnamese/Chinese; IND; BLR Belarusian/Russian; B; Eu Serbian/Croatian; ME; ME; ME; NPL Nepali; NPL Nepali; PAK Urdu; 1 Eu; CHN; CHN; Eu,As; Eu,As Eu Polish; Eu; Eu

Legend: Arabic · Chinese · English · French · German · Hindi · Indonesian · Japanese · Mandarin · Portuguese · Russian · Spanish · Other

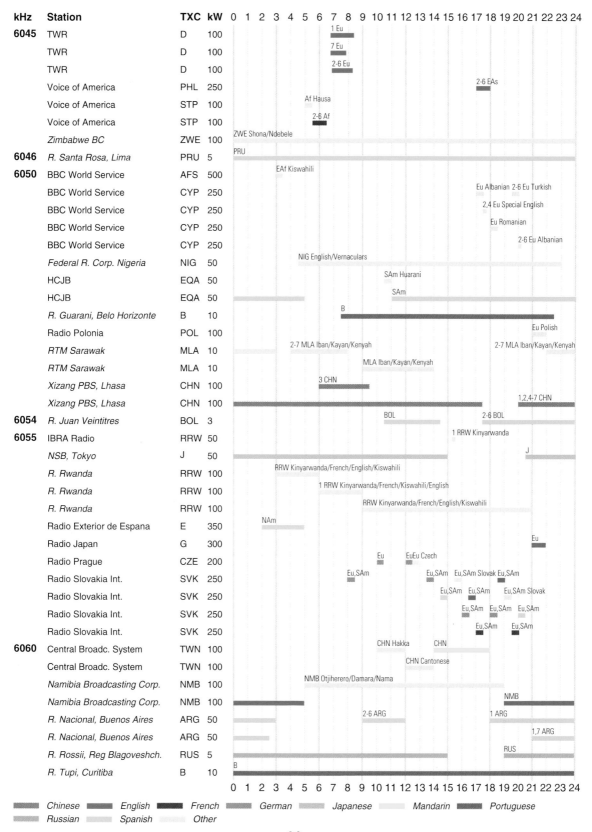

kHz	Station	TXC	kW
6045	TWR	D	100
	TWR	D	100
	TWR	D	100
	Voice of America	PHL	250
	Voice of America	STP	100
	Voice of America	STP	100
	Zimbabwe BC	ZWE	100
6046	R. Santa Rosa, Lima	PRU	5
6050	BBC World Service	AFS	500
	BBC World Service	CYP	250
	BBC World Service	CYP	250
	BBC World Service	CYP	250
	BBC World Service	CYP	250
	Federal R. Corp. Nigeria	NIG	50
	HCJB	EQA	50
	HCJB	EQA	50
	R. Guarani, Belo Horizonte	B	10
	Radio Polonia	POL	100
	RTM Sarawak	MLA	10
	RTM Sarawak	MLA	10
	Xizang PBS, Lhasa	CHN	100
	Xizang PBS, Lhasa	CHN	100
6054	R. Juan Veintitres	BOL	3
6055	IBRA Radio	RRW	50
	NSB, Tokyo	J	50
	R. Rwanda	RRW	100
	R. Rwanda	RRW	100
	R. Rwanda	RRW	100
	Radio Exterior de Espana	E	350
	Radio Japan	G	300
	Radio Prague	CZE	200
	Radio Slovakia Int.	SVK	250
	Radio Slovakia Int.	SVK	250
	Radio Slovakia Int.	SVK	250
	Radio Slovakia Int.	SVK	250
6060	Central Broadc. System	TWN	100
	Central Broadc. System	TWN	100
	Namibia Broadcasting Corp.	NMB	100
	Namibia Broadcasting Corp.	NMB	100
	R. Nacional, Buenos Aires	ARG	50
	R. Nacional, Buenos Aires	ARG	50
	R. Rossii, Reg Blagoveshch.	RUS	5
	R. Tupi, Curitiba	B	10

Programme annotations (by time, 0–24):

- 6045 TWR D 100: 1 Eu
- 6045 TWR D 100: 7 Eu
- 6045 TWR D 100: 2-6 Eu
- Voice of America PHL 250: 2-6 EAs
- Voice of America STP 100: Af Hausa
- Voice of America STP 100: 2-6 Af
- Zimbabwe BC ZWE 100: ZWE Shona/Ndebele
- 6046 R. Santa Rosa, Lima PRU 5: PRU
- BBC World Service AFS 500: EAf Kiswahili
- BBC World Service CYP 250: Eu Albanian 2-6 Eu Turkish
- BBC World Service CYP 250: 2,4 Eu Special English
- BBC World Service CYP 250: Eu Romanian
- BBC World Service CYP 250: 2-6 Eu Albanian
- Federal R. Corp. Nigeria NIG 50: NIG English/Vernaculars
- HCJB EQA 50: SAm Huarani
- HCJB EQA 50: SAm
- R. Guarani, Belo Horizonte B 10: B
- Radio Polonia POL 100: Eu Polish
- RTM Sarawak MLA 10: 2-7 MLA Iban/Kayan/Kenyah ... 2-7 MLA Iban/Kayan/Kenyah
- RTM Sarawak MLA 10: MLA Iban/Kayan/Kenyah
- Xizang PBS, Lhasa CHN 100: 3 CHN
- Xizang PBS, Lhasa CHN 100: 1,2,4-7 CHN
- R. Juan Veintitres BOL 3: BOL / 2-6 BOL
- IBRA Radio RRW 50: 1 RRW Kinyarwanda
- NSB, Tokyo J 50: J
- R. Rwanda RRW 100: RRW Kinyarwanda/French/English/Kiswahili
- R. Rwanda RRW 100: 1 RRW Kinyarwanda/French/Kiswahili/English
- R. Rwanda RRW 100: RRW Kinyarwanda/French/English/Kiswahili
- Radio Exterior de Espana E 350: NAm
- Radio Japan G 300: Eu
- Radio Prague CZE 200: Eu / EuEu Czech
- Radio Slovakia Int. SVK 250: Eu,SAm / Eu,SAm Eu,SAm Slovak Eu,SAm
- Radio Slovakia Int. SVK 250: Eu,SAm Eu,SAm Eu,SAm Slovak
- Radio Slovakia Int. SVK 250: Eu,SAm Eu,SAm Eu,SAm
- Radio Slovakia Int. SVK 250: Eu,SAm Eu,SAm
- Central Broadc. System TWN 100: CHN Hakka CHN
- Central Broadc. System TWN 100: CHN Cantonese
- Namibia Broadcasting Corp. NMB 100: NMB Otjiherero/Damara/Nama
- Namibia Broadcasting Corp. NMB 100: NMB
- R. Nacional, Buenos Aires ARG 50: 2-6 ARG / 1 ARG
- R. Nacional, Buenos Aires ARG 50: 1,7 ARG
- R. Rossii, Reg Blagoveshch. RUS 5: RUS
- R. Tupi, Curitiba B 10: B

Legend:

Chinese	English	French
German	Japanese	Mandarin	Portuguese
Russian	Spanish	Other

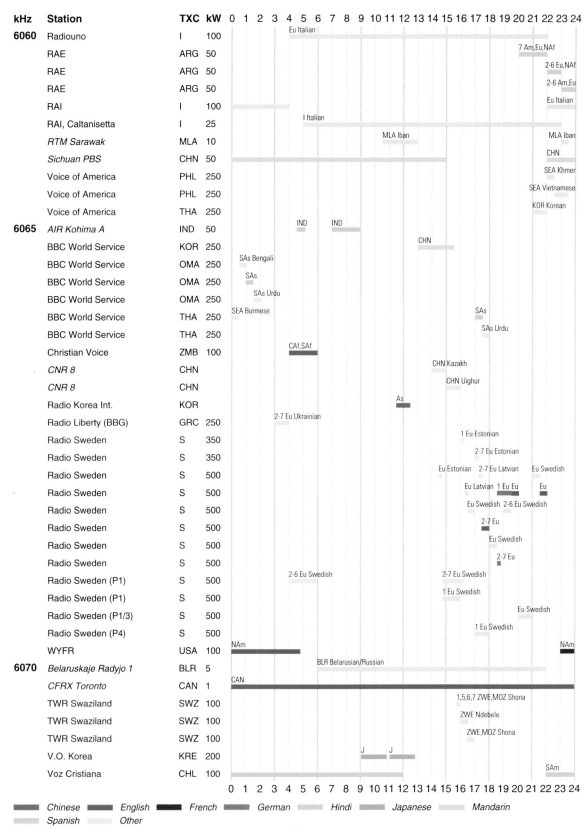

kHz	Station	TXC	kW	0 1 2 3 4 5 6 7 8 9 10 11 12 13 14 15 16 17 18 19 20 21 22 23 24
6060	Radiouno	I	100	Eu Italian
	RAE	ARG	50	7 Am,Eu,NAf
	RAE	ARG	50	2-6 Eu,NAf
	RAE	ARG	50	2-6 Am,Eu
	RAI	I	100	Eu Italian
	RAI, Caltanisetta	I	25	I Italian
	RTM Sarawak	MLA	10	MLA Iban MLA Iban
	Sichuan PBS	CHN	50	CHN
	Voice of America	PHL	250	SEA Khmer
	Voice of America	PHL	250	SEA Vietnamese
	Voice of America	THA	250	KOR Korean
6065	AIR Kohima A	IND	50	IND IND
	BBC World Service	KOR	250	CHN
	BBC World Service	OMA	250	SAs Bengali
	BBC World Service	OMA	250	SAs
	BBC World Service	OMA	250	SAs Urdu
	BBC World Service	THA	250	SEA Burmese SAs
	BBC World Service	THA	250	SAs Urdu
	Christian Voice	ZMB	100	CAf,SAf
	CNR 8	CHN		CHN Kazakh
	CNR 8	CHN		CHN Uighur
	Radio Korea Int.	KOR		As
	Radio Liberty (BBG)	GRC	250	2-7 Eu Ukrainian
	Radio Sweden	S	350	1 Eu Estonian
	Radio Sweden	S	350	2-7 Eu Estonian
	Radio Sweden	S	500	Eu Estonian 2-7 Eu Latvian Eu Swedish
	Radio Sweden	S	500	Eu Latvian 1 Eu Eu Eu
	Radio Sweden	S	500	Eu Swedish 2-6 Eu Swedish
	Radio Sweden	S	500	2-7 Eu
	Radio Sweden	S	500	Eu Swedish
	Radio Sweden	S	500	2-7 Eu
	Radio Sweden (P1)	S	500	2-6 Eu Swedish 2-7 Eu Swedish
	Radio Sweden (P1)	S	500	1 Eu Swedish
	Radio Sweden (P1/3)	S	500	Eu Swedish
	Radio Sweden (P4)	S	500	1 Eu Swedish
	WYFR	USA	100	NAm NAm
6070	Belaruskaje Radyjo 1	BLR	5	BLR Belarusian/Russian
	CFRX Toronto	CAN	1	CAN
	TWR Swaziland	SWZ	100	1,5,6,7 ZWE,MOZ Shona
	TWR Swaziland	SWZ	100	ZWE Ndebele
	TWR Swaziland	SWZ	100	ZWE,MOZ Shona
	V.O. Korea	KRE	200	J J
	Voz Cristiana	CHL	100	SAm

0 1 2 3 4 5 6 7 8 9 10 11 12 13 14 15 16 17 18 19 20 21 22 23 24

Chinese	English	French	German	Hindi	Japanese	Mandarin
Spanish	Other			

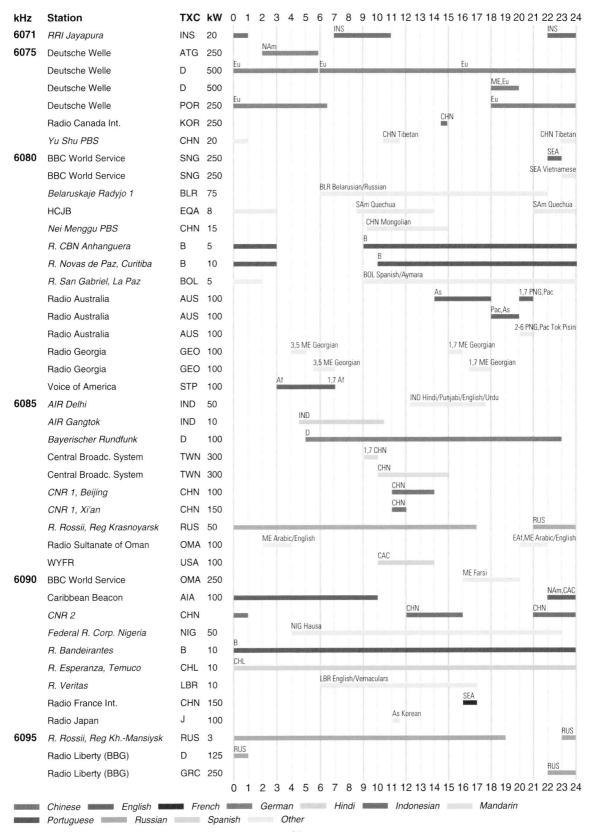

kHz	Station	TXC	kW
6071	*RRI Jayapura*	INS	20
6075	Deutsche Welle	ATG	250
	Deutsche Welle	D	500
	Deutsche Welle	D	500
	Deutsche Welle	POR	250
	Radio Canada Int.	KOR	250
	Yu Shu PBS	CHN	20
6080	BBC World Service	SNG	250
	BBC World Service	SNG	250
	Belaruskaje Radyjo 1	BLR	75
	HCJB	EQA	8
	Nei Menggu PBS	CHN	15
	R. CBN Anhanguera	B	5
	R. Novas de Paz, Curitiba	B	10
	R. San Gabriel, La Paz	BOL	5
	Radio Australia	AUS	100
	Radio Australia	AUS	100
	Radio Australia	AUS	100
	Radio Georgia	GEO	100
	Radio Georgia	GEO	100
	Voice of America	STP	100
6085	*AIR Delhi*	IND	50
	AIR Gangtok	IND	10
	Bayerischer Rundfunk	D	100
	Central Broadc. System	TWN	300
	Central Broadc. System	TWN	300
	CNR 1, Beijing	CHN	100
	CNR 1, Xi'an	CHN	150
	R. Rossii, Reg Krasnoyarsk	RUS	50
	Radio Sultanate of Oman	OMA	100
	WYFR	USA	100
6090	BBC World Service	OMA	250
	Caribbean Beacon	AIA	100
	CNR 2	CHN	
	Federal R. Corp. Nigeria	NIG	50
	R. Bandeirantes	B	10
	R. Esperanza, Temuco	CHL	10
	R. Veritas	LBR	10
	Radio France Int.	CHN	150
	Radio Japan	J	100
6095	*R. Rossii, Reg Kh.-Mansiysk*	RUS	3
	Radio Liberty (BBG)	D	125
	Radio Liberty (BBG)	GRC	250

Chinese   English   French   German   Hindi   Indonesian   Mandarin
Portuguese   Russian   Spanish   Other

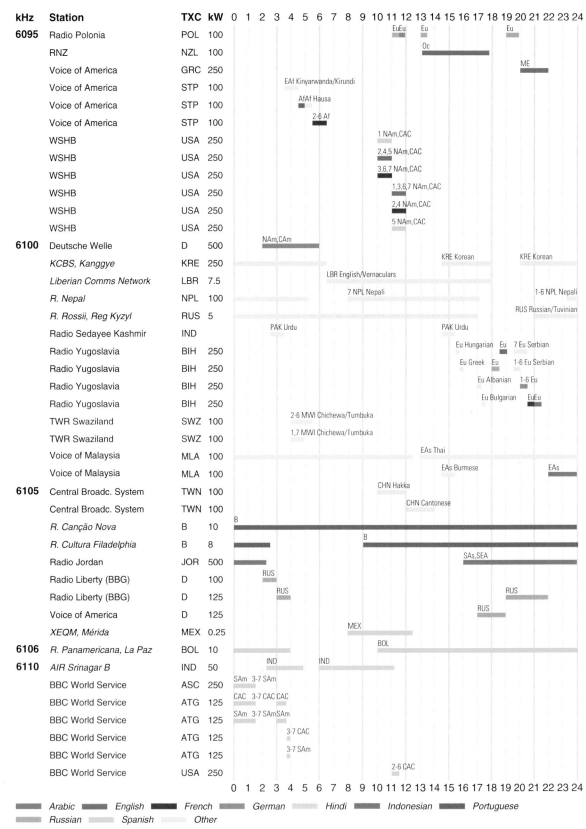

kHz	Station	TXC	kW	Time chart (0–24) annotations
6095	Radio Polonia	POL	100	EuEu Eu / Eu
	RNZ	NZL	100	Oc
	Voice of America	GRC	250	ME
	Voice of America	STP	100	EAf Kinyarwanda/Kirundi
	Voice of America	STP	100	AfAf Hausa
	Voice of America	STP	100	2-6 Af
	WSHB	USA	250	1 NAm,CAC
	WSHB	USA	250	2,4,5 NAm,CAC
	WSHB	USA	250	3,6,7 NAm,CAC
	WSHB	USA	250	1,3,6,7 NAm,CAC
	WSHB	USA	250	2,4 NAm,CAC
	WSHB	USA	250	5 NAm,CAC
6100	Deutsche Welle	D	500	NAm,CAm
	KCBS, Kanggye	KRE	250	KRE Korean / KRE Korean
	Liberian Comms Network	LBR	7.5	LBR English/Vernaculars
	R. Nepal	NPL	100	7 NPL Nepali / 1-6 NPL Nepali
	R. Rossii, Reg Kyzyl	RUS	5	RUS Russian/Tuvinian
	Radio Sedayee Kashmir	IND		PAK Urdu / PAK Urdu
	Radio Yugoslavia	BIH	250	Eu Hungarian Eu 7 Eu Serbian
	Radio Yugoslavia	BIH	250	Eu Greek Eu 1-6 Eu Serbian
	Radio Yugoslavia	BIH	250	Eu Albanian 1-6 Eu
	Radio Yugoslavia	BIH	250	Eu Bulgarian EuEu
	TWR Swaziland	SWZ	100	2-6 MWI Chichewa/Tumbuka
	TWR Swaziland	SWZ	100	1,7 MWI Chichewa/Tumbuka
	Voice of Malaysia	MLA	100	EAs Thai
	Voice of Malaysia	MLA	100	EAs Burmese EAs
6105	Central Broadc. System	TWN	100	CHN Hakka
	Central Broadc. System	TWN	100	CHN Cantonese
	R. Canção Nova	B	10	B
	R. Cultura Filadelphia	B	8	B
	Radio Jordan	JOR	500	SAs,SEA
	Radio Liberty (BBG)	D	100	RUS
	Radio Liberty (BBG)	D	125	RUS RUS
	Voice of America	D	125	RUS
	XEQM, Mérida	MEX	0.25	MEX
6106	R. Panamericana, La Paz	BOL	10	BOL
6110	AIR Srinagar B	IND	50	IND IND
	BBC World Service	ASC	250	SAm 3-7 SAm
	BBC World Service	ATG	125	CAC 3-7 CAC CAC
	BBC World Service	ATG	125	SAm 3-7 SAmSAm
	BBC World Service	ATG	125	3-7 CAC
	BBC World Service	ATG	125	3-7 SAm
	BBC World Service	USA	250	2-6 CAC

Arabic English French German Hindi Indonesian Portuguese
Russian Spanish Other

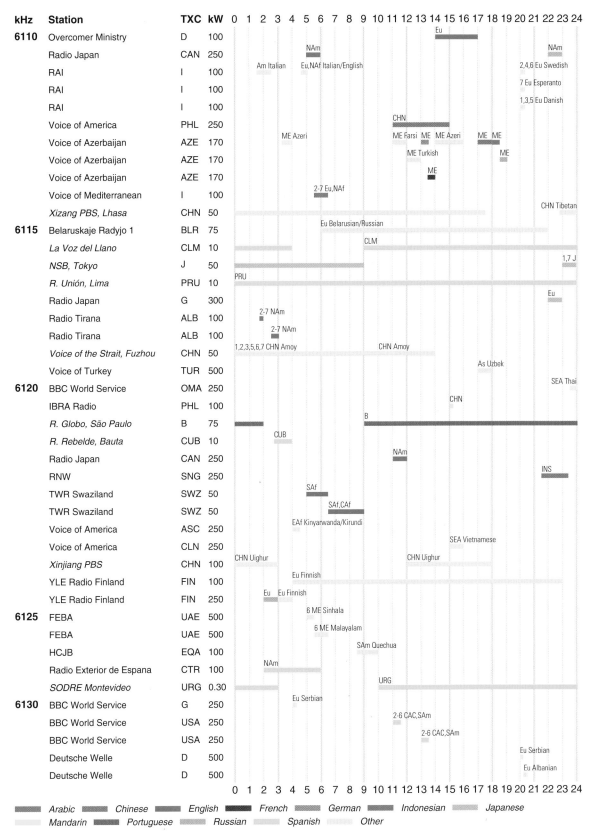

kHz	Station	TXC	kW
6110	Overcomer Ministry	D	100
	Radio Japan	CAN	250
	RAI	I	100
	RAI	I	100
	RAI	I	100
	Voice of America	PHL	250
	Voice of Azerbaijan	AZE	170
	Voice of Azerbaijan	AZE	170
	Voice of Azerbaijan	AZE	170
	Voice of Mediterranean	I	100
	Xizang PBS, Lhasa	CHN	50
6115	Belaruskaje Radyjo 1	BLR	75
	La Voz del Llano	CLM	10
	NSB, Tokyo	J	50
	R. Unión, Lima	PRU	10
	Radio Japan	G	300
	Radio Tirana	ALB	100
	Radio Tirana	ALB	100
	Voice of the Strait, Fuzhou	CHN	50
	Voice of Turkey	TUR	500
6120	BBC World Service	OMA	250
	IBRA Radio	PHL	100
	R. Globo, São Paulo	B	75
	R. Rebelde, Bauta	CUB	10
	Radio Japan	CAN	250
	RNW	SNG	250
	TWR Swaziland	SWZ	50
	TWR Swaziland	SWZ	50
	Voice of America	ASC	250
	Voice of America	CLN	250
	Xinjiang PBS	CHN	100
	YLE Radio Finland	FIN	100
	YLE Radio Finland	FIN	250
6125	FEBA	UAE	500
	FEBA	UAE	500
	HCJB	EQA	100
	Radio Exterior de Espana	CTR	100
	SODRE Montevideo	URG	0.30
6130	BBC World Service	G	250
	BBC World Service	USA	250
	BBC World Service	USA	250
	Deutsche Welle	D	500
	Deutsche Welle	D	500

Labels on chart: Eu; NAm; NAm; Am Italian; Eu,NAf Italian/English; 2,4,6 Eu Swedish; 7 Eu Esperanto; 1,3,5 Eu Danish; CHN; ME Azeri; ME Farsi; ME; ME Azeri; ME; ME; ME Turkish; ME; ME; 2-7 Eu,NAf; CHN Tibetan; Eu Belarusian/Russian; CLM; 1,7 J; PRU; Eu; 2-7 NAm; 2-7 NAm; 1,2,3,5,6,7 CHN Amoy; CHN Amoy; As Uzbek; SEA Thai; CHN; B; CUB; NAm; INS; SAf; SAf,CAf; EAf Kinyarwanda/Kirundi; SEA Vietnamese; CHN Uighur; CHN Uighur; Eu Finnish; Eu; Eu Finnish; 6 ME Sinhala; 6 ME Malayalam; SAm Quechua; NAm; URG; Eu Serbian; 2-6 CAC,SAm; 2-6 CAC,SAm; Eu Serbian; Eu Albanian

Legend: Arabic Chinese English French German Indonesian Japanese Mandarin Portuguese Russian Spanish Other

kHz	Station	TXC	kW	Time / Notes
6130	*Ghana Broadcasting Corp.*	GHA	35	GHA
	R. Nationale Lao	LAO	25	LAO Lao/H'mong/Khmu · LAO Lao · LAO Khmu/Lao
	Radio Free Europe (BBG)	D	100	Eu Serbian/Croatian
	Radio Free Europe (BBG)	D	100	Eu Serbian/Croatian
	Radio Santec	XXX		6 Eu,Af
	Radio Tirana	ALB	100	2-7 ME Turkish
	Radio Tirana	ALB	100	2-7 Eu Greek
	RAI	I	100	Eu Czech Eu,NAf
	RAI	I	100	Eu Slovak Eu,NAf
	RAI	I	100	Eu Polish Eu,NAf
	RAI	I	100	Eu Serbian Eu,NAf
	RAI	I	100	Eu Hungarian
	TWR Swaziland	SWZ	100	Af Chokwe
	TWR Swaziland	SWZ	100	AGL Umbundu
	TWR Swaziland	SWZ	100	1 AGL Luvale
	TWR Swaziland	SWZ	100	2-7 AGL Kikongo
	TWR Swaziland	SWZ	100	7 Af Lunyaneka
	TWR Swaziland	SWZ	100	2,3,5 Af
	TWR Swaziland	SWZ	100	1 Af Kuanyama
	TWR Swaziland	SWZ	100	4,6 Af Luchazi
	TWR Swaziland	SWZ	100	AGL
	TWR Swaziland	SWZ	100	AGL Kimbundu
	Voice of America	D	100	Eu Croatian
	Voice of America	USA	250	3-7 CAC
	Xizang PBS, Lhasa	CHN	150	CHN Tibetan
6135	BBC World Service	AFS	250	SAf SAf
	BBC World Service	THA	250	FE
	KBS, Hwasong	KOR	10	KOR Korean · KOR Korean
	R. Aparecida	B	25	B
	R. Madigasikara	MDG	30	MDG Malagasy/French
	R. Santa Cruz	BOL	10	BOL
	Radio Japan	ASC	250	Af Kiswahili
	Radio Korea Int.	KOR	10	RUS J As
	Radio Tirana	ALB	100	2-7 Eu Serbian
	TWR Swaziland	SWZ	100	MWI Chichewa/Tumbuka
	TWR Swaziland	SWZ	100	7 ZMB Bemba
	TWR Swaziland	SWZ	50	2-6 ZMB Chichewa
6139	*R. UNAMSIL (UN)*	SRL	10	SRL English/Vernaculars
6140	BBC World Service	OMA	250	SAs SAs Sinhala
	BBC World Service	OMA	250	1 SAs
	BBC World Service	OMA	250	2-7 SAs Special English
	BBC World Service	OMA	250	SAs Tamil
	Central Broadc. System	TWN	300	1,7 CHN

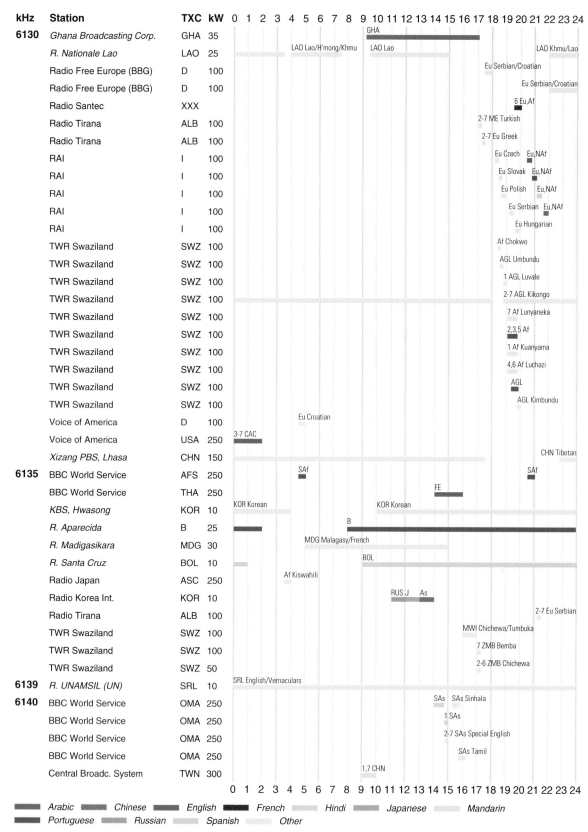

Legend: Arabic · Chinese · English · French · Hindi · Japanese · Mandarin · Portuguese · Russian · Spanish · Other

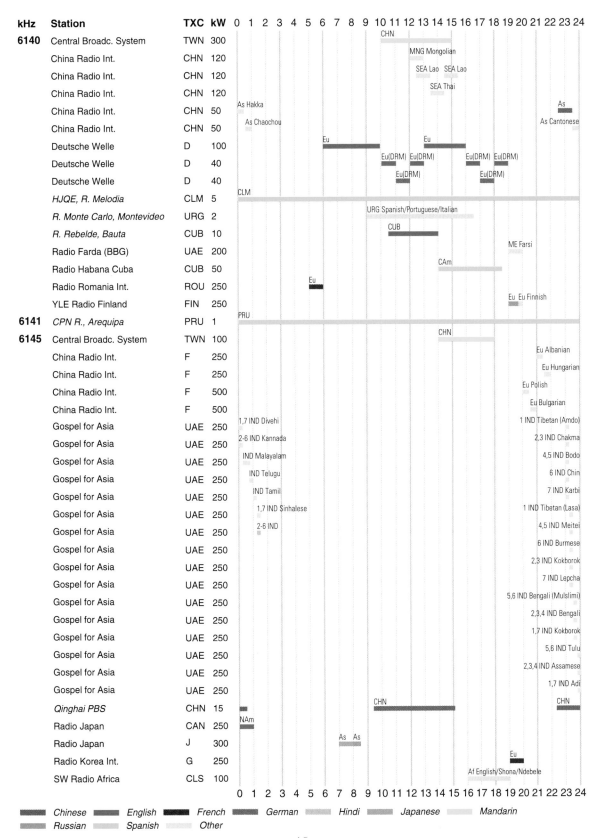

kHz	Station	TXC	kW	Schedule
6140	Central Broadc. System	TWN	300	CHN (10-11)
	China Radio Int.	CHN	120	MNG Mongolian (12-13)
	China Radio Int.	CHN	120	SEA Lao (13-14) SEA Lao (14-15)
	China Radio Int.	CHN	120	SEA Thai (14-15)
	China Radio Int.	CHN	50	As Hakka (0); As (22-23)
	China Radio Int.	CHN	50	As Chaochou (0-1); As Cantonese (23-24)
	Deutsche Welle	D	100	Eu (6-10); Eu (13-15)
	Deutsche Welle	D	40	Eu(DRM) (10-11) Eu(DRM) (12) Eu(DRM) (16-17) Eu(DRM) (18)
	Deutsche Welle	D	40	Eu(DRM) (11-12); Eu(DRM) (17-18)
	HJQE, R. Melodia	CLM	5	CLM (0-24)
	R. Monte Carlo, Montevideo	URG	2	URG Spanish/Portuguese/Italian (10-15)
	R. Rebelde, Bauta	CUB	10	CUB (10-13)
	Radio Farda (BBG)	UAE	200	ME Farsi (18)
	Radio Habana Cuba	CUB	50	CAm (13-18)
	Radio Romania Int.	ROU	250	Eu (5-6)
	YLE Radio Finland	FIN	250	Eu Eu Finnish (18-19)
6141	*CPN R., Arequipa*	PRU	1	PRU (0-24)
6145	Central Broadc. System	TWN	100	CHN (12-16)
	China Radio Int.	F	250	Eu Albanian (21)
	China Radio Int.	F	250	Eu Hungarian (21)
	China Radio Int.	F	500	Eu Polish (20)
	China Radio Int.	F	500	Eu Bulgarian (21)
	Gospel for Asia	UAE	250	1,7 IND Divehi (0); 1 IND Tibetan (Amdo) (22)
	Gospel for Asia	UAE	250	2-6 IND Kannada (0); 2,3 IND Chakma (23)
	Gospel for Asia	UAE	250	IND Malayalam (1); 4,5 IND Bodo (23)
	Gospel for Asia	UAE	250	IND Telugu (1); 6 IND Chin (23)
	Gospel for Asia	UAE	250	IND Tamil (1); 7 IND Karbi (23)
	Gospel for Asia	UAE	250	1,7 IND Sinhalese (2); 1 IND Tibetan (Lasa) (22)
	Gospel for Asia	UAE	250	2-6 IND (2); 4,5 IND Meitei (23)
	Gospel for Asia	UAE	250	6 IND Burmese (23)
	Gospel for Asia	UAE	250	2,3 IND Kokborok (23)
	Gospel for Asia	UAE	250	7 IND Lepcha (23)
	Gospel for Asia	UAE	250	5,6 IND Bengali (Mulslimi) (22)
	Gospel for Asia	UAE	250	2,3,4 IND Bengali (23)
	Gospel for Asia	UAE	250	1,7 IND Kokborok (22)
	Gospel for Asia	UAE	250	5,6 IND Tulu (22)
	Gospel for Asia	UAE	250	2,3,4 IND Assamese (23)
	Gospel for Asia	UAE	250	1,7 IND Adi (22)
	Qinghai PBS	CHN	15	CHN (0); CHN (10-15); CHN (22-23)
	Radio Japan	CAN	250	NAm (0-1)
	Radio Japan	J	300	As As (7-8)
	Radio Korea Int.	G	250	Eu (19)
	SW Radio Africa	CLS	100	Af English/Shona/Ndebele (16-19)

Legend:
■ Chinese   ■ English   ■ French   ▨ German   ▨ Hindi   ▨ Japanese   ▨ Mandarin
▨ Russian   ▨ Spanish   ▨ Other

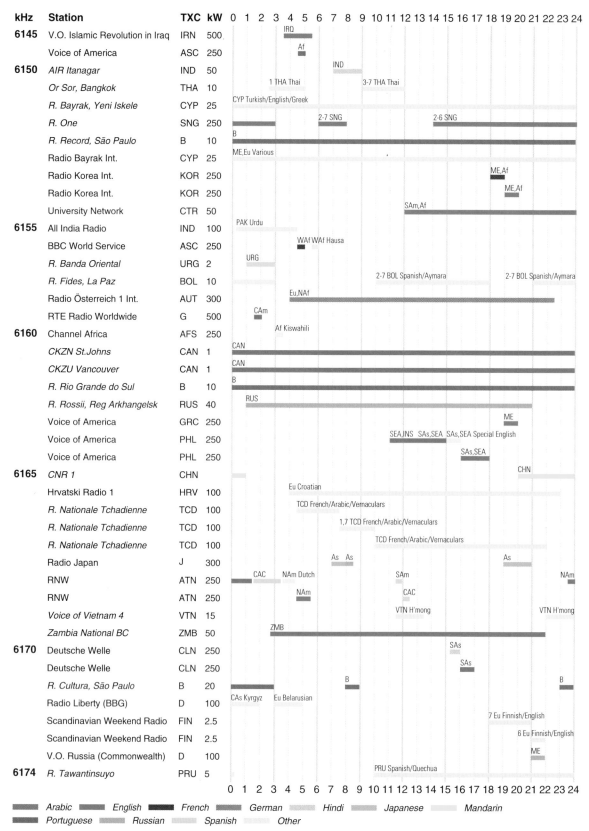

kHz	Station	TXC	kW
6145	V.O. Islamic Revolution in Iraq	IRN	500
	Voice of America	ASC	250
6150	AIR Itanagar	IND	50
	Or Sor, Bangkok	THA	10
	R. Bayrak, Yeni Iskele	CYP	25
	R. One	SNG	250
	R. Record, São Paulo	B	10
	Radio Bayrak Int.	CYP	25
	Radio Korea Int.	KOR	250
	Radio Korea Int.	KOR	250
	University Network	CTR	50
6155	All India Radio	IND	100
	BBC World Service	ASC	250
	R. Banda Oriental	URG	2
	R. Fides, La Paz	BOL	10
	Radio Österreich 1 Int.	AUT	300
	RTE Radio Worldwide	G	500
6160	Channel Africa	AFS	250
	CKZN St.Johns	CAN	1
	CKZU Vancouver	CAN	1
	R. Rio Grande do Sul	B	10
	R. Rossii, Reg Arkhangelsk	RUS	40
	Voice of America	GRC	250
	Voice of America	PHL	250
	Voice of America	PHL	250
6165	CNR 1	CHN	
	Hrvatski Radio 1	HRV	100
	R. Nationale Tchadienne	TCD	100
	R. Nationale Tchadienne	TCD	100
	R. Nationale Tchadienne	TCD	100
	Radio Japan	J	300
	RNW	ATN	250
	RNW	ATN	250
	Voice of Vietnam 4	VTN	15
	Zambia National BC	ZMB	50
6170	Deutsche Welle	CLN	250
	Deutsche Welle	CLN	250
	R. Cultura, São Paulo	B	20
	Radio Liberty (BBG)	D	100
	Scandinavian Weekend Radio	FIN	2.5
	Scandinavian Weekend Radio	FIN	2.5
	V.O. Russia (Commonwealth)	D	100
6174	R. Tawantinsuyo	PRU	5

Legend: Arabic · English · French · German · Hindi · Japanese · Mandarin · Portuguese · Russian · Spanish · Other

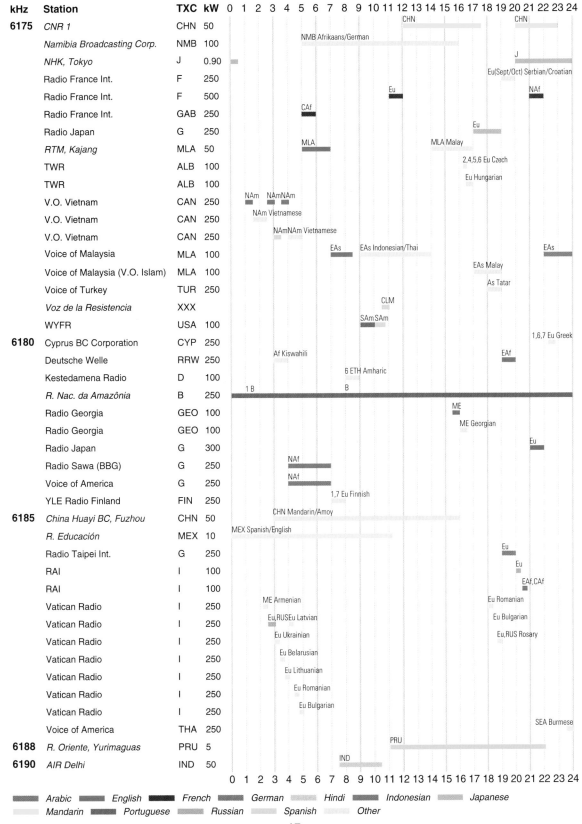

kHz	Station	TXC	kW
6175	*CNR 1*	CHN	50
	Namibia Broadcasting Corp.	NMB	100
	NHK, Tokyo	J	0.90
	Radio France Int.	F	250
	Radio France Int.	F	500
	Radio France Int.	GAB	250
	Radio Japan	G	250
	RTM, Kajang	MLA	50
	TWR	ALB	100
	TWR	ALB	100
	V.O. Vietnam	CAN	250
	V.O. Vietnam	CAN	250
	V.O. Vietnam	CAN	250
	Voice of Malaysia	MLA	100
	Voice of Malaysia (V.O. Islam)	MLA	100
	Voice of Turkey	TUR	250
	Voz de la Resistencia	XXX	
	WYFR	USA	100
6180	Cyprus BC Corporation	CYP	250
	Deutsche Welle	RRW	250
	Kestedamena Radio	D	100
	R. Nac. da Amazônia	B	250
	Radio Georgia	GEO	100
	Radio Georgia	GEO	100
	Radio Japan	G	300
	Radio Sawa (BBG)	G	250
	Voice of America	G	250
	YLE Radio Finland	FIN	250
6185	*China Huayi BC, Fuzhou*	CHN	50
	R. Educación	MEX	10
	Radio Taipei Int.	G	250
	RAI	I	100
	RAI	I	100
	Vatican Radio	I	250
	Vatican Radio	I	250
	Vatican Radio	I	250
	Vatican Radio	I	250
	Vatican Radio	I	250
	Vatican Radio	I	250
	Vatican Radio	I	250
	Voice of America	THA	250
6188	*R. Oriente, Yurimaguas*	PRU	5
6190	*AIR Delhi*	IND	50

Legend: Arabic, English, French, German, Hindi, Indonesian, Japanese, Mandarin, Portuguese, Russian, Spanish, Other

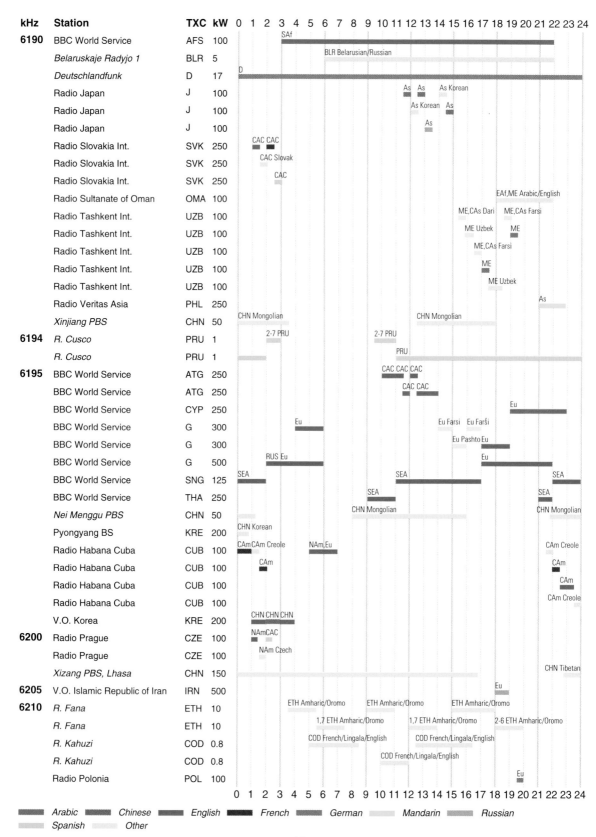

kHz	Station	TXC	kW	Schedule (0–24)
6190	BBC World Service	AFS	100	SAf (3–22)
	Belaruskaje Radyjo 1	BLR	5	BLR Belarusian/Russian (6–22)
	Deutschlandfunk	D	17	D (0–24)
	Radio Japan	J	100	As / As / As Korean (11–15)
	Radio Japan	J	100	As Korean / As (12–15)
	Radio Japan	J	100	As (13–14)
	Radio Slovakia Int.	SVK	250	CAC CAC (1–2)
	Radio Slovakia Int.	SVK	250	CAC Slovak (1–2)
	Radio Slovakia Int.	SVK	250	CAC (2–3)
	Radio Sultanate of Oman	OMA	100	EAf,ME Arabic/English (17–22)
	Radio Tashkent Int.	UZB	100	ME,CAs Dari / ME,CAs Farsi (14–18)
	Radio Tashkent Int.	UZB	100	ME Uzbek / ME (14–18)
	Radio Tashkent Int.	UZB	100	ME,CAs Farsi (14–15)
	Radio Tashkent Int.	UZB	100	ME (15–16)
	Radio Tashkent Int.	UZB	100	ME Uzbek (15–16)
	Radio Veritas Asia	PHL	250	As (21–23)
	Xinjiang PBS	CHN	50	CHN Mongolian (0–2 / 11–15)
6194	*R. Cusco*	PRU	1	2-7 PRU (2–3 / 10–11)
	R. Cusco	PRU	1	PRU (0–2 / 11–12)
6195	BBC World Service	ATG	250	CAC CAC CAC (11–13)
	BBC World Service	ATG	250	CAC CAC (12–13)
	BBC World Service	CYP	250	Eu (18–22)
	BBC World Service	G	300	Eu (4–6)
	BBC World Service	G	300	Eu Farsi / Eu Farši (13–16)
	BBC World Service	G	300	Eu Pashto Eu (16–19)
	BBC World Service	G	500	RUS Eu (2–6) / Eu (16–22)
	BBC World Service	SNG	125	SEA (0–1 / 9–15 / 18–22)
	BBC World Service	THA	250	SEA (9–10 / 18–19)
	Nei Menggu PBS	CHN	50	CHN Mongolian (9–12 / 18–24)
	Pyongyang BS	KRE	200	CHN Korean (0–1)
	Radio Habana Cuba	CUB	100	CAm CAm Creole (0–2) / NAm,Eu (5–7) / CAm Creole (21–23)
	Radio Habana Cuba	CUB	100	CAm (2–3) / CAm (21–22)
	Radio Habana Cuba	CUB	100	CAm (21–23)
	Radio Habana Cuba	CUB	100	CAm Creole (22–24)
	V.O. Korea	KRE	200	CHN CHN CHN (0–3)
6200	Radio Prague	CZE	100	NAmCAC (0–1)
	Radio Prague	CZE	100	NAm Czech (1–2)
	Xizang PBS, Lhasa	CHN	150	CHN Tibetan (0–16 / 22–24)
6205	V.O. Islamic Republic of Iran	IRN	500	Eu (18–20)
6210	*R. Fana*	ETH	10	ETH Amharic/Oromo (4–17)
	R. Fana	ETH	10	1,7 ETH Amharic/Oromo (5–18)
	R. Kahuzi	COD	0.8	COD French/Lingala/English (4–13)
	R. Kahuzi	COD	0.8	COD French/Lingala/English (9–12)
	Radio Polonia	POL	100	Eu (18–20)

Legend:
- Arabic
- Chinese
- English
- French
- German
- Mandarin
- Russian
- Spanish
- Other

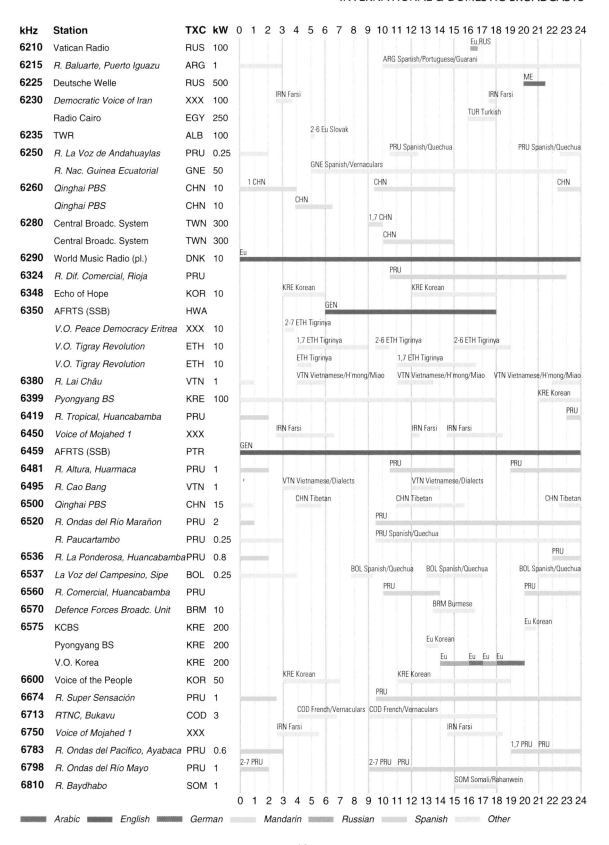

kHz	Station	TXC	kW	Schedule (0–24)
6210	Vatican Radio	RUS	100	Eu,RUS
6215	*R. Baluarte, Puerto Iguazu*	ARG	1	ARG Spanish/Portuguese/Guarani
6225	Deutsche Welle	RUS	500	ME
6230	*Democratic Voice of Iran*	XXX	100	IRN Farsi / IRN Farsi
	Radio Cairo	EGY	250	TUR Turkish
6235	TWR	ALB	100	2-6 Eu Slovak
6250	*R. La Voz de Andahuaylas*	PRU	0.25	PRU Spanish/Quechua / PRU Spanish/Quechua
	R. Nac. Guinea Ecuatorial	GNE	50	GNE Spanish/Vernaculars
6260	*Qinghai PBS*	CHN	10	1 CHN / CHN / CHN
	Qinghai PBS	CHN	10	CHN
6280	Central Broadc. System	TWN	300	1,7 CHN
	Central Broadc. System	TWN	300	CHN
6290	World Music Radio (pl.)	DNK	10	Eu
6324	*R. Dif. Comercial, Rioja*	PRU		PRU
6348	Echo of Hope	KOR	10	KRE Korean / KRE Korean
6350	AFRTS (SSB)	HWA		GEN
	V.O. Peace Democracy Eritrea	XXX	10	2-7 ETH Tigrinya
	V.O. Tigray Revolution	ETH	10	1,7 ETH Tigrinya / 2-6 ETH Tigrinya / 2-6 ETH Tigrinya
	V.O. Tigray Revolution	ETH	10	ETH Tigrinya / 1,7 ETH Tigrinya
6380	*R. Lai Châu*	VTN	1	VTN Vietnamese/H'mong/Miao / VTN Vietnamese/H'mong/Miao / VTN Vietnamese/H'mong/Miao
6399	*Pyongyang BS*	KRE	100	KRE Korean
6419	*R. Tropical, Huancabamba*	PRU		PRU
6450	*Voice of Mojahed 1*	XXX		IRN Farsi / IRN Farsi IRN Farsi
6459	AFRTS (SSB)	PTR		GEN
6481	*R. Altura, Huarmaca*	PRU	1	PRU / PRU
6495	*R. Cao Bang*	VTN	1	VTN Vietnamese/Dialects / VTN Vietnamese/Dialects
6500	*Qinghai PBS*	CHN	15	CHN Tibetan / CHN Tibetan / CHN Tibetan
6520	*R. Ondas del Río Marañon*	PRU	2	PRU
	R. Paucartambo	PRU	0.25	PRU Spanish/Quechua
6536	*R. La Ponderosa, Huancabamba*	PRU	0.8	PRU
6537	*La Voz del Campesino, Sipe*	BOL	0.25	BOL Spanish/Quechua / BOL Spanish/Quechua / BOL Spanish/Quechua
6560	*R. Comercial, Huancabamba*	PRU		PRU / PRU
6570	*Defence Forces Broadc. Unit*	BRM	10	BRM Burmese
6575	KCBS	KRE	200	Eu Korean
	Pyongyang BS	KRE	200	Eu Korean
	V.O. Korea	KRE	200	Eu Eu Eu Eu
6600	Voice of the People	KOR	50	KRE Korean / KRE Korean
6674	*R. Super Sensación*	PRU	1	PRU
6713	RTNC, Bukavu	COD	3	COD French/Vernaculars COD French/Vernaculars
6750	*Voice of Mojahed 1*	XXX		IRN Farsi / IRN Farsi
6783	*R. Ondas del Pacifico, Ayabaca*	PRU	0.6	1,7 PRU PRU
6798	*R. Ondas del Río Mayo*	PRU	1	2-7 PRU / 2-7 PRU PRU
6810	*R. Baydhabo*	SOM	1	SOM Somali/Rahanwein

Legend: ▨ Arabic ▨ English ▨ German ▨ Mandarin ▨ Russian ▨ Spanish ▨ Other

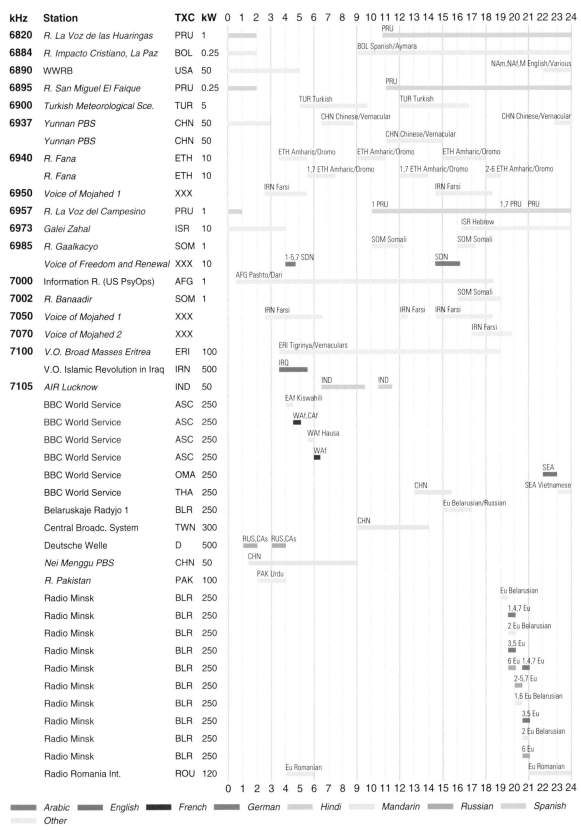

kHz	Station	TXC	kW	Schedule (0–24)
6820	*R. La Voz de las Huaringas*	PRU	1	PRU
6884	*R. Impacto Cristiano, La Paz*	BOL	0.25	BOL Spanish/Aymara
6890	*WWRB*	USA	50	NAm,NAf,M English/Various
6895	*R. San Miguel El Faique*	PRU	0.25	PRU
6900	*Turkish Meteorological Sce.*	TUR	5	TUR Turkish / TUR Turkish
6937	*Yunnan PBS*	CHN	50	CHN Chinese/Vernacular / CHN Chinese/Vernacular
	Yunnan PBS	CHN	50	CHN Chinese/Vernacular
6940	*R. Fana*	ETH	10	ETH Amharic/Oromo / ETH Amharic/Oromo / ETH Amharic/Oromo
	R. Fana	ETH	10	1,7 ETH Amharic/Oromo / 1,7 ETH Amharic/Oromo / 2-6 ETH Amharic/Oromo
6950	*Voice of Mojahed 1*	XXX		IRN Farsi / IRN Farsi
6957	*R. La Voz del Campesino*	PRU	1	1 PRU / 1,7 PRU PRU
6973	*Galei Zahal*	ISR	10	ISR Hebrew
6985	*R. Gaalkacyo*	SOM	1	SOM Somali / SOM Somali
	Voice of Freedom and Renewal	XXX	10	1-5,7 SDN / SDN
7000	Information R. (US PsyOps)	AFG	1	AFG Pashto/Dari
7002	*R. Banaadir*	SOM	1	SOM Somali
7050	*Voice of Mojahed 1*	XXX		IRN Farsi / IRN Farsi IRN Farsi
7070	*Voice of Mojahed 2*	XXX		IRN Farsi
7100	*V.O. Broad Masses Eritrea*	ERI	100	ERI Tigrinya/Vernaculars
	V.O. Islamic Revolution in Iraq	IRN	500	IRQ
7105	*AIR Lucknow*	IND	50	IND / IND
	BBC World Service	ASC	250	EAf Kiswahili
	BBC World Service	ASC	250	WAf,CAf
	BBC World Service	ASC	250	WAf Hausa
	BBC World Service	ASC	250	WAf
	BBC World Service	OMA	250	SEA
	BBC World Service	THA	250	CHN / SEA Vietnamese
	Belaruskaje Radyjo 1	BLR	250	Eu Belarusian/Russian
	Central Broadc. System	TWN	300	CHN
	Deutsche Welle	D	500	RUS,CAs RUS,CAs
	Nei Menggu PBS	CHN	50	CHN
	R. Pakistan	PAK	100	PAK Urdu
	Radio Minsk	BLR	250	Eu Belarusian
	Radio Minsk	BLR	250	1,4,7 Eu
	Radio Minsk	BLR	250	2 Eu Belarusian
	Radio Minsk	BLR	250	3,5 Eu
	Radio Minsk	BLR	250	6 Eu 1,4,7 Eu
	Radio Minsk	BLR	250	2-5,7 Eu
	Radio Minsk	BLR	250	1,6 Eu Belarusian
	Radio Minsk	BLR	250	3,5 Eu
	Radio Minsk	BLR	250	2 Eu Belarusian
	Radio Minsk	BLR	250	6 Eu
	Radio Romania Int.	ROU	120	Eu Romanian / Eu Romanian

Legend: Arabic — English — French — German — Hindi — Mandarin — Russian — Spanish — Other

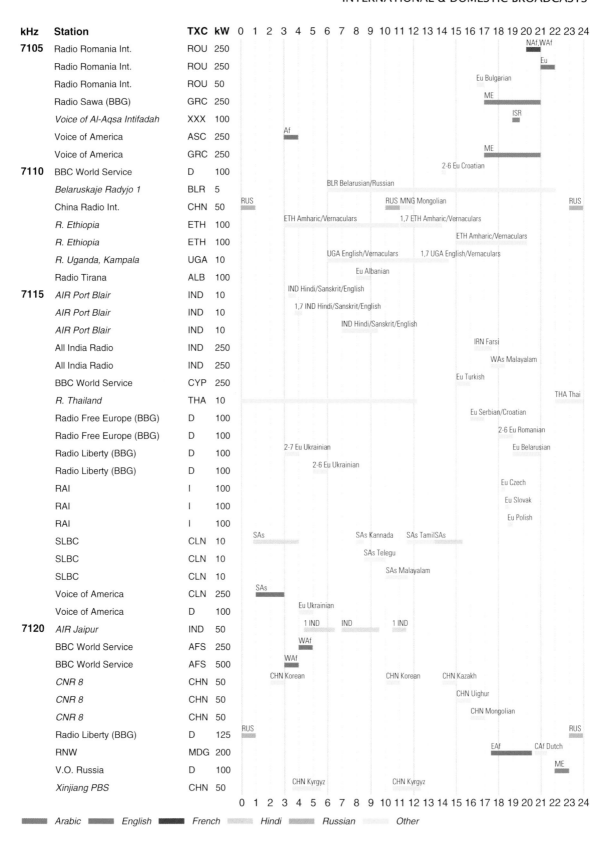

kHz	Station	TXC	kW	Broadcast schedule
7105	Radio Romania Int.	ROU	250	NAf,WAf
	Radio Romania Int.	ROU	250	Eu
	Radio Romania Int.	ROU	50	Eu Bulgarian
	Radio Sawa (BBG)	GRC	250	ME
	Voice of Al-Aqsa Intifadah	XXX	100	ISR
	Voice of America	ASC	250	Af
	Voice of America	GRC	250	ME
7110	BBC World Service	D	100	2-6 Eu Croatian
	Belaruskaje Radyjo 1	BLR	5	BLR Belarusian/Russian
	China Radio Int.	CHN	50	RUS / RUS MNG Mongolian / RUS
	R. Ethiopia	ETH	100	ETH Amharic/Vernaculars / 1,7 ETH Amharic/Vernaculars
	R. Ethiopia	ETH	100	ETH Amharic/Vernaculars
	R. Uganda, Kampala	UGA	10	UGA English/Vernaculars / 1,7 UGA English/Vernaculars
	Radio Tirana	ALB	100	Eu Albanian
7115	*AIR Port Blair*	IND	10	IND Hindi/Sanskrit/English
	AIR Port Blair	IND	10	1,7 IND Hindi/Sanskrit/English
	AIR Port Blair	IND	10	IND Hindi/Sanskrit/English
	All India Radio	IND	250	IRN Farsi
	All India Radio	IND	250	WAs Malayalam
	BBC World Service	CYP	250	Eu Turkish
	R. Thailand	THA	10	THA Thai
	Radio Free Europe (BBG)	D	100	Eu Serbian/Croatian
	Radio Free Europe (BBG)	D	100	2-6 Eu Romanian
	Radio Liberty (BBG)	D	100	2-7 Eu Ukrainian / Eu Belarusian
	Radio Liberty (BBG)	D	100	2-6 Eu Ukrainian
	RAI	I	100	Eu Czech
	RAI	I	100	Eu Slovak
	RAI	I	100	Eu Polish
	SLBC	CLN	10	SAs / SAs Kannada / SAs TamilSAs
	SLBC	CLN	10	SAs Telegu
	SLBC	CLN	10	SAs Malayalam
	Voice of America	CLN	250	SAs
	Voice of America	D	100	Eu Ukrainian
7120	*AIR Jaipur*	IND	50	1 IND / IND / 1 IND
	BBC World Service	AFS	250	WAf
	BBC World Service	AFS	500	WAf
	CNR 8	CHN	50	CHN Korean / CHN Korean / CHN Kazakh
	CNR 8	CHN	50	CHN Uighur
	CNR 8	CHN	50	CHN Mongolian
	Radio Liberty (BBG)	D	125	RUS / RUS
	RNW	MDG	200	EAf / CAf Dutch
	V.O. Russia	D	100	ME
	Xinjiang PBS	CHN	50	CHN Kyrgyz / CHN Kyrgyz

Arabic   English   French   Hindi   Russian   Other

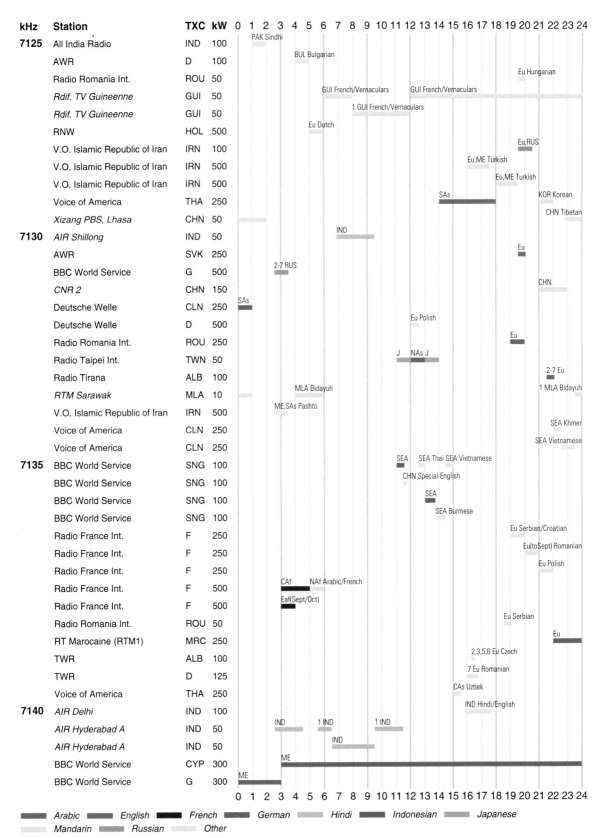

kHz	Station	TXC	kW	Schedule (0–24h)
7125	All India Radio	IND	100	PAK Sindhi
	AWR	D	100	BUL Bulgarian
	Radio Romania Int.	ROU	50	Eu Hungarian
	Rdif. TV Guineenne	GUI	50	GUI French/Vernaculars · GUI French/Vernaculars
	Rdif. TV Guineenne	GUI	50	1 GUI French/Vernaculars
	RNW	HOL	500	Eu Dutch
	V.O. Islamic Republic of Iran	IRN	100	Eu,RUS
	V.O. Islamic Republic of Iran	IRN	500	Eu,ME Turkish
	V.O. Islamic Republic of Iran	IRN	500	Eu,ME Turkish
	Voice of America	THA	250	SAs · KOR Korean
	Xizang PBS, Lhasa	CHN	50	CHN Tibetan
7130	*AIR Shillong*	IND	50	IND
	AWR	SVK	250	Eu
	BBC World Service	G	500	2-7 RUS
	CNR 2	CHN	150	CHN
	Deutsche Welle	CLN	250	SAs
	Deutsche Welle	D	500	Eu Polish
	Radio Romania Int.	ROU	250	Eu
	Radio Taipei Int.	TWN	50	J · NAs J
	Radio Tirana	ALB	100	2-7 Eu
	RTM Sarawak	MLA	10	MLA Bidayuh · 1 MLA Bidayuh
	V.O. Islamic Republic of Iran	IRN	500	ME,SAs Pashto
	Voice of America	CLN	250	SEA Khmer
	Voice of America	CLN	250	SEA Vietnamese
7135	BBC World Service	SNG	100	SEA · SEA Thai SEA Vietnamese
	BBC World Service	SNG	100	CHN Special English
	BBC World Service	SNG	100	SEA
	BBC World Service	SNG	100	SEA Burmese
	Radio France Int.	F	250	Eu Serbian/Croatian
	Radio France Int.	F	250	Eu(to$ept) Romanian
	Radio France Int.	F	250	Eu Polish
	Radio France Int.	F	500	CAf · NAf Arabic/French
	Radio France Int.	F	500	Eaf(Sept/Oct)
	Radio Romania Int.	ROU	50	Eu Serbian
	RT Marocaine (RTM1)	MRC	250	Eu
	TWR	ALB	100	2,3,5,6 Eu Czech
	TWR	D	125	7 Eu Romanian
	Voice of America	THA	250	CAs Uzbek
7140	*AIR Delhi*	IND	100	IND Hindi/English
	AIR Hyderabad A	IND	50	IND · 1 IND · 1 IND
	AIR Hyderabad A	IND	50	IND
	BBC World Service	CYP	300	ME
	BBC World Service	G	300	ME

Legend: Arabic · English · French · German · Hindi · Indonesian · Japanese · Mandarin · Russian · Other

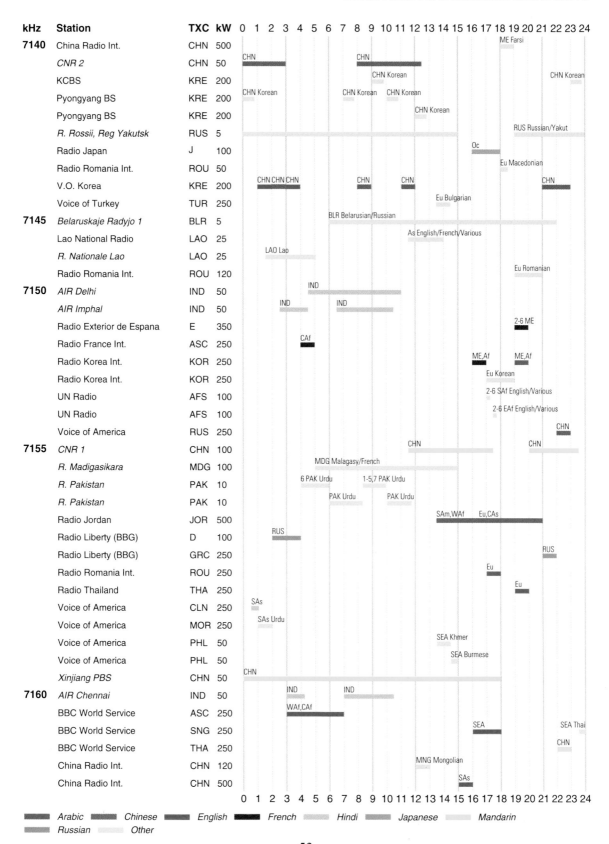

kHz	Station	TXC	kW
7140	China Radio Int.	CHN	500
	CNR 2	CHN	50
	KCBS	KRE	200
	Pyongyang BS	KRE	200
	Pyongyang BS	KRE	200
	R. Rossii, Reg Yakutsk	RUS	5
	Radio Japan	J	100
	Radio Romania Int.	ROU	50
	V.O. Korea	KRE	200
	Voice of Turkey	TUR	250
7145	Belaruskaje Radyjo 1	BLR	5
	Lao National Radio	LAO	25
	R. Nationale Lao	LAO	25
	Radio Romania Int.	ROU	120
7150	AIR Delhi	IND	50
	AIR Imphal	IND	50
	Radio Exterior de Espana	E	350
	Radio France Int.	ASC	250
	Radio Korea Int.	KOR	250
	Radio Korea Int.	KOR	250
	UN Radio	AFS	100
	UN Radio	AFS	100
	Voice of America	RUS	250
7155	CNR 1	CHN	100
	R. Madigasikara	MDG	100
	R. Pakistan	PAK	10
	R. Pakistan	PAK	10
	Radio Jordan	JOR	500
	Radio Liberty (BBG)	D	100
	Radio Liberty (BBG)	GRC	250
	Radio Romania Int.	ROU	250
	Radio Thailand	THA	250
	Voice of America	CLN	250
	Voice of America	MOR	250
	Voice of America	PHL	50
	Voice of America	PHL	50
	Xinjiang PBS	CHN	50
7160	AIR Chennai	IND	50
	BBC World Service	ASC	250
	BBC World Service	SNG	250
	BBC World Service	THA	250
	China Radio Int.	CHN	120
	China Radio Int.	CHN	500

Legend: Arabic Chinese English French Hindi Japanese Mandarin Russian Other

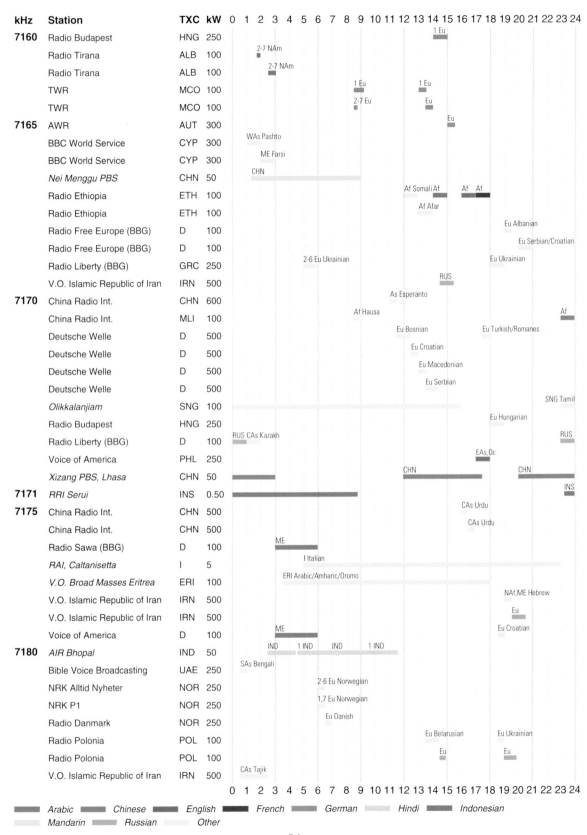

kHz	Station	TXC	kW	Schedule (0–24h)
7160	Radio Budapest	HNG	250	1 Eu (13-15)
	Radio Tirana	ALB	100	2-7 NAm
	Radio Tirana	ALB	100	2-7 NAm
	TWR	MCO	100	1 Eu (8-9); 1 Eu (13-14)
	TWR	MCO	100	2-7 Eu (8-9); Eu (13-14)
7165	AWR	AUT	300	Eu (15-16)
	BBC World Service	CYP	300	WAs Pashto
	BBC World Service	CYP	300	ME Farsi
	Nei Menggu PBS	CHN	50	CHN
	Radio Ethiopia	ETH	100	Af Somali Af; Af; Af
	Radio Ethiopia	ETH	100	Af Afar
	Radio Free Europe (BBG)	D	100	Eu Albanian
	Radio Free Europe (BBG)	D	100	Eu Serbian/Croatian
	Radio Liberty (BBG)	GRC	250	2-6 Eu Ukrainian; Eu Ukrainian
	V.O. Islamic Republic of Iran	IRN	500	RUS
7170	China Radio Int.	CHN	600	As Esperanto
	China Radio Int.	MLI	100	Af Hausa; Af
	Deutsche Welle	D	500	Eu Bosnian; Eu Turkish/Romanes
	Deutsche Welle	D	500	Eu Croatian
	Deutsche Welle	D	500	Eu Macedonian
	Deutsche Welle	D	500	Eu Serbian
	Olikkalanjiam	SNG	100	SNG Tamil
	Radio Budapest	HNG	250	Eu Hungarian
	Radio Liberty (BBG)	D	100	RUS CAs Kazakh; RUS
	Voice of America	PHL	250	EAs,Oc
	Xizang PBS, Lhasa	CHN	50	CHN; CHN
7171	RRI Serui	INS	0.50	INS
7175	China Radio Int.	CHN	500	CAs Urdu
	China Radio Int.	CHN	500	CAs Urdu
	Radio Sawa (BBG)	D	100	ME
	RAI, Caltanisetta	I	5	I Italian
	V.O. Broad Masses Eritrea	ERI	100	ERI Arabic/Amharic/Oromo
	V.O. Islamic Republic of Iran	IRN	500	NAf,ME Hebrew
	V.O. Islamic Republic of Iran	IRN	500	Eu
	Voice of America	D	100	ME; Eu Croatian
7180	AIR Bhopal	IND	50	IND; 1 IND; IND; 1 IND
	Bible Voice Broadcasting	UAE	250	SAs Bengali
	NRK Alltid Nyheter	NOR	250	2-6 Eu Norwegian
	NRK P1	NOR	250	1,7 Eu Norwegian
	Radio Danmark	NOR	250	Eu Danish
	Radio Polonia	POL	100	Eu Belarusian; Eu Ukrainian
	Radio Polonia	POL	100	Eu; Eu
	V.O. Islamic Republic of Iran	IRN	500	CAs Tajik

Legend: Arabic · Chinese · English · French · German · Hindi · Indonesian · Mandarin · Russian · Other

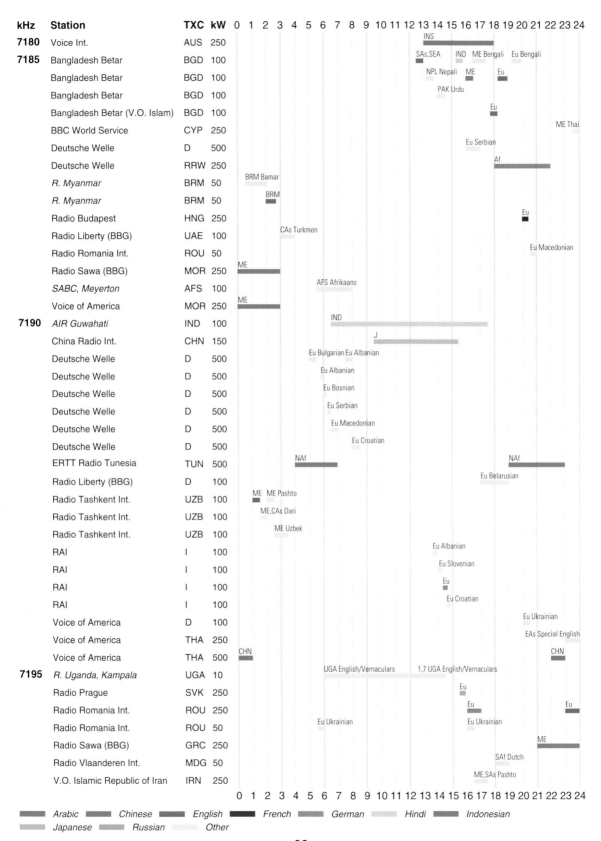

kHz	Station	TXC	kW
7180	Voice Int.	AUS	250
7185	Bangladesh Betar	BGD	100
	Bangladesh Betar	BGD	100
	Bangladesh Betar	BGD	100
	Bangladesh Betar (V.O. Islam)	BGD	100
	BBC World Service	CYP	250
	Deutsche Welle	D	500
	Deutsche Welle	RRW	250
	R. Myanmar	BRM	50
	R. Myanmar	BRM	50
	Radio Budapest	HNG	250
	Radio Liberty (BBG)	UAE	100
	Radio Romania Int.	ROU	50
	Radio Sawa (BBG)	MOR	250
	SABC, Meyerton	AFS	100
	Voice of America	MOR	250
7190	AIR Guwahati	IND	100
	China Radio Int.	CHN	150
	Deutsche Welle	D	500
	Deutsche Welle	D	500
	Deutsche Welle	D	500
	Deutsche Welle	D	500
	Deutsche Welle	D	500
	Deutsche Welle	D	500
	ERTT Radio Tunesia	TUN	500
	Radio Liberty (BBG)	D	100
	Radio Tashkent Int.	UZB	100
	Radio Tashkent Int.	UZB	100
	Radio Tashkent Int.	UZB	100
	RAI	I	100
	RAI	I	100
	RAI	I	100
	RAI	I	100
	Voice of America	D	100
	Voice of America	THA	250
	Voice of America	THA	500
7195	R. Uganda, Kampala	UGA	10
	Radio Prague	SVK	250
	Radio Romania Int.	ROU	250
	Radio Romania Int.	ROU	50
	Radio Sawa (BBG)	GRC	250
	Radio Vlaanderen Int.	MDG	50
	V.O. Islamic Republic of Iran	IRN	250

Chart labels (by row):
- Voice Int.: INS
- Bangladesh Betar: SAs,SEA / IND / ME Bengali / Eu Bengali
- Bangladesh Betar: NPL Nepali / ME / Eu
- Bangladesh Betar: PAK Urdu
- Bangladesh Betar (V.O. Islam): Eu
- BBC World Service: ME Thai
- Deutsche Welle: Eu Serbian
- Deutsche Welle: Af
- R. Myanmar: BRM Bamar
- R. Myanmar: BRM
- Radio Budapest: Eu
- Radio Liberty (BBG): CAs Turkmen
- Radio Romania Int.: Eu Macedonian
- Radio Sawa (BBG): ME
- SABC, Meyerton: AFS Afrikaans
- Voice of America: ME
- AIR Guwahati: IND
- China Radio Int.: J
- Deutsche Welle: Eu Bulgarian Eu Albanian
- Deutsche Welle: Eu Albanian
- Deutsche Welle: Eu Bosnian
- Deutsche Welle: Eu Serbian
- Deutsche Welle: Eu Macedonian
- Deutsche Welle: Eu Croatian
- ERTT Radio Tunesia: NAf / NAf
- Radio Liberty (BBG): Eu Belarusian
- Radio Tashkent Int.: ME ME Pashto
- Radio Tashkent Int.: ME,CAs Dari
- Radio Tashkent Int.: ME Uzbek
- RAI: Eu Albanian
- RAI: Eu Slovenian
- RAI: Eu
- RAI: Eu Croatian
- Voice of America: Eu Ukrainian
- Voice of America: EAs Special English
- Voice of America: CHN / CHN
- R. Uganda, Kampala: UGA English/Vernaculars / 1,7 UGA English/Vernaculars
- Radio Prague: Eu
- Radio Romania Int.: Eu / Eu
- Radio Romania Int.: Eu Ukrainian / Eu Ukrainian
- Radio Sawa (BBG): ME
- Radio Vlaanderen Int.: SAf Dutch
- V.O. Islamic Republic of Iran: ME,SAs Pashto

Legend: Arabic — Chinese — English — French — German — Hindi — Indonesian — Japanese — Russian — Other

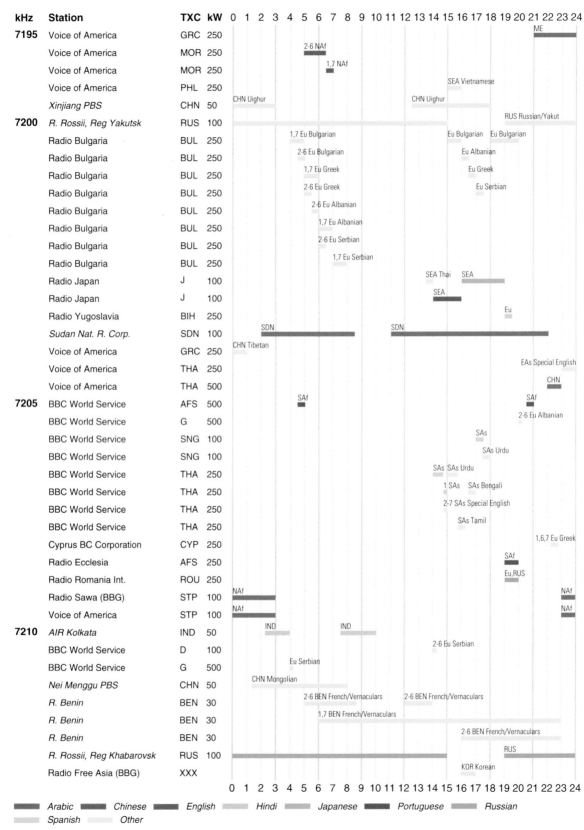

kHz	Station	TXC	kW	Schedule notes
7195	Voice of America	GRC	250	ME
	Voice of America	MOR	250	2-6 NAf
	Voice of America	MOR	250	1,7 NAf
	Voice of America	PHL	250	SEA Vietnamese
	Xinjiang PBS	CHN	50	CHN Uighur / CHN Uighur
7200	*R. Rossii, Reg Yakutsk*	RUS	100	RUS Russian/Yakut
	Radio Bulgaria	BUL	250	1,7 Eu Bulgarian / Eu Bulgarian / Eu Bulgarian
	Radio Bulgaria	BUL	250	2-6 Eu Bulgarian / Eu Albanian
	Radio Bulgaria	BUL	250	1,7 Eu Greek / Eu Greek
	Radio Bulgaria	BUL	250	2-6 Eu Greek / Eu Serbian
	Radio Bulgaria	BUL	250	2-6 Eu Albanian
	Radio Bulgaria	BUL	250	1,7 Eu Albanian
	Radio Bulgaria	BUL	250	2-6 Eu Serbian
	Radio Bulgaria	BUL	250	1,7 Eu Serbian
	Radio Japan	J	100	SEA Thai / SEA
	Radio Japan	J	100	SEA
	Radio Yugoslavia	BIH	250	Eu
	Sudan Nat. R. Corp.	SDN	100	SDN / SDN
	Voice of America	GRC	250	CHN Tibetan
	Voice of America	THA	250	EAs Special English
	Voice of America	THA	500	CHN
7205	BBC World Service	AFS	500	SAf / SAf
	BBC World Service	G	500	2-6 Eu Albanian
	BBC World Service	SNG	100	SAs
	BBC World Service	SNG	100	SAs Urdu
	BBC World Service	THA	250	SAs SAs Urdu
	BBC World Service	THA	250	1 SAs SAs Bengali
	BBC World Service	THA	250	2-7 SAs Special English
	BBC World Service	THA	250	SAs Tamil
	Cyprus BC Corporation	CYP	250	1,6,7 Eu Greek
	Radio Ecclesia	AFS	250	SAf
	Radio Romania Int.	ROU	250	Eu,RUS
	Radio Sawa (BBG)	STP	100	NAf / NAf
	Voice of America	STP	100	NAf / NAf
7210	*AIR Kolkata*	IND	50	IND / IND
	BBC World Service	D	100	2-6 Eu Serbian
	BBC World Service	G	500	Eu Serbian
	Nei Menggu PBS	CHN	50	CHN Mongolian
	R. Benin	BEN	30	2-6 BEN French/Vernaculars / 2-6 BEN French/Vernaculars
	R. Benin	BEN	30	1,7 BEN French/Vernaculars
	R. Benin	BEN	30	2-6 BEN French/Vernaculars
	R. Rossii, Reg Khabarovsk	RUS	100	RUS
	Radio Free Asia (BBG)	XXX		KOR Korean

Legend: *Arabic* · *Chinese* · *English* · *Hindi* · *Japanese* · *Portuguese* · *Russian* · *Spanish* · *Other*

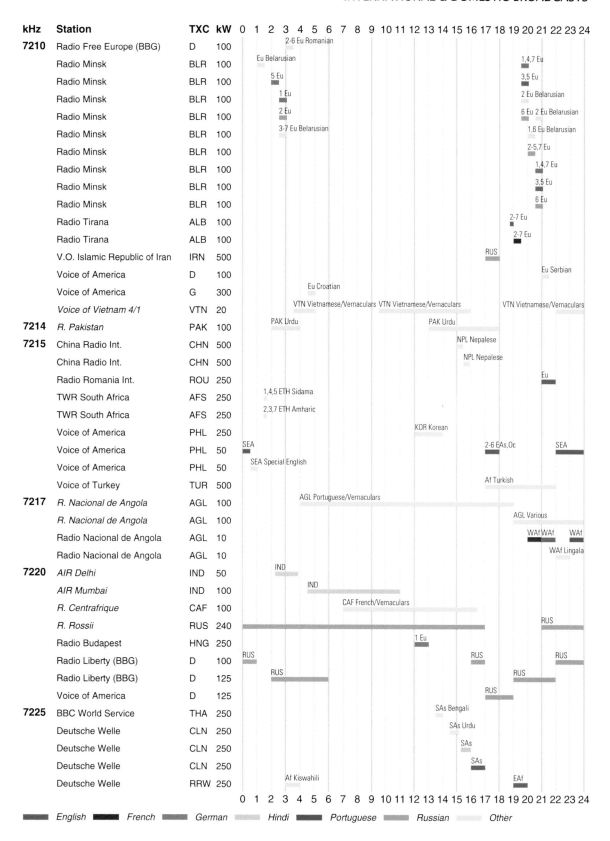

kHz	Station	TXC	kW		
7210	Radio Free Europe (BBG)	D	100		2-6 Eu Romanian
	Radio Minsk	BLR	100		Eu Belarusian / 1,4,7 Eu
	Radio Minsk	BLR	100		5 Eu / 3,5 Eu
	Radio Minsk	BLR	100		1 Eu / 2 Eu Belarusian
	Radio Minsk	BLR	100		2 Eu / 6 Eu 2 Eu Belarusian
	Radio Minsk	BLR	100		3-7 Eu Belarusian / 1,6 Eu Belarusian
	Radio Minsk	BLR	100		2-5,7 Eu
	Radio Minsk	BLR	100		1,4,7 Eu
	Radio Minsk	BLR	100		3,5 Eu
	Radio Minsk	BLR	100		6 Eu
	Radio Tirana	ALB	100		2-7 Eu
	Radio Tirana	ALB	100		2-7 Eu
	V.O. Islamic Republic of Iran	IRN	500		RUS
	Voice of America	D	100		Eu Serbian
	Voice of America	G	300		Eu Croatian
	Voice of Vietnam 4/1	VTN	20		VTN Vietnamese/Vernaculars VTN Vietnamese/Vernaculars VTN Vietnamese/Vernaculars
7214	*R. Pakistan*	PAK	100		PAK Urdu PAK Urdu
7215	China Radio Int.	CHN	500		NPL Nepalese
	China Radio Int.	CHN	500		NPL Nepalese
	Radio Romania Int.	ROU	250		Eu
	TWR South Africa	AFS	250		1,4,5 ETH Sidama
	TWR South Africa	AFS	250		2,3,7 ETH Amharic
	Voice of America	PHL	250		KOR Korean
	Voice of America	PHL	50		SEA / 2-6 EAs,Oc / SEA
	Voice of America	PHL	50		SEA Special English
	Voice of Turkey	TUR	500		Af Turkish
7217	*R. Nacional de Angola*	AGL	100		AGL Portuguese/Vernaculars
	R. Nacional de Angola	AGL	100		AGL Various
	Radio Nacional de Angola	AGL	10		WAf WAf WAf
	Radio Nacional de Angola	AGL	10		WAf Lingala
7220	*AIR Delhi*	IND	50		IND
	AIR Mumbai	IND	100		IND
	R. Centrafrique	CAF	100		CAF French/Vernaculars
	R. Rossii	RUS	240		RUS
	Radio Budapest	HNG	250		1 Eu
	Radio Liberty (BBG)	D	100		RUS / RUS / RUS
	Radio Liberty (BBG)	D	125		RUS / RUS
	Voice of America	D	125		RUS
7225	BBC World Service	THA	250		SAs Bengali
	Deutsche Welle	CLN	250		SAs Urdu
	Deutsche Welle	CLN	250		SAs
	Deutsche Welle	CLN	250		SAs
	Deutsche Welle	RRW	250		Af Kiswahili / EAf

0 1 2 3 4 5 6 7 8 9 10 11 12 13 14 15 16 17 18 19 20 21 22 23 24

English   French   German   Hindi   Portuguese   Russian   Other

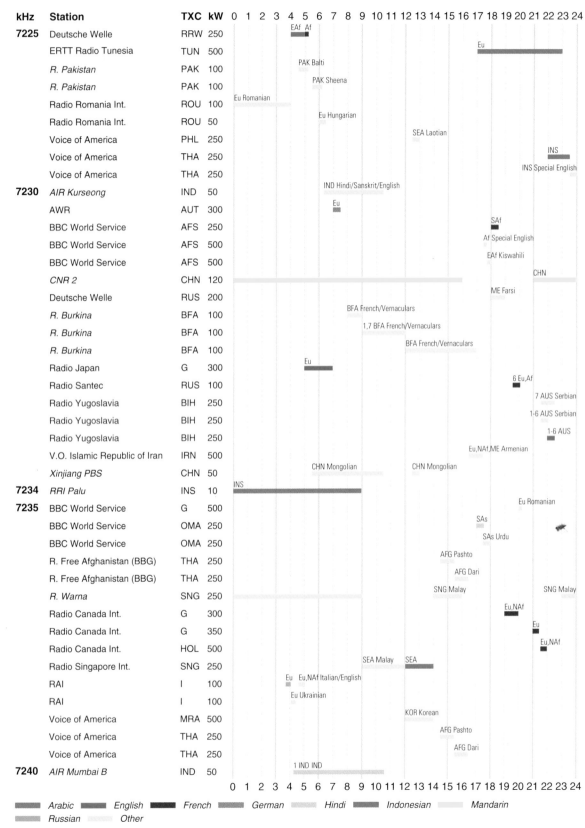

kHz	Station	TXC	kW	Schedule (0-24)
7225	Deutsche Welle	RRW	250	EAf Af
	ERTT Radio Tunisia	TUN	500	Eu
	R. Pakistan	PAK	100	PAK Balti
	R. Pakistan	PAK	100	PAK Sheena
	Radio Romania Int.	ROU	100	Eu Romanian
	Radio Romania Int.	ROU	50	Eu Hungarian
	Voice of America	PHL	250	SEA Laotian
	Voice of America	THA	250	INS
	Voice of America	THA	250	INS Special English
7230	AIR Kurseong	IND	50	IND Hindi/Sanskrit/English
	AWR	AUT	300	Eu
	BBC World Service	AFS	250	SAf
	BBC World Service	AFS	500	Af Special English
	BBC World Service	AFS	500	EAf Kiswahili
	CNR 2	CHN	120	CHN
	Deutsche Welle	RUS	200	ME Farsi
	R. Burkina	BFA	100	BFA French/Vernaculars
	R. Burkina	BFA	100	1,7 BFA French/Vernaculars
	R. Burkina	BFA	100	BFA French/Vernaculars
	Radio Japan	G	300	Eu
	Radio Santec	RUS	100	6 Eu,Af
	Radio Yugoslavia	BIH	250	7 AUS Serbian
	Radio Yugoslavia	BIH	250	1-6 AUS Serbian
	Radio Yugoslavia	BIH	250	1-6 AUS
	V.O. Islamic Republic of Iran	IRN	500	Eu,NAf,ME Armenian
	Xinjiang PBS	CHN	50	CHN Mongolian CHN Mongolian
7234	RRI Palu	INS	10	INS
7235	BBC World Service	G	500	Eu Romanian
	BBC World Service	OMA	250	SAs
	BBC World Service	OMA	250	SAs Urdu
	R. Free Afghanistan (BBG)	THA	250	AFG Pashto
	R. Free Afghanistan (BBG)	THA	250	AFG Dari
	R. Warna	SNG	250	SNG Malay SNG Malay
	Radio Canada Int.	G	300	Eu,NAf
	Radio Canada Int.	G	350	Eu
	Radio Canada Int.	HOL	500	Eu,NAf
	Radio Singapore Int.	SNG	250	SEA Malay SEA
	RAI	I	100	Eu Eu,NAf Italian/English
	RAI	I	100	Eu Ukrainian
	Voice of America	MRA	500	KOR Korean
	Voice of America	THA	250	AFG Pashto
	Voice of America	THA	250	AFG Dari
7240	AIR Mumbai B	IND	50	1 IND IND

Legend: Arabic · English · French · German · Hindi · Indonesian · Mandarin · Russian · Other

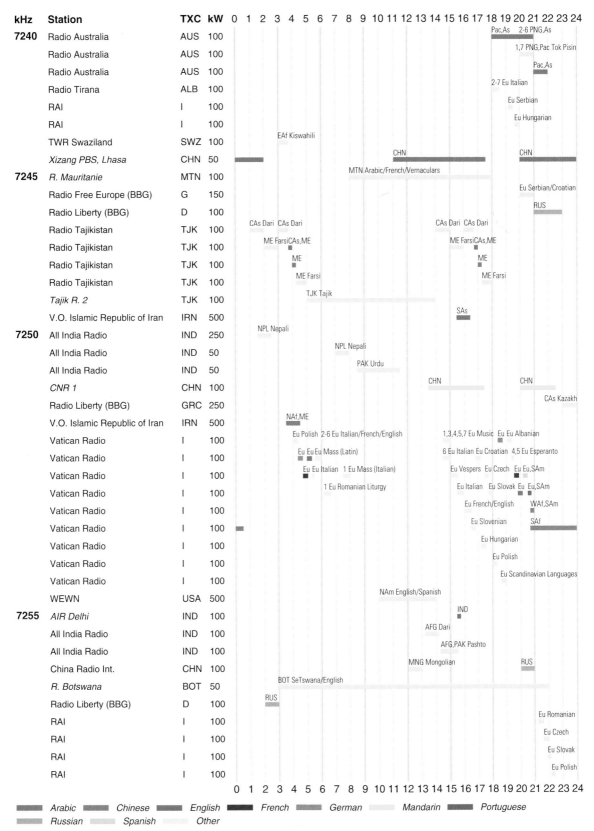

kHz	Station	TXC	kW	0 1 2 3 4 5 6 7 8 9 10 11 12 13 14 15 16 17 18 19 20 21 22 23 24
7240	Radio Australia	AUS	100	Pac,As 2-6 PNG,As
	Radio Australia	AUS	100	1,7 PNG,Pac Tok Pisin
	Radio Australia	AUS	100	Pac,As
	Radio Tirana	ALB	100	2-7 Eu Italian
	RAI	I	100	Eu Serbian
	RAI	I	100	Eu Hungarian
	TWR Swaziland	SWZ	100	EAf Kiswahili
	Xizang PBS, Lhasa	CHN	50	CHN CHN
7245	*R. Mauritanie*	MTN	100	MTN Arabic/French/Vernaculars
	Radio Free Europe (BBG)	G	150	Eu Serbian/Croatian
	Radio Liberty (BBG)	D	100	RUS
	Radio Tajikistan	TJK	100	CAs Dari CAs Dari CAs Dari CAs Dari
	Radio Tajikistan	TJK	100	ME FarsiCAs,ME ME FarsiCAs,ME
	Radio Tajikistan	TJK	100	ME ME
	Radio Tajikistan	TJK	100	ME Farsi ME Farsi
	Tajik R. 2	TJK	100	TJK Tajik
	V.O. Islamic Republic of Iran	IRN	500	SAs
7250	All India Radio	IND	250	NPL Nepali
	All India Radio	IND	50	NPL Nepali
	All India Radio	IND	50	PAK Urdu
	CNR 1	CHN	100	CHN CHN
	Radio Liberty (BBG)	GRC	250	CAs Kazakh
	V.O. Islamic Republic of Iran	IRN	500	NAf,ME
	Vatican Radio	I	100	Eu Polish 2-6 Eu Italian/French/English 1,3,4,5,7 Eu Music Eu Eu Albanian
	Vatican Radio	I	100	Eu Eu Eu Mass (Latin) 6 Eu Italian Eu Croatian 4,5 Eu Esperanto
	Vatican Radio	I	100	Eu Eu Italian 1 Eu Mass (Italian) Eu Vespers Eu Czech Eu Eu,SAm
	Vatican Radio	I	100	1 Eu Romanian Liturgy Eu Italian Eu Slovak Eu Eu,SAm
	Vatican Radio	I	100	Eu French/English WAf,SAm
	Vatican Radio	I	100	Eu Slovenian SAf
	Vatican Radio	I	100	Eu Hungarian
	Vatican Radio	I	100	Eu Polish
	Vatican Radio	I	100	Eu Scandinavian Languages
	WEWN	USA	500	NAm English/Spanish
7255	*AIR Delhi*	IND	100	IND
	All India Radio	IND	100	AFG Dari
	All India Radio	IND	100	AFG,PAK Pashto
	China Radio Int.	CHN	100	MNG Mongolian RUS
	R. Botswana	BOT	50	BOT SeTswana/English
	Radio Liberty (BBG)	D	100	RUS
	RAI	I	100	Eu Romanian
	RAI	I	100	Eu Czech
	RAI	I	100	Eu Slovak
	RAI	I	100	Eu Polish

0 1 2 3 4 5 6 7 8 9 10 11 12 13 14 15 16 17 18 19 20 21 22 23 24

Arabic　*Chinese*　*English*　*French*　*German*　*Mandarin*　*Portuguese*
Russian　*Spanish*　*Other*

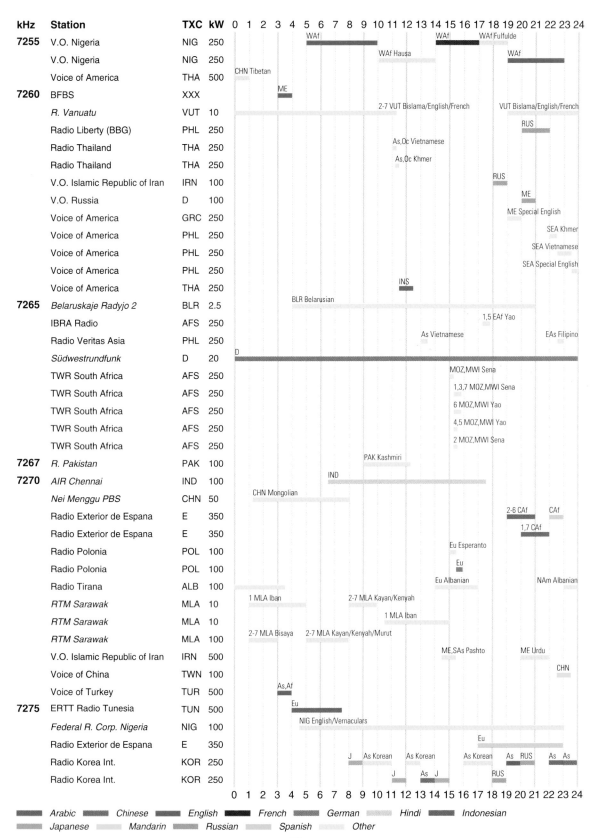

kHz	Station	TXC	kW	Notes
7255	V.O. Nigeria	NIG	250	WAf / WAf / WAf Fulfulde
	V.O. Nigeria	NIG	250	WAf Hausa / WAf
	Voice of America	THA	500	CHN Tibetan
7260	BFBS	XXX		ME
	R. Vanuatu	VUT	10	2-7 VUT Bislama/English/French / VUT Bislama/English/French
	Radio Liberty (BBG)	PHL	250	RUS
	Radio Thailand	THA	250	As,Oc Vietnamese
	Radio Thailand	THA	250	As,Oc Khmer
	V.O. Islamic Republic of Iran	IRN	100	RUS
	V.O. Russia	D	100	ME
	Voice of America	GRC	250	ME Special English
	Voice of America	PHL	250	SEA Khmer
	Voice of America	PHL	250	SEA Vietnamese
	Voice of America	PHL	250	SEA Special English
	Voice of America	THA	250	INS
7265	*Belaruskaje Radyjo 2*	BLR	2.5	BLR Belarusian
	IBRA Radio	AFS	250	1,5 EAf Yao
	Radio Veritas Asia	PHL	250	As Vietnamese / EAs Filipino
	Südwestrundfunk	D	20	D
	TWR South Africa	AFS	250	MOZ,MWI Sena
	TWR South Africa	AFS	250	1,3,7 MOZ,MWI Sena
	TWR South Africa	AFS	250	6 MOZ,MWI Yao
	TWR South Africa	AFS	250	4,5 MOZ,MWI Yao
	TWR South Africa	AFS	250	2 MOZ,MWI Sena
7267	*R. Pakistan*	PAK	100	PAK Kashmiri
7270	*AIR Chennai*	IND	100	IND
	Nei Menggu PBS	CHN	50	CHN Mongolian
	Radio Exterior de Espana	E	350	2-6 CAf / CAf
	Radio Exterior de Espana	E	350	1,7 CAf
	Radio Polonia	POL	100	Eu Esperanto
	Radio Polonia	POL	100	Eu
	Radio Tirana	ALB	100	Eu Albanian / NAm Albanian
	RTM Sarawak	MLA	10	1 MLA Iban / 2-7 MLA Kayan/Kenyah
	RTM Sarawak	MLA	10	1 MLA Iban
	RTM Sarawak	MLA	100	2-7 MLA Bisaya / 2-7 MLA Kayan/Kenyah/Murut
	V.O. Islamic Republic of Iran	IRN	500	ME,SAs Pashto / ME Urdu
	Voice of China	TWN	100	CHN
	Voice of Turkey	TUR	500	As,Af
7275	ERTT Radio Tunisia	TUN	500	Eu
	Federal R. Corp. Nigeria	NIG	100	NIG English/Vernaculars
	Radio Exterior de Espana	E	350	Eu
	Radio Korea Int.	KOR	250	J / As Korean / As Korean / As Korean / As RUS / As As
	Radio Korea Int.	KOR	250	J / As J / RUS

Legend:
- Arabic
- Chinese
- English
- French
- German
- Hindi
- Indonesian
- Japanese
- Mandarin
- Russian
- Spanish
- Other

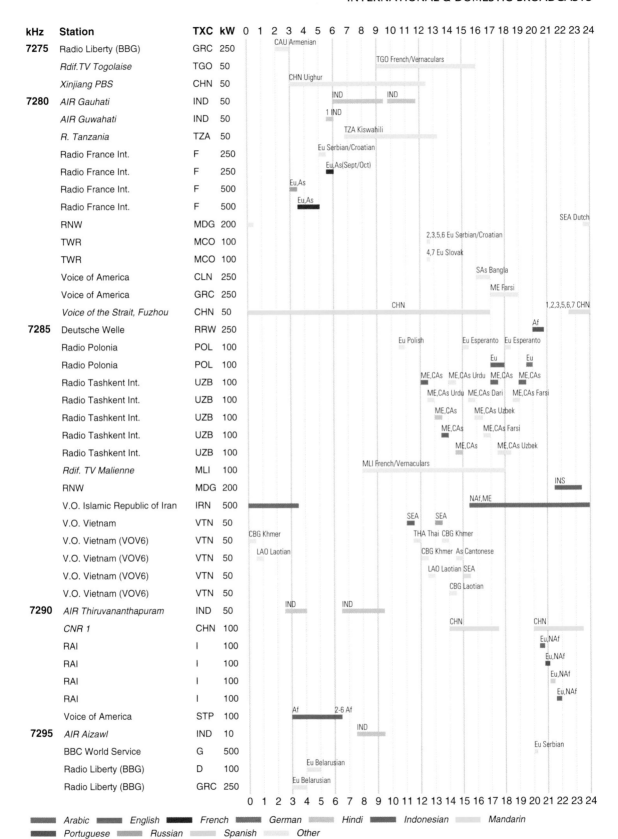

kHz	Station	TXC	kW
7275	Radio Liberty (BBG)	GRC	250
	Rdif.TV Togolaise	TGO	50
	Xinjiang PBS	CHN	50
7280	AIR Gauhati	IND	50
	AIR Guwahati	IND	50
	R. Tanzania	TZA	50
	Radio France Int.	F	250
	Radio France Int.	F	250
	Radio France Int.	F	500
	Radio France Int.	F	500
	RNW	MDG	200
	TWR	MCO	100
	TWR	MCO	100
	Voice of America	CLN	250
	Voice of America	GRC	250
	Voice of the Strait, Fuzhou	CHN	50
7285	Deutsche Welle	RRW	250
	Radio Polonia	POL	100
	Radio Polonia	POL	100
	Radio Tashkent Int.	UZB	100
	Radio Tashkent Int.	UZB	100
	Radio Tashkent Int.	UZB	100
	Radio Tashkent Int.	UZB	100
	Radio Tashkent Int.	UZB	100
	Rdif. TV Malienne	MLI	100
	RNW	MDG	200
	V.O. Islamic Republic of Iran	IRN	500
	V.O. Vietnam	VTN	50
	V.O. Vietnam (VOV6)	VTN	50
	V.O. Vietnam (VOV6)	VTN	50
	V.O. Vietnam (VOV6)	VTN	50
	V.O. Vietnam (VOV6)	VTN	50
	V.O. Vietnam (VOV6)	VTN	50
7290	AIR Thiruvananthapuram	IND	50
	CNR 1	CHN	100
	RAI	I	100
	RAI	I	100
	RAI	I	100
	RAI	I	100
	Voice of America	STP	100
7295	AIR Aizawl	IND	10
	BBC World Service	G	500
	Radio Liberty (BBG)	D	100
	Radio Liberty (BBG)	GRC	250

Broadcast timeline annotations (0–24 hours):

- Radio Liberty (BBG) GRC 250: CAU Armenian
- Rdif.TV Togolaise TGO 50: TGO French/Vernaculars
- Xinjiang PBS CHN 50: CHN Uighur
- AIR Gauhati IND 50: IND, IND
- AIR Guwahati IND 50: 1 IND
- R. Tanzania TZA 50: TZA Kiswahili
- Radio France Int. F 250: Eu Serbian/Croatian
- Radio France Int. F 250: Eu,As(Sept/Oct)
- Radio France Int. F 500: Eu,As
- Radio France Int. F 500: Eu,As
- RNW MDG 200: SEA Dutch
- TWR MCO 100: 2,3,5,6 Eu Serbian/Croatian
- TWR MCO 100: 4,7 Eu Slovak
- Voice of America CLN 250: SAs Bangla
- Voice of America GRC 250: ME Farsi
- Voice of the Strait, Fuzhou CHN 50: CHN, 1,2,3,5,6,7 CHN
- Deutsche Welle RRW 250: Af
- Radio Polonia POL 100: Eu Polish, Eu Esperanto, Eu Esperanto
- Radio Polonia POL 100: Eu, Eu
- Radio Tashkent Int. UZB 100: ME,CAs, ME,CAs Urdu, ME,CAs, ME,CAs
- Radio Tashkent Int. UZB 100: ME,CAs Urdu, ME,CAs Dari, ME,CAs Farsi
- Radio Tashkent Int. UZB 100: ME,CAs, ME,CAs Uzbek
- Radio Tashkent Int. UZB 100: ME,CAs, ME,CAs Farsi
- Radio Tashkent Int. UZB 100: ME,CAs, ME,CAs Uzbek
- Rdif. TV Malienne MLI 100: MLI French/Vernaculars
- RNW MDG 200: INS
- V.O. Islamic Republic of Iran IRN 500: NAf,ME
- V.O. Vietnam VTN 50: SEA, SEA
- V.O. Vietnam (VOV6) VTN 50: CBG Khmer, THA Thai, CBG Khmer
- V.O. Vietnam (VOV6) VTN 50: LAO Laotian, CBG Khmer, As Cantonese
- V.O. Vietnam (VOV6) VTN 50: LAO Laotian SEA
- V.O. Vietnam (VOV6) VTN 50: CBG Laotian
- AIR Thiruvananthapuram IND 50: IND, IND
- CNR 1 CHN 100: CHN, CHN
- RAI I 100: Eu,NAf
- RAI I 100: Eu,NAf
- RAI I 100: Eu,NAf
- RAI I 100: Eu,NAf
- Voice of America STP 100: Af, 2-6 Af
- AIR Aizawl IND 10: IND
- BBC World Service G 500: Eu Serbian
- Radio Liberty (BBG) D 100: Eu Belarusian
- Radio Liberty (BBG) GRC 250: Eu Belarusian

Legend:

Arabic — English — French — German — Hindi — Indonesian — Mandarin — Portuguese — Russian — Spanish — Other

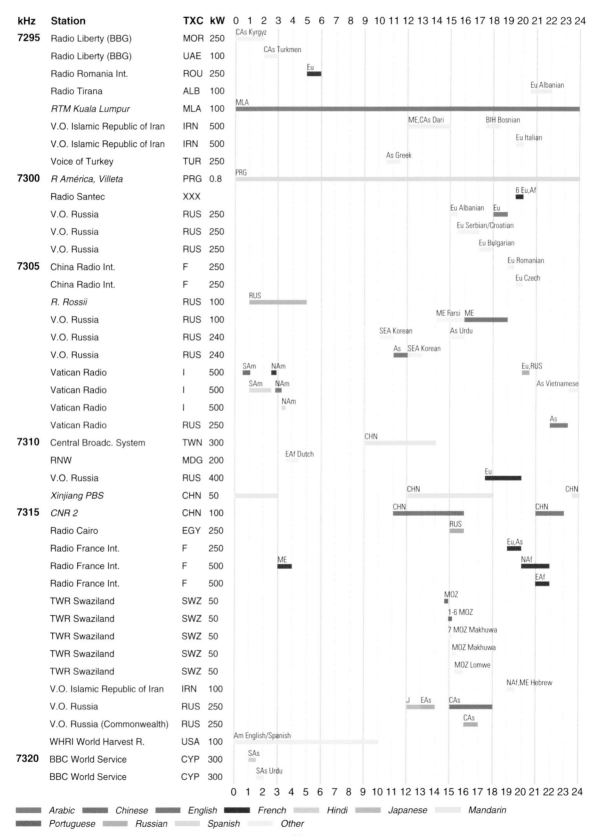

kHz	Station	TXC	kW
7295	Radio Liberty (BBG)	MOR	250
	Radio Liberty (BBG)	UAE	100
	Radio Romania Int.	ROU	250
	Radio Tirana	ALB	100
	RTM Kuala Lumpur	MLA	100
	V.O. Islamic Republic of Iran	IRN	500
	V.O. Islamic Republic of Iran	IRN	500
	Voice of Turkey	TUR	250
7300	R América, Villeta	PRG	0.8
	Radio Santec	XXX	
	V.O. Russia	RUS	250
	V.O. Russia	RUS	250
	V.O. Russia	RUS	250
7305	China Radio Int.	F	250
	China Radio Int.	F	250
	R. Rossii	RUS	100
	V.O. Russia	RUS	100
	V.O. Russia	RUS	240
	V.O. Russia	RUS	240
	Vatican Radio	I	500
	Vatican Radio	I	500
	Vatican Radio	I	500
	Vatican Radio	RUS	250
7310	Central Broadc. System	TWN	300
	RNW	MDG	200
	V.O. Russia	RUS	400
	Xinjiang PBS	CHN	50
7315	CNR 2	CHN	100
	Radio Cairo	EGY	250
	Radio France Int.	F	250
	Radio France Int.	F	500
	Radio France Int.	F	500
	TWR Swaziland	SWZ	50
	TWR Swaziland	SWZ	50
	TWR Swaziland	SWZ	50
	TWR Swaziland	SWZ	50
	TWR Swaziland	SWZ	50
	V.O. Islamic Republic of Iran	IRN	100
	V.O. Russia	RUS	250
	V.O. Russia (Commonwealth)	RUS	250
	WHRI World Harvest R.	USA	100
7320	BBC World Service	CYP	300
	BBC World Service	CYP	300

Arabic　Chinese　English　French　Hindi　Japanese　Mandarin
Portuguese　Russian　Spanish　Other

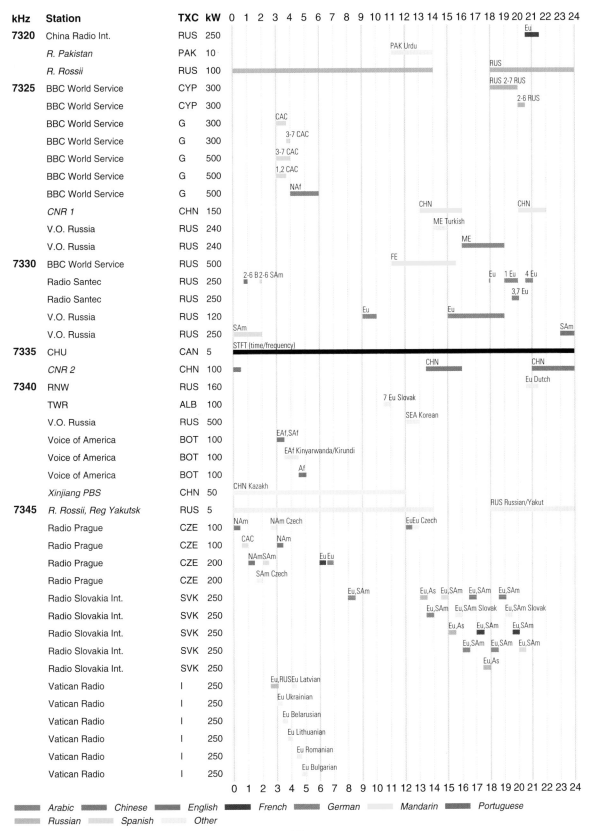

kHz	Station	TXC	kW
7320	China Radio Int.	RUS	250
	R. Pakistan	PAK	10
	R. Rossii	RUS	100
7325	BBC World Service	CYP	300
	BBC World Service	CYP	300
	BBC World Service	G	300
	BBC World Service	G	300
	BBC World Service	G	500
	BBC World Service	G	500
	BBC World Service	G	500
	CNR 1	CHN	150
	V.O. Russia	RUS	240
	V.O. Russia	RUS	240
7330	BBC World Service	RUS	500
	Radio Santec	RUS	250
	Radio Santec	RUS	250
	V.O. Russia	RUS	120
	V.O. Russia	RUS	250
7335	CHU	CAN	5
	CNR 2	CHN	100
7340	RNW	RUS	160
	TWR	ALB	100
	V.O. Russia	RUS	500
	Voice of America	BOT	100
	Voice of America	BOT	100
	Voice of America	BOT	100
	Xinjiang PBS	CHN	50
7345	*R. Rossii, Reg Yakutsk*	RUS	5
	Radio Prague	CZE	100
	Radio Prague	CZE	100
	Radio Prague	CZE	200
	Radio Prague	CZE	200
	Radio Slovakia Int.	SVK	250
	Radio Slovakia Int.	SVK	250
	Radio Slovakia Int.	SVK	250
	Radio Slovakia Int.	SVK	250
	Radio Slovakia Int.	SVK	250
	Vatican Radio	I	250
	Vatican Radio	I	250
	Vatican Radio	I	250
	Vatican Radio	I	250
	Vatican Radio	I	250
	Vatican Radio	I	250

Legend: Arabic, Chinese, English, French, German, Mandarin, Portuguese, Russian, Spanish, Other

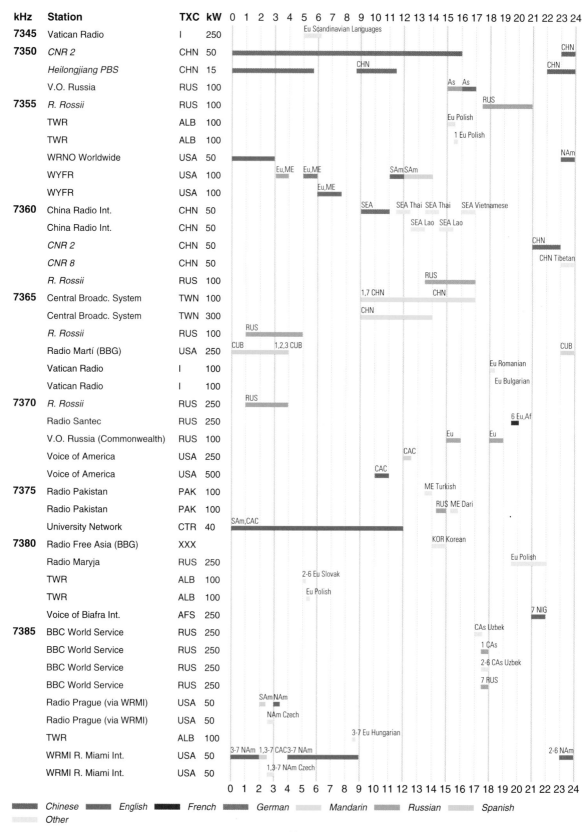

kHz	Station	TXC	kW
7345	Vatican Radio	I	250
7350	*CNR 2*	CHN	50
	Heilongjiang PBS	CHN	15
	V.O. Russia	RUS	100
7355	*R. Rossii*	RUS	100
	TWR	ALB	100
	TWR	ALB	100
	WRNO Worldwide	USA	50
	WYFR	USA	100
	WYFR	USA	100
7360	China Radio Int.	CHN	50
	China Radio Int.	CHN	50
	CNR 2	CHN	50
	CNR 8	CHN	50
	R. Rossii	RUS	100
7365	Central Broadc. System	TWN	100
	Central Broadc. System	TWN	300
	R. Rossii	RUS	100
	Radio Martí (BBG)	USA	250
	Vatican Radio	I	100
	Vatican Radio	I	100
7370	*R. Rossii*	RUS	250
	Radio Santec	RUS	250
	V.O. Russia (Commonwealth)	RUS	100
	Voice of America	USA	250
	Voice of America	USA	500
7375	Radio Pakistan	PAK	100
	Radio Pakistan	PAK	100
	University Network	CTR	40
7380	Radio Free Asia (BBG)	XXX	
	Radio Maryja	RUS	250
	TWR	ALB	100
	TWR	ALB	100
	Voice of Biafra Int.	AFS	250
7385	BBC World Service	RUS	250
	BBC World Service	RUS	250
	BBC World Service	RUS	250
	BBC World Service	RUS	250
	Radio Prague (via WRMI)	USA	50
	Radio Prague (via WRMI)	USA	50
	TWR	ALB	100
	WRMI R. Miami Int.	USA	50
	WRMI R. Miami Int.	USA	50

Legend: Chinese, English, French, German, Mandarin, Russian, Spanish, Other

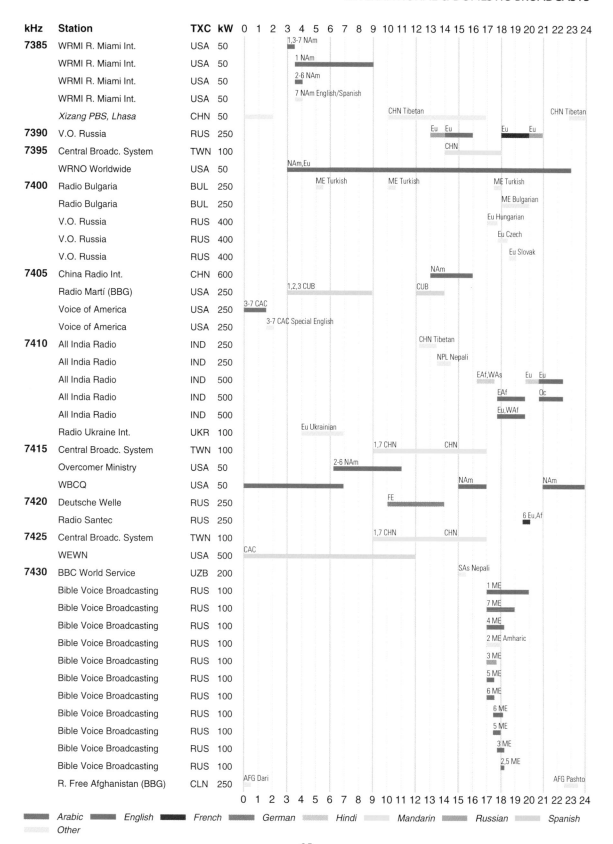

kHz	Station	TXC	kW	
7385	WRMI R. Miami Int.	USA	50	1,3-7 NAm
	WRMI R. Miami Int.	USA	50	1 NAm
	WRMI R. Miami Int.	USA	50	2-6 NAm
	WRMI R. Miami Int.	USA	50	7 NAm English/Spanish
	Xizang PBS, Lhasa	CHN	50	CHN Tibetan / CHN Tibetan
7390	V.O. Russia	RUS	250	Eu Eu Eu Eu
7395	Central Broadc. System	TWN	100	CHN
	WRNO Worldwide	USA	50	NAm,Eu
7400	Radio Bulgaria	BUL	250	ME Turkish ME Turkish ME Turkish
	Radio Bulgaria	BUL	250	ME Bulgarian
	V.O. Russia	RUS	400	Eu Hungarian
	V.O. Russia	RUS	400	Eu Czech
	V.O. Russia	RUS	400	Eu Slovak
7405	China Radio Int.	CHN	600	NAm
	Radio Martí (BBG)	USA	250	1,2,3 CUB CUB
	Voice of America	USA	250	3-7 CAC
	Voice of America	USA	250	3-7 CAC Special English
7410	All India Radio	IND	250	CHN Tibetan
	All India Radio	IND	250	NPL Nepali
	All India Radio	IND	500	EAf,WAs Eu Eu
	All India Radio	IND	500	EAf Oc
	All India Radio	IND	500	Eu,WAf
	Radio Ukraine Int.	UKR	100	Eu Ukrainian
7415	Central Broadc. System	TWN	100	1,7 CHN CHN
	Overcomer Ministry	USA	50	2-6 NAm
	WBCQ	USA	50	NAm NAm
7420	Deutsche Welle	RUS	250	FE
	Radio Santec	RUS	250	6 Eu,Af
7425	Central Broadc. System	TWN	100	1,7 CHN CHN
	WEWN	USA	500	CAC
7430	BBC World Service	UZB	200	SAs Nepali
	Bible Voice Broadcasting	RUS	100	1 ME
	Bible Voice Broadcasting	RUS	100	7 ME
	Bible Voice Broadcasting	RUS	100	4 ME
	Bible Voice Broadcasting	RUS	100	2 ME Amharic
	Bible Voice Broadcasting	RUS	100	3 ME
	Bible Voice Broadcasting	RUS	100	5 ME
	Bible Voice Broadcasting	RUS	100	6 ME
	Bible Voice Broadcasting	RUS	100	6 ME
	Bible Voice Broadcasting	RUS	100	5 ME
	Bible Voice Broadcasting	RUS	100	3 ME
	Bible Voice Broadcasting	RUS	100	2,5 ME
	R. Free Afghanistan (BBG)	CLN	250	AFG Dari AFG Pashto

Legend: *Arabic* *English* *French* *German* *Hindi* *Mandarin* *Russian* *Spanish* *Other*

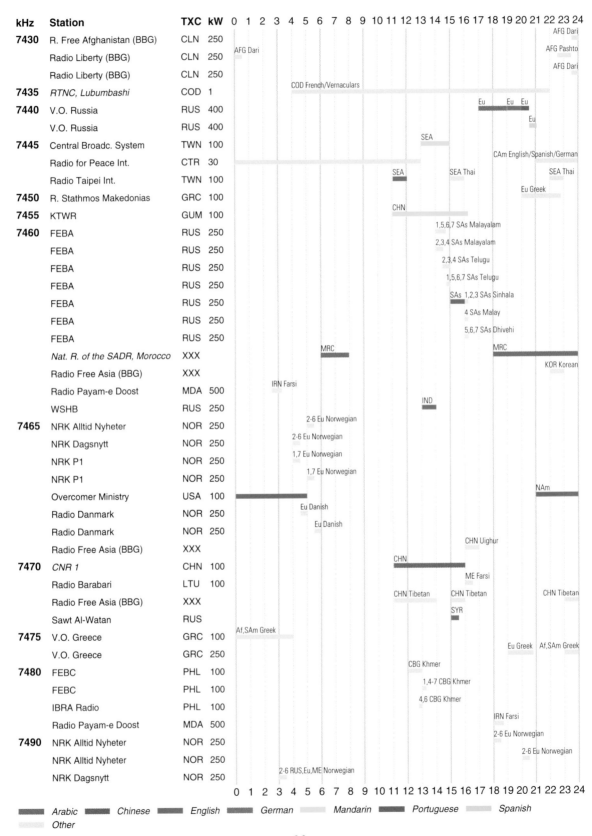

kHz	Station	TXC	kW
7430	R. Free Afghanistan (BBG)	CLN	250
	Radio Liberty (BBG)	CLN	250
	Radio Liberty (BBG)	CLN	250
7435	*RTNC, Lubumbashi*	COD	1
7440	V.O. Russia	RUS	400
	V.O. Russia	RUS	400
7445	Central Broadc. System	TWN	100
	Radio for Peace Int.	CTR	30
	Radio Taipei Int.	TWN	100
7450	R. Stathmos Makedonias	GRC	100
7455	KTWR	GUM	100
7460	FEBA	RUS	250
	FEBA	RUS	250
	FEBA	RUS	250
	FEBA	RUS	250
	FEBA	RUS	250
	FEBA	RUS	250
	FEBA	RUS	250
	Nat. R. of the SADR, Morocco	XXX	
	Radio Free Asia (BBG)	XXX	
	Radio Payam-e Doost	MDA	500
	WSHB	RUS	250
7465	NRK Alltid Nyheter	NOR	250
	NRK Dagsnytt	NOR	250
	NRK P1	NOR	250
	NRK P1	NOR	250
	Overcomer Ministry	USA	100
	Radio Danmark	NOR	250
	Radio Danmark	NOR	250
	Radio Free Asia (BBG)	XXX	
7470	*CNR 1*	CHN	100
	Radio Barabari	LTU	100
	Radio Free Asia (BBG)	XXX	
	Sawt Al-Watan	RUS	
7475	V.O. Greece	GRC	100
	V.O. Greece	GRC	250
7480	FEBC	PHL	100
	FEBC	PHL	100
	IBRA Radio	PHL	100
	Radio Payam-e Doost	MDA	500
7490	NRK Alltid Nyheter	NOR	250
	NRK Alltid Nyheter	NOR	250
	NRK Dagsnytt	NOR	250

Legend: Arabic, Chinese, English, German, Mandarin, Portuguese, Spanish, Other

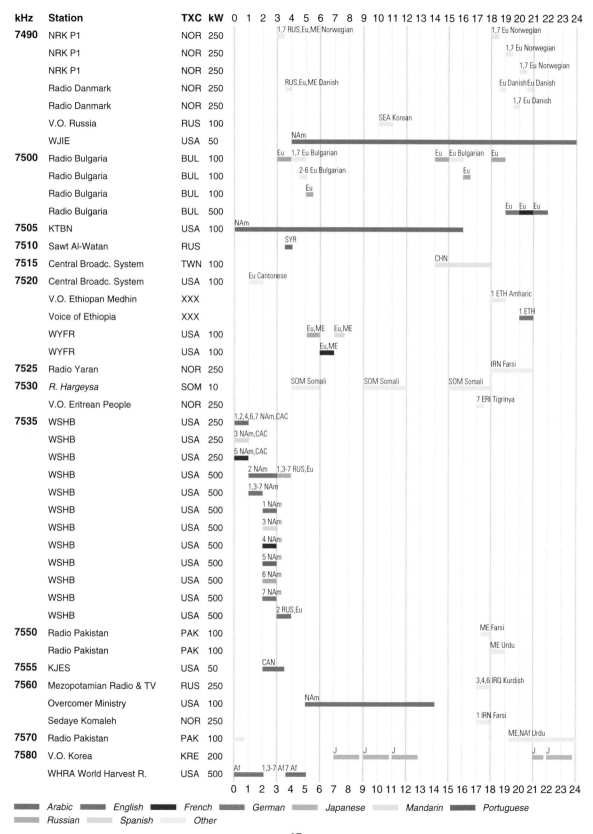

kHz	Station	TXC	kW
7490	NRK P1	NOR	250
	NRK P1	NOR	250
	NRK P1	NOR	250
	Radio Danmark	NOR	250
	Radio Danmark	NOR	250
	V.O. Russia	RUS	100
	WJIE	USA	50
7500	Radio Bulgaria	BUL	100
	Radio Bulgaria	BUL	100
	Radio Bulgaria	BUL	100
	Radio Bulgaria	BUL	500
7505	KTBN	USA	100
7510	Sawt Al-Watan	RUS	
7515	Central Broadc. System	TWN	100
7520	Central Broadc. System	USA	100
	V.O. Ethiopan Medhin	XXX	
	Voice of Ethiopia	XXX	
	WYFR	USA	100
	WYFR	USA	100
7525	Radio Yaran	NOR	250
7530	R. Hargeysa	SOM	10
	V.O. Eritrean People	NOR	250
7535	WSHB	USA	250
	WSHB	USA	250
	WSHB	USA	250
	WSHB	USA	500
	WSHB	USA	500
	WSHB	USA	500
	WSHB	USA	500
	WSHB	USA	500
	WSHB	USA	500
	WSHB	USA	500
	WSHB	USA	500
	WSHB	USA	500
7550	Radio Pakistan	PAK	100
	Radio Pakistan	PAK	100
7555	KJES	USA	50
7560	Mezopotamian Radio & TV	RUS	250
	Overcomer Ministry	USA	100
	Sedaye Komaleh	NOR	250
7570	Radio Pakistan	PAK	100
7580	V.O. Korea	KRE	200
	WHRA World Harvest R.	USA	500

Arabic   English   French   German   Japanese   Mandarin   Portuguese
Russian   Spanish   Other

67

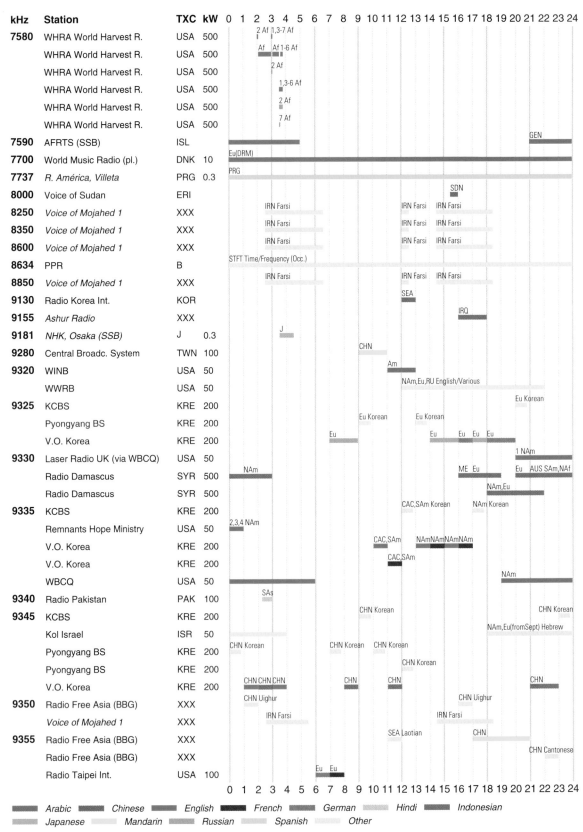

kHz	Station	TXC	kW
7580	WHRA World Harvest R.	USA	500
	WHRA World Harvest R.	USA	500
	WHRA World Harvest R.	USA	500
	WHRA World Harvest R.	USA	500
	WHRA World Harvest R.	USA	500
	WHRA World Harvest R.	USA	500
7590	AFRTS (SSB)	ISL	
7700	World Music Radio (pl.)	DNK	10
7737	R. América, Villeta	PRG	0.3
8000	Voice of Sudan	ERI	
8250	Voice of Mojahed 1	XXX	
8350	Voice of Mojahed 1	XXX	
8600	Voice of Mojahed 1	XXX	
8634	PPR	B	
8850	Voice of Mojahed 1	XXX	
9130	Radio Korea Int.	KOR	
9155	Ashur Radio	XXX	
9181	NHK, Osaka (SSB)	J	0.3
9280	Central Broadc. System	TWN	100
9320	WINB	USA	50
	WWRB	USA	50
9325	KCBS	KRE	200
	Pyongyang BS	KRE	200
	V.O. Korea	KRE	200
9330	Laser Radio UK (via WBCQ)	USA	50
	Radio Damascus	SYR	500
	Radio Damascus	SYR	500
9335	KCBS	KRE	200
	Remnants Hope Ministry	USA	50
	V.O. Korea	KRE	200
	V.O. Korea	KRE	200
	WBCQ	USA	50
9340	Radio Pakistan	PAK	100
9345	KCBS	KRE	200
	Kol Israel	ISR	50
	Pyongyang BS	KRE	200
	Pyongyang BS	KRE	200
	V.O. Korea	KRE	200
9350	Radio Free Asia (BBG)	XXX	
	Voice of Mojahed 1	XXX	
9355	Radio Free Asia (BBG)	XXX	
	Radio Free Asia (BBG)	XXX	
	Radio Taipei Int.	USA	100

Legend: Arabic · Chinese · English · French · German · Hindi · Indonesian · Japanese · Mandarin · Russian · Spanish · Other

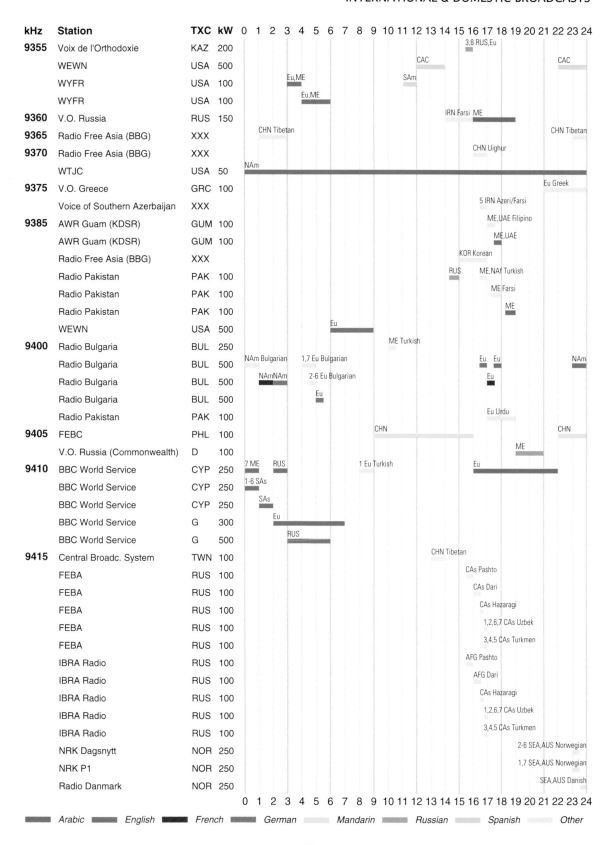

kHz	Station	TXC	kW	Notes
9355	Voix de l'Orthodoxie	KAZ	200	3,6 RUS,Eu
	WEWN	USA	500	CAC / CAC
	WYFR	USA	100	Eu,ME / SAm
	WYFR	USA	100	Eu,ME
9360	V.O. Russia	RUS	150	IRN Farsi / ME
9365	Radio Free Asia (BBG)	XXX		CHN Tibetan / CHN Tibetan
9370	Radio Free Asia (BBG)	XXX		CHN Uighur
	WTJC	USA	50	NAm
9375	V.O. Greece	GRC	100	Eu Greek
	Voice of Southern Azerbaijan	XXX		5 IRN Azeri/Farsi
9385	AWR Guam (KDSR)	GUM	100	ME,UAE Filipino
	AWR Guam (KDSR)	GUM	100	ME,UAE
	Radio Free Asia (BBG)	XXX		KOR Korean
	Radio Pakistan	PAK	100	RUS / ME,NAf Turkish
	Radio Pakistan	PAK	100	ME Farsi
	Radio Pakistan	PAK	100	ME
	WEWN	USA	500	Eu
9400	Radio Bulgaria	BUL	250	ME Turkish
	Radio Bulgaria	BUL	500	NAm Bulgarian / 1,7 Eu Bulgarian / Eu Eu / NAm
	Radio Bulgaria	BUL	500	NAmNAm / 2-6 Eu Bulgarian / Eu
	Radio Bulgaria	BUL	500	Eu
	Radio Pakistan	PAK	100	Eu Urdu
9405	FEBC	PHL	100	CHN / CHN
	V.O. Russia (Commonwealth)	D	100	ME
9410	BBC World Service	CYP	250	7 ME / RUS / 1 Eu Turkish / Eu
	BBC World Service	CYP	250	1-6 SAs
	BBC World Service	CYP	250	SAs
	BBC World Service	G	300	Eu
	BBC World Service	G	500	RUS
9415	Central Broadc. System	TWN	100	CHN Tibetan
	FEBA	RUS	100	CAs Pashto
	FEBA	RUS	100	CAs Dari
	FEBA	RUS	100	CAs Hazaragi
	FEBA	RUS	100	1,2,6,7 CAs Uzbek
	FEBA	RUS	100	3,4,5 CAs Turkmen
	IBRA Radio	RUS	100	AFG Pashto
	IBRA Radio	RUS	100	AFG Dari
	IBRA Radio	RUS	100	CAs Hazaragi
	IBRA Radio	RUS	100	1,2,6,7 CAs Uzbek
	IBRA Radio	RUS	100	3,4,5 CAs Turkmen
	NRK Dagsnytt	NOR	250	2-6 SEA,AUS Norwegian
	NRK P1	NOR	250	1,7 SEA,AUS Norwegian
	Radio Danmark	NOR	250	SEA,AUS Danish

Legend: *Arabic* · *English* · *French* · *German* · *Mandarin* · *Russian* · *Spanish* · *Other*

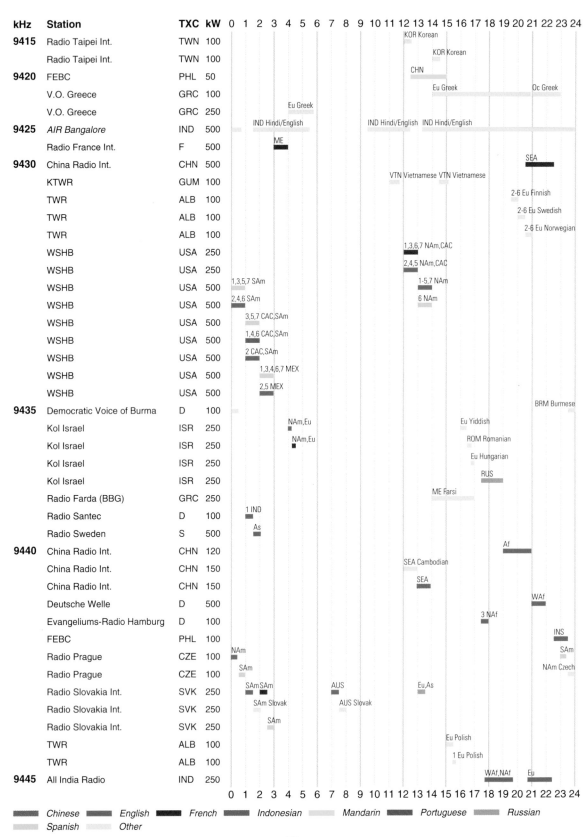

kHz	Station	TXC	kW
9415	Radio Taipei Int.	TWN	100
	Radio Taipei Int.	TWN	100
9420	FEBC	PHL	50
	V.O. Greece	GRC	100
	V.O. Greece	GRC	250
9425	*AIR Bangalore*	IND	500
	Radio France Int.	F	500
9430	China Radio Int.	CHN	500
	KTWR	GUM	100
	TWR	ALB	100
	TWR	ALB	100
	TWR	ALB	100
	WSHB	USA	250
	WSHB	USA	250
	WSHB	USA	500
	WSHB	USA	500
	WSHB	USA	500
	WSHB	USA	500
	WSHB	USA	500
	WSHB	USA	500
	WSHB	USA	500
9435	Democratic Voice of Burma	D	100
	Kol Israel	ISR	250
	Kol Israel	ISR	250
	Kol Israel	ISR	250
	Kol Israel	ISR	250
	Radio Farda (BBG)	GRC	250
	Radio Santec	D	100
	Radio Sweden	S	500
9440	China Radio Int.	CHN	120
	China Radio Int.	CHN	150
	China Radio Int.	CHN	150
	Deutsche Welle	D	500
	Evangeliums-Radio Hamburg	D	100
	FEBC	PHL	100
	Radio Prague	CZE	100
	Radio Prague	CZE	100
	Radio Slovakia Int.	SVK	250
	Radio Slovakia Int.	SVK	250
	Radio Slovakia Int.	SVK	250
	TWR	ALB	100
	TWR	ALB	100
9445	All India Radio	IND	250

Programme markings on the chart (by time band 0–24):

- 9415 Radio Taipei Int.: KOR Korean
- 9415 Radio Taipei Int.: KOR Korean
- 9420 FEBC: CHN
- 9420 V.O. Greece: Eu Greek / Oc Greek
- 9420 V.O. Greece: Eu Greek
- 9425 AIR Bangalore: IND Hindi/English; IND Hindi/English; IND Hindi/English
- 9425 Radio France Int.: ME
- 9430 China Radio Int.: SEA
- 9430 KTWR: VTN Vietnamese; VTN Vietnamese
- 9430 TWR: 2-6 Eu Finnish
- 9430 TWR: 2-6 Eu Swedish
- 9430 TWR: 2-6 Eu Norwegian
- 9430 WSHB: 1,3,6,7 NAm,CAC
- 9430 WSHB: 2,4,5 NAm,CAC
- 9430 WSHB: 1,3,5,7 SAm; 1-5,7 NAm
- 9430 WSHB: 2,4,6 SAm; 6 NAm
- 9430 WSHB: 3,5,7 CAC,SAm
- 9430 WSHB: 1,4,6 CAC,SAm
- 9430 WSHB: 2 CAC,SAm
- 9430 WSHB: 1,3,4,6,7 MEX
- 9430 WSHB: 2,5 MEX
- 9435 Democratic Voice of Burma: BRM Burmese
- 9435 Kol Israel: NAm,Eu; Eu Yiddish
- 9435 Kol Israel: NAm,Eu; ROM Romanian
- 9435 Kol Israel: Eu Hungarian
- 9435 Kol Israel: RUS
- 9435 Radio Farda (BBG): ME Farsi
- 9435 Radio Santec: 1 IND
- 9435 Radio Sweden: As
- 9440 China Radio Int.: Af
- 9440 China Radio Int.: SEA Cambodian
- 9440 China Radio Int.: SEA
- 9440 Deutsche Welle: WAf
- 9440 Evangeliums-Radio Hamburg: 3 NAf
- 9440 FEBC: INS
- 9440 Radio Prague: NAm; SAm
- 9440 Radio Prague: SAm; NAm Czech
- 9440 Radio Slovakia Int.: SAm SAm; AUS; Eu,As
- 9440 Radio Slovakia Int.: SAm Slovak; AUS Slovak
- 9440 Radio Slovakia Int.: SAm
- 9440 TWR: Eu Polish
- 9440 TWR: 1 Eu Polish
- 9445 All India Radio: WAf,NAf; Eu

Legend: Chinese · English · French · Indonesian · Mandarin · Portuguese · Russian · Spanish · Other

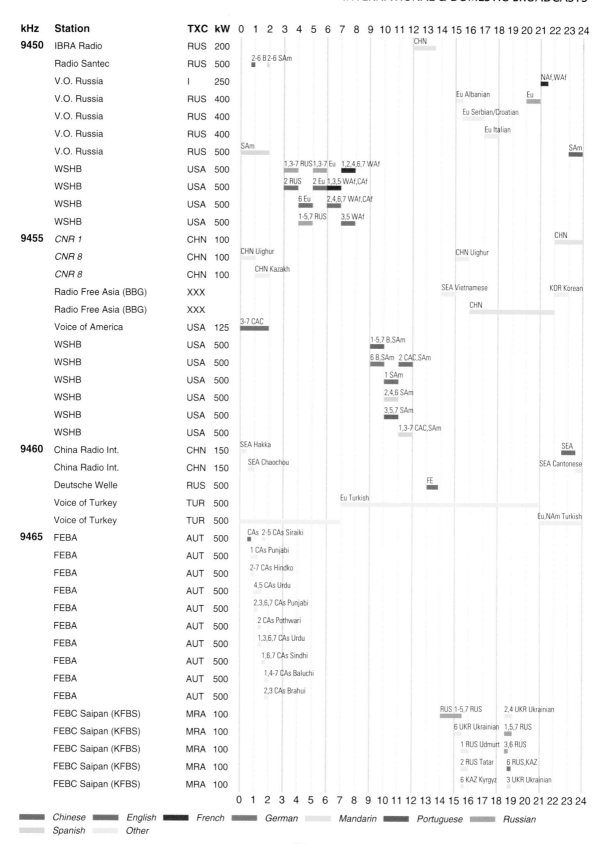

kHz	Station	TXC	kW																								
9450	IBRA Radio	RUS	200														CHN									
	Radio Santec	RUS	500		2-6 B 2-6 SAm																					
	V.O. Russia	I	250																		NAf,WAf				
	V.O. Russia	RUS	400													Eu Albanian				Eu				
	V.O. Russia	RUS	400													Eu Serbian/Croatian								
	V.O. Russia	RUS	400													Eu Italian							
	V.O. Russia	RUS	500	SAm																			SAm
	WSHB	USA	500				1,3-7 RUS 1,3-7 Eu		1,2,4,6,7 WAf																
	WSHB	USA	500				2 RUS		2 Eu 1,3,5 WAf,CAf															
	WSHB	USA	500				6 Eu		2,4,6,7 WAf,CAf															
	WSHB	USA	500				1-5,7 RUS		3,5 WAf														
9455	*CNR 1*	CHN	100																			CHN	
	CNR 8	CHN	100	CHN Uighur											CHN Uighur								
	CNR 8	CHN	100		CHN Kazakh																		
	Radio Free Asia (BBG)	XXX													SEA Vietnamese				KOR Korean			
	Radio Free Asia (BBG)	XXX												CHN								
	Voice of America	USA	125	3-7 CAC																		
	WSHB	USA	500								1-5,7 B,SAm										
	WSHB	USA	500							6 B,SAm	2 CAC,SAm										
	WSHB	USA	500								1 SAm										
	WSHB	USA	500								2,4,6 SAm										
	WSHB	USA	500								3,5,7 SAm										
	WSHB	USA	500									1,3-7 CAC,SAm								
9460	China Radio Int.	CHN	150	SEA Hakka																SEA
	China Radio Int.	CHN	150		SEA Chaochou															SEA Cantonese
	Deutsche Welle	RUS	500											FE						
	Voice of Turkey	TUR	500							Eu Turkish										
	Voice of Turkey	TUR	500																	Eu,NAm Turkish
9465	FEBA	AUT	500		CAs 2-5 CAs Siraiki															
	FEBA	AUT	500		1 CAs Punjabi															
	FEBA	AUT	500		2-7 CAs Hindko															
	FEBA	AUT	500			4,5 CAs Urdu														
	FEBA	AUT	500			2,3,6,7 CAs Punjabi														
	FEBA	AUT	500			2 CAs Pothwari														
	FEBA	AUT	500			1,3,6,7 CAs Urdu														
	FEBA	AUT	500				1,6,7 CAs Sindhi													
	FEBA	AUT	500				1,4-7 CAs Baluchi													
	FEBA	AUT	500				2,3 CAs Brahui													
	FEBC Saipan (KFBS)	MRA	100												RUS 1-5,7 RUS		2,4 UKR Ukrainian		
	FEBC Saipan (KFBS)	MRA	100												6 UKR Ukrainian	1,5,7 RUS		
	FEBC Saipan (KFBS)	MRA	100												1 RUS Udmurt	3,6 RUS		
	FEBC Saipan (KFBS)	MRA	100												2 RUS Tatar	6 RUS,KAZ		
	FEBC Saipan (KFBS)	MRA	100												6 KAZ Kyrgyz	3 UKR Ukrainian		

▬▬ *Chinese*	▬▬ *English*	▬ *French*	▬▬ *German*	▬▬ *Mandarin*	▬▬ *Portuguese*	▬▬ *Russian*
▬▬ *Spanish*	▬▬ *Other*					

THE SHORTWAVE GUIDE

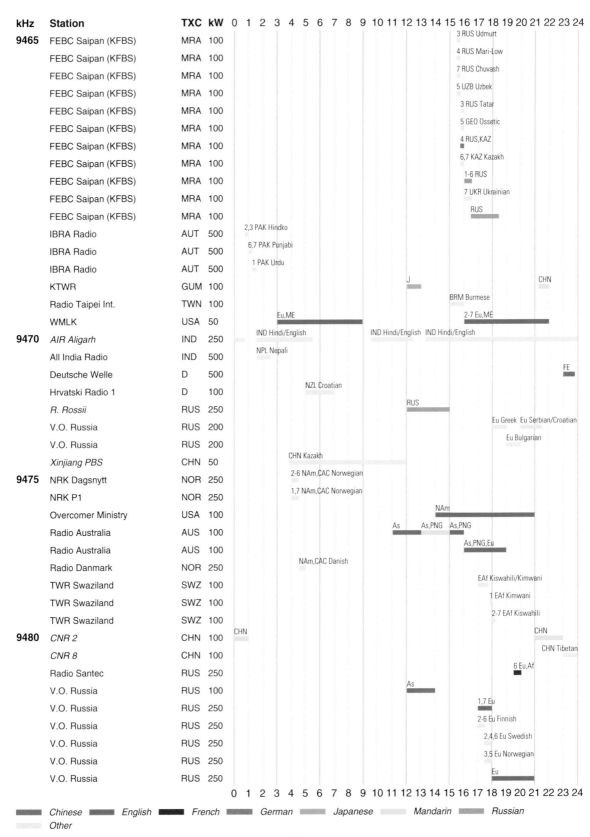

kHz	Station	TXC	kW	Schedule / Notes
9465	FEBC Saipan (KFBS)	MRA	100	3 RUS Udmurt
	FEBC Saipan (KFBS)	MRA	100	4 RUS Mari-Low
	FEBC Saipan (KFBS)	MRA	100	7 RUS Chuvash
	FEBC Saipan (KFBS)	MRA	100	5 UZB Uzbek
	FEBC Saipan (KFBS)	MRA	100	3 RUS Tatar
	FEBC Saipan (KFBS)	MRA	100	5 GEO Ossetic
	FEBC Saipan (KFBS)	MRA	100	4 RUS,KAZ
	FEBC Saipan (KFBS)	MRA	100	6,7 KAZ Kazakh
	FEBC Saipan (KFBS)	MRA	100	1-6 RUS
	FEBC Saipan (KFBS)	MRA	100	7 UKR Ukrainian
	FEBC Saipan (KFBS)	MRA	100	RUS
	IBRA Radio	AUT	500	2,3 PAK Hindko
	IBRA Radio	AUT	500	6,7 PAK Punjabi
	IBRA Radio	AUT	500	1 PAK Urdu
	KTWR	GUM	100	J / CHN
	Radio Taipei Int.	TWN	100	BRM Burmese
	WMLK	USA	50	Eu,ME / 2-7 Eu,ME
9470	*AIR Aligarh*	IND	250	IND Hindi/English / IND Hindi/English / IND Hindi/English
	All India Radio	IND	500	NPL Nepali
	Deutsche Welle	D	500	FE
	Hrvatski Radio 1	D	100	NZL Croatian
	R. Rossii	RUS	250	RUS
	V.O. Russia	RUS	200	Eu Greek / Eu Serbian/Croatian
	V.O. Russia	RUS	200	Eu Bulgarian
	Xinjiang PBS	CHN	50	CHN Kazakh
9475	NRK Dagsnytt	NOR	250	2-6 NAm,CAC Norwegian
	NRK P1	NOR	250	1,7 NAm,CAC Norwegian
	Overcomer Ministry	USA	100	NAm
	Radio Australia	AUS	100	As / As,PNG / As,PNG
	Radio Australia	AUS	100	As,PNG,Eu
	Radio Danmark	NOR	250	NAm,CAC Danish
	TWR Swaziland	SWZ	100	EAf Kiswahili/Kimwani
	TWR Swaziland	SWZ	100	1 EAf Kimwani
	TWR Swaziland	SWZ	100	2-7 EAf Kiswahili
9480	*CNR 2*	CHN	100	CHN / CHN
	CNR 8	CHN	100	CHN Tibetan
	Radio Santec	RUS	250	6 Eu,Af
	V.O. Russia	RUS	100	As
	V.O. Russia	RUS	250	1,7 Eu
	V.O. Russia	RUS	250	2-6 Eu Finnish
	V.O. Russia	RUS	250	2,4,6 Eu Swedish
	V.O. Russia	RUS	250	3,5 Eu Norwegian
	V.O. Russia	RUS	250	Eu

Legend: Chinese · English · French · German · Japanese · Mandarin · Russian · Other

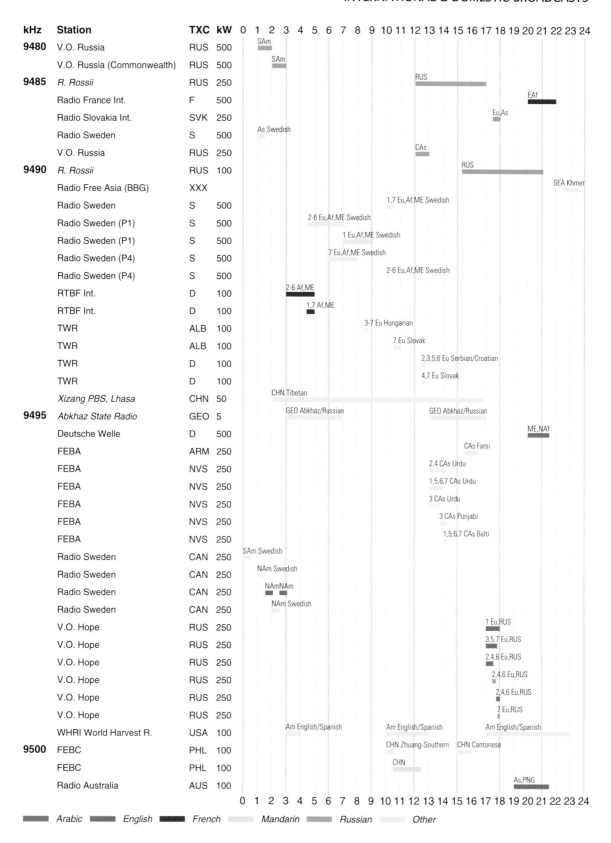

kHz	Station	TXC	kW	Schedule (0–24)
9480	V.O. Russia	RUS	500	SAm
	V.O. Russia (Commonwealth)	RUS	500	SAm
9485	*R. Rossii*	RUS	250	RUS
	Radio France Int.	F	500	EAf
	Radio Slovakia Int.	SVK	250	Eu,As
	Radio Sweden	S	500	As Swedish
	V.O. Russia	RUS	250	CAs
9490	*R. Rossii*	RUS	100	RUS
	Radio Free Asia (BBG)	XXX		SEA Khmer
	Radio Sweden	S	500	1,7 Eu,Af,ME Swedish
	Radio Sweden (P1)	S	500	2-6 Eu,Af,ME Swedish
	Radio Sweden (P1)	S	500	1 Eu,Af,ME Swedish
	Radio Sweden (P4)	S	500	7 Eu,Af,ME Swedish
	Radio Sweden (P4)	S	500	2-6 Eu,Af,ME Swedish
	RTBF Int.	D	100	2-6 Af,ME
	RTBF Int.	D	100	1,7 Af,ME
	TWR	ALB	100	3-7 Eu Hungarian
	TWR	ALB	100	7 Eu Slovak
	TWR	D	100	2,3,5,6 Eu Serbian/Croatian
	TWR	D	100	4,7 Eu Slovak
	Xizang PBS, Lhasa	CHN	50	CHN Tibetan
9495	*Abkhaz State Radio*	GEO	5	GEO Abkhaz/Russian ... GEO Abkhaz/Russian
	Deutsche Welle	D	500	ME,NAf
	FEBA	ARM	250	CAs Farsi
	FEBA	NVS	250	2,4 CAs Urdu
	FEBA	NVS	250	1,5,6,7 CAs Urdu
	FEBA	NVS	250	3 CAs Urdu
	FEBA	NVS	250	3 CAs Punjabi
	FEBA	NVS	250	1,5,6,7 CAs Balti
	Radio Sweden	CAN	250	SAm Swedish
	Radio Sweden	CAN	250	NAm Swedish
	Radio Sweden	CAN	250	NAmNAm
	Radio Sweden	CAN	250	NAm Swedish
	V.O. Hope	RUS	250	1 Eu,RUS
	V.O. Hope	RUS	250	3,5,7 Eu,RUS
	V.O. Hope	RUS	250	2,4,6 Eu,RUS
	V.O. Hope	RUS	250	2,4,6 Eu,RUS
	V.O. Hope	RUS	250	2,4,6 Eu,RUS
	V.O. Hope	RUS	250	7 Eu,RUS
	WHRI World Harvest R.	USA	100	Am English/Spanish ... Am English/Spanish ... Am English/Spanish
9500	FEBC	PHL	100	CHN Zhuang-Southern ... CHN Cantonese
	FEBC	PHL	100	CHN
	Radio Australia	AUS	100	As,PNG

Legend: Arabic — English — French — Mandarin — Russian — Other

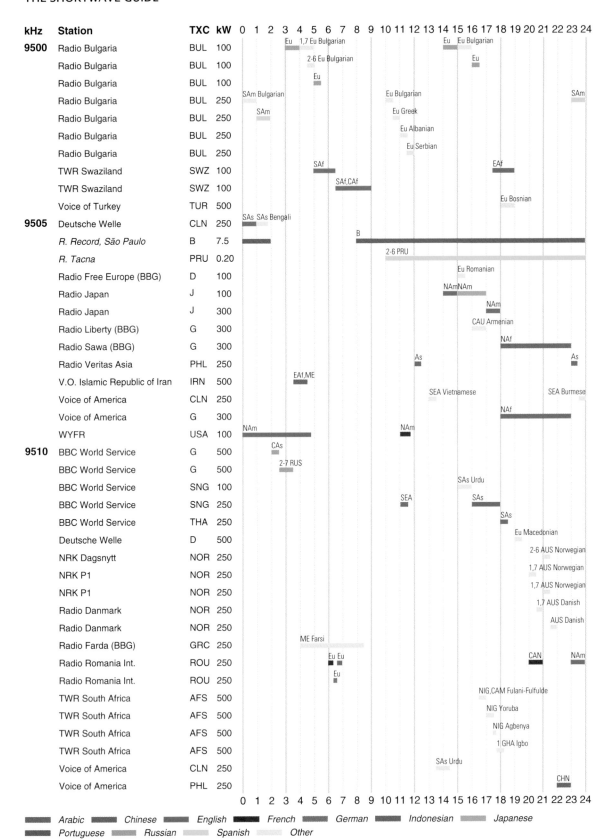

kHz	Station	TXC	kW	Schedule (0–24)
9500	Radio Bulgaria	BUL	100	Eu 1,7 Eu Bulgarian · Eu Eu Bulgarian
	Radio Bulgaria	BUL	100	2-6 Eu Bulgarian · Eu
	Radio Bulgaria	BUL	100	Eu
	Radio Bulgaria	BUL	250	SAm Bulgarian · Eu Bulgarian · SAm
	Radio Bulgaria	BUL	250	SAm · Eu Greek
	Radio Bulgaria	BUL	250	Eu Albanian
	Radio Bulgaria	BUL	250	Eu Serbian
	TWR Swaziland	SWZ	100	SAf · EAf
	TWR Swaziland	SWZ	100	SAf,CAf
	Voice of Turkey	TUR	500	Eu Bosnian
9505	Deutsche Welle	CLN	250	SAs SAs Bengali
	R. Record, São Paulo	B	7.5	B
	R. Tacna	PRU	0.20	2-6 PRU
	Radio Free Europe (BBG)	D	100	Eu Romanian
	Radio Japan	J	100	NAmNAm
	Radio Japan	J	300	NAm
	Radio Liberty (BBG)	G	300	CAU Armenian
	Radio Sawa (BBG)	G	300	NAf
	Radio Veritas Asia	PHL	250	As · As
	V.O. Islamic Republic of Iran	IRN	500	EAf,ME
	Voice of America	CLN	250	SEA Vietnamese · SEA Burmese
	Voice of America	G	300	NAf
	WYFR	USA	100	NAm · NAm
9510	BBC World Service	G	500	CAs
	BBC World Service	G	500	2-7 RUS
	BBC World Service	SNG	100	SAs Urdu
	BBC World Service	SNG	250	SEA · SAs
	BBC World Service	THA	250	SAs
	Deutsche Welle	D	500	Eu Macedonian
	NRK Dagsnytt	NOR	250	2-6 AUS Norwegian
	NRK P1	NOR	250	1,7 AUS Norwegian
	NRK P1	NOR	250	1,7 AUS Norwegian
	Radio Danmark	NOR	250	1,7 AUS Danish
	Radio Danmark	NOR	250	AUS Danish
	Radio Farda (BBG)	GRC	250	ME Farsi
	Radio Romania Int.	ROU	250	Eu Eu · CAN · NAm
	Radio Romania Int.	ROU	250	Eu
	TWR South Africa	AFS	500	NIG,CAM Fulani-Fulfulde
	TWR South Africa	AFS	500	NIG Yoruba
	TWR South Africa	AFS	500	NIG Agbenya
	TWR South Africa	AFS	500	1 GHA Igbo
	Voice of America	CLN	250	SAs Urdu
	Voice of America	PHL	250	CHN

Legend: Arabic · Chinese · English · French · German · Indonesian · Japanese · Portuguese · Russian · Spanish · Other

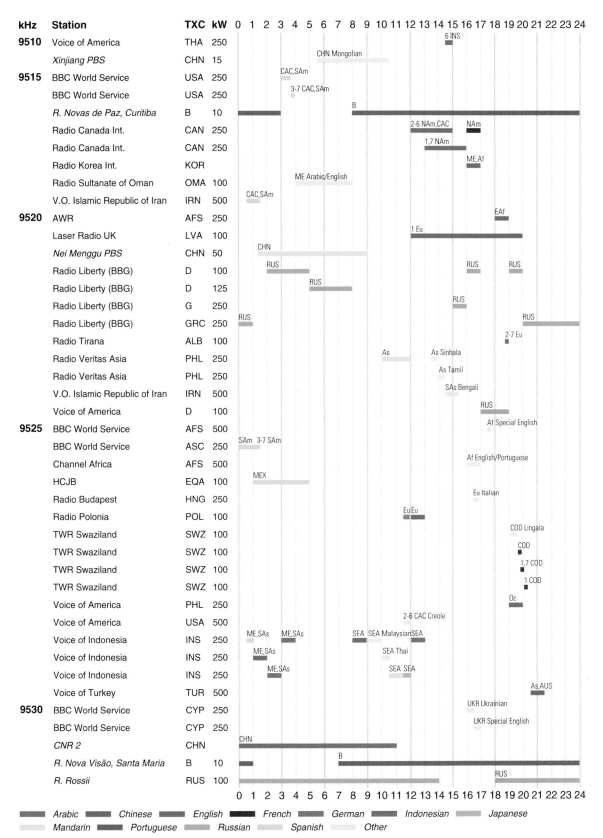

kHz	Station	TXC	kW	0	1	2	3	4	5	6	7	8	9	10	11	12	13	14	15	16	17	18	19	20	21	22	23	24
9510	Voice of America	THA	250																6 INS									
	Xinjiang PBS	CHN	15							CHN Mongolian																		
9515	BBC World Service	USA	250				CAC,SAm																					
	BBC World Service	USA	250				3-7 CAC,SAm																					
	R. Novas de Paz, Curitiba	B	10									B																
	Radio Canada Int.	CAN	250													2-6 NAm,CAC		NAm									
	Radio Canada Int.	CAN	250													1,7 NAm											
	Radio Korea Int.	KOR																		ME,Af							
	Radio Sultanate of Oman	OMA	100							ME Arabic/English																	
	V.O. Islamic Republic of Iran	IRN	500		CAC,SAm																						
9520	AWR	AFS	250																	EAf							
	Laser Radio UK	LVA	100													1 Eu											
	Nei Menggu PBS	CHN	50			CHN																					
	Radio Liberty (BBG)	D	100			RUS													RUS		RUS						
	Radio Liberty (BBG)	D	125					RUS																			
	Radio Liberty (BBG)	G	250														RUS										
	Radio Liberty (BBG)	GRC	250	RUS																		RUS					
	Radio Tirana	ALB	100																		2-7 Eu						
	Radio Veritas Asia	PHL	250									As			As Sinhala											
	Radio Veritas Asia	PHL	250												As Tamil											
	V.O. Islamic Republic of Iran	IRN	500												SAs Bengali											
	Voice of America	D	100																RUS								
9525	BBC World Service	AFS	500																	Af Special English						
	BBC World Service	ASC	250	SAm	3-7 SAm																						
	Channel Africa	AFS	500																Af English/Portuguese							
	HCJB	EQA	100		MEX																						
	Radio Budapest	HNG	250																	Eu Italian							
	Radio Polonia	POL	100													Eu Eu											
	TWR Swaziland	SWZ	100																			COD Lingala					
	TWR Swaziland	SWZ	100																			COD					
	TWR Swaziland	SWZ	100																			1,7 COD					
	TWR Swaziland	SWZ	100																			1 COD					
	Voice of America	PHL	250																			Oc					
	Voice of America	USA	500												2-6 CAC Creole												
	Voice of Indonesia	INS	250	ME,SAs		ME,SAs						SEA	SEA Malaysian	SEA												
	Voice of Indonesia	INS	250		ME,SAs									SEA Thai												
	Voice of Indonesia	INS	250			ME,SAs							SEA	SEA												
	Voice of Turkey	TUR	500																			As,AUS					
9530	BBC World Service	CYP	250													UKR Ukrainian											
	BBC World Service	CYP	250													UKR Special English											
	CNR 2	CHN			CHN																						
	R. Nova Visão, Santa Maria	B	10								B																	
	R. Rossii	RUS	100																			RUS					

Arabic ▬ *Chinese* ▬ *English* ▬ *French* ▬ *German* ▬ *Indonesian* ▬ *Japanese* ▬ *Mandarin* ▬ *Portuguese* ▬ *Russian* ▬ *Spanish* ▬ *Other* ▬

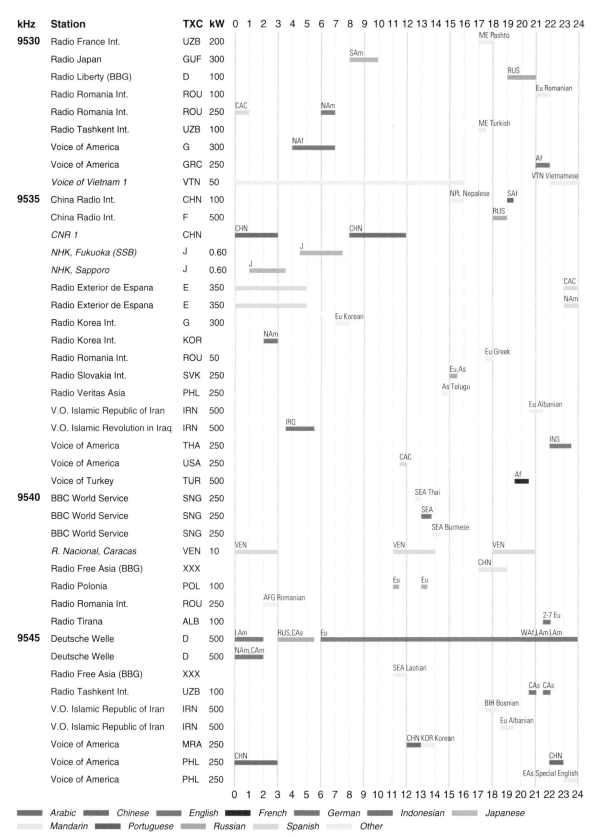

kHz	Station	TXC	kW	Schedule
9530	Radio France Int.	UZB	200	ME Pashto (17-18)
	Radio Japan	GUF	300	SAm (8-10)
	Radio Liberty (BBG)	D	100	RUS (19-21)
	Radio Romania Int.	ROU	100	Eu Romanian (21-23)
	Radio Romania Int.	ROU	250	CAC (0-1); NAm (6-7)
	Radio Tashkent Int.	UZB	100	ME Turkish (17-18)
	Voice of America	G	300	NAf (4-7)
	Voice of America	GRC	250	Af (21-22)
	Voice of Vietnam 1	VTN	50	VTN Vietnamese (0-24)
9535	China Radio Int.	CHN	100	NPL Nepalese (15-17); SAf (19-20)
	China Radio Int.	F	500	RUS (18-19)
	CNR 1	CHN		CHN (0-3); CHN (7-12)
	NHK, Fukuoka (SSB)	J	0.60	J (5-7)
	NHK, Sapporo	J	0.60	J (1-3)
	Radio Exterior de Espana	E	350	CAC (22-23)
	Radio Exterior de Espana	E	350	NAm (22-23)
	Radio Korea Int.	G	300	Eu Korean (6-8)
	Radio Korea Int.	KOR		NAm (2-3)
	Radio Romania Int.	ROU	50	Eu Greek (17-18)
	Radio Slovakia Int.	SVK	250	Eu,As (13-14)
	Radio Veritas Asia	PHL	250	As Telugu (13-14)
	V.O. Islamic Republic of Iran	IRN	500	Eu Albanian (21-22)
	V.O. Islamic Revolution in Iraq	IRN	500	IRQ (4-6)
	Voice of America	THA	250	INS (22-23)
	Voice of America	USA	250	CAC (11-12)
	Voice of Turkey	TUR	500	Af (18-19)
9540	BBC World Service	SNG	250	SEA Thai (12-13)
	BBC World Service	SNG	250	SEA (12-13)
	BBC World Service	SNG	250	SEA Burmese (13-14)
	R. Nacional, Caracas	VEN	10	VEN (0-1); VEN (11-12); VEN (18-19)
	Radio Free Asia (BBG)	XXX		CHN (17-19)
	Radio Polonia	POL	100	Eu (11-12); Eu (13-14)
	Radio Romania Int.	ROU	250	AFG Romanian (3-5)
	Radio Tirana	ALB	100	2-7 Eu (22-23)
9545	Deutsche Welle	D	500	LAm (0-1); RUS,CAs (4-5); Eu (6-7); WAf,LAm LAm (18-24)
	Deutsche Welle	D	500	NAm,CAm (0-3)
	Radio Free Asia (BBG)	XXX		SEA Laotian (12-13)
	Radio Tashkent Int.	UZB	100	CAs (20-21); CAs (21-22)
	V.O. Islamic Republic of Iran	IRN	500	BIH Bosnian (18-19)
	V.O. Islamic Republic of Iran	IRN	500	Eu Albanian (19-20)
	Voice of America	MRA	250	CHN KOR Korean (12-14)
	Voice of America	PHL	250	CHN (0-3); CHN (22-23)
	Voice of America	PHL	250	EAs Special English (23-24)

Arabic ▪ Chinese ▪ English ▪ French ▪ German ▪ Indonesian ▪ Japanese
Mandarin ▪ Portuguese ▪ Russian ▪ Spanish ▪ Other

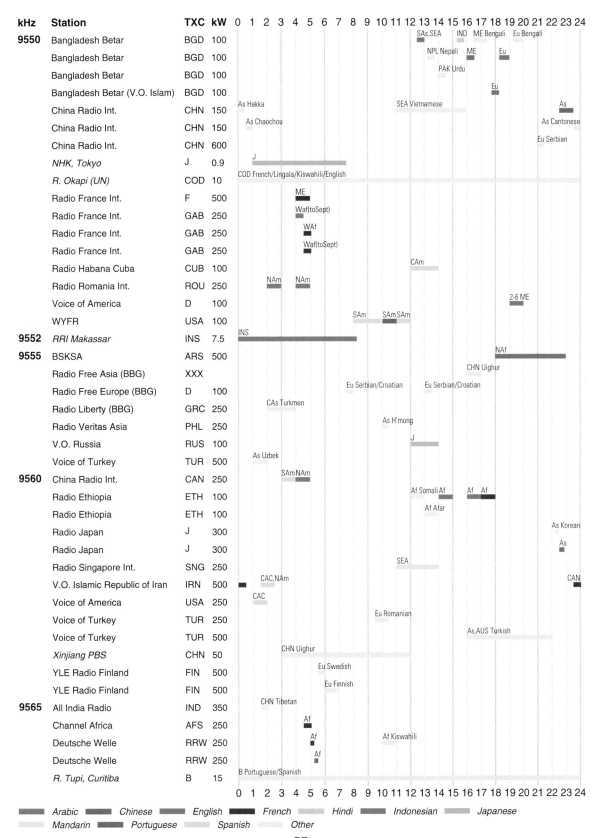

kHz	Station	TXC	kW
9550	Bangladesh Betar	BGD	100
	Bangladesh Betar	BGD	100
	Bangladesh Betar	BGD	100
	Bangladesh Betar (V.O. Islam)	BGD	100
	China Radio Int.	CHN	150
	China Radio Int.	CHN	150
	China Radio Int.	CHN	600
	NHK, Tokyo	J	0.9
	R. Okapi (UN)	COD	10
	Radio France Int.	F	500
	Radio France Int.	GAB	250
	Radio France Int.	GAB	250
	Radio France Int.	GAB	250
	Radio Habana Cuba	CUB	100
	Radio Romania Int.	ROU	250
	Voice of America	D	100
	WYFR	USA	100
9552	RRI Makassar	INS	7.5
9555	BSKSA	ARS	500
	Radio Free Asia (BBG)	XXX	
	Radio Free Europe (BBG)	D	100
	Radio Liberty (BBG)	GRC	250
	Radio Veritas Asia	PHL	250
	V.O. Russia	RUS	100
	Voice of Turkey	TUR	500
9560	China Radio Int.	CAN	250
	Radio Ethiopia	ETH	100
	Radio Ethiopia	ETH	100
	Radio Japan	J	300
	Radio Japan	J	300
	Radio Singapore Int.	SNG	250
	V.O. Islamic Republic of Iran	IRN	500
	Voice of America	USA	250
	Voice of Turkey	TUR	250
	Voice of Turkey	TUR	500
	Xinjiang PBS	CHN	50
	YLE Radio Finland	FIN	500
	YLE Radio Finland	FIN	500
9565	All India Radio	IND	350
	Channel Africa	AFS	250
	Deutsche Welle	RRW	250
	Deutsche Welle	RRW	250
	R. Tupi, Curitiba	B	15

Legend: Arabic · Chinese · English · French · Hindi · Indonesian · Japanese · Mandarin · Portuguese · Spanish · Other

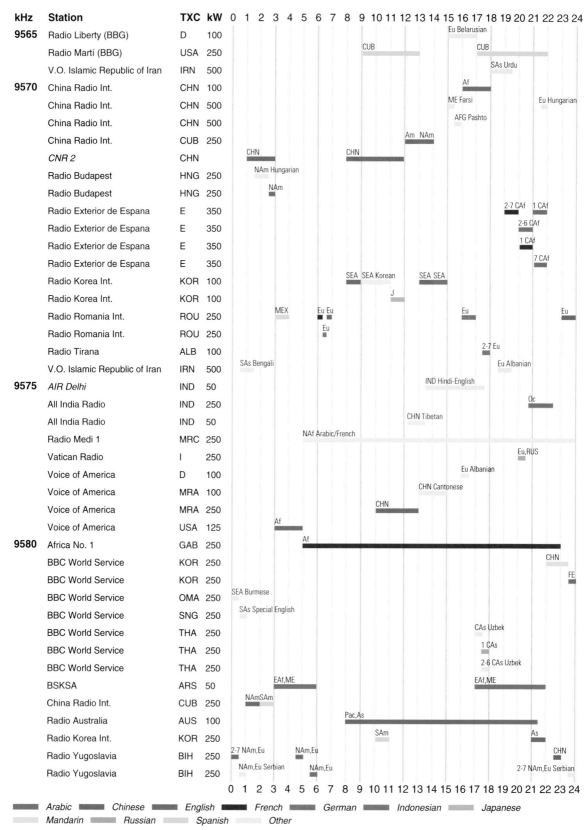

kHz	Station	TXC	kW
9565	Radio Liberty (BBG)	D	100
	Radio Martí (BBG)	USA	250
	V.O. Islamic Republic of Iran	IRN	500
9570	China Radio Int.	CHN	100
	China Radio Int.	CHN	500
	China Radio Int.	CHN	500
	China Radio Int.	CUB	250
	CNR 2	CHN	
	Radio Budapest	HNG	250
	Radio Budapest	HNG	250
	Radio Exterior de Espana	E	350
	Radio Exterior de Espana	E	350
	Radio Exterior de Espana	E	350
	Radio Exterior de Espana	E	350
	Radio Korea Int.	KOR	100
	Radio Korea Int.	KOR	100
	Radio Romania Int.	ROU	250
	Radio Romania Int.	ROU	250
	Radio Tirana	ALB	100
	V.O. Islamic Republic of Iran	IRN	500
9575	*AIR Delhi*	IND	50
	All India Radio	IND	250
	All India Radio	IND	50
	Radio Medi 1	MRC	250
	Vatican Radio	I	250
	Voice of America	D	100
	Voice of America	MRA	100
	Voice of America	MRA	250
	Voice of America	USA	125
9580	Africa No. 1	GAB	250
	BBC World Service	KOR	250
	BBC World Service	KOR	250
	BBC World Service	OMA	250
	BBC World Service	SNG	250
	BBC World Service	THA	250
	BBC World Service	THA	250
	BBC World Service	THA	250
	BSKSA	ARS	50
	China Radio Int.	CUB	250
	Radio Australia	AUS	100
	Radio Korea Int.	KOR	250
	Radio Yugoslavia	BIH	250
	Radio Yugoslavia	BIH	250

Legend: *Arabic* · *Chinese* · *English* · *French* · *German* · *Indonesian* · *Japanese* · *Mandarin* · *Russian* · *Spanish* · *Other*

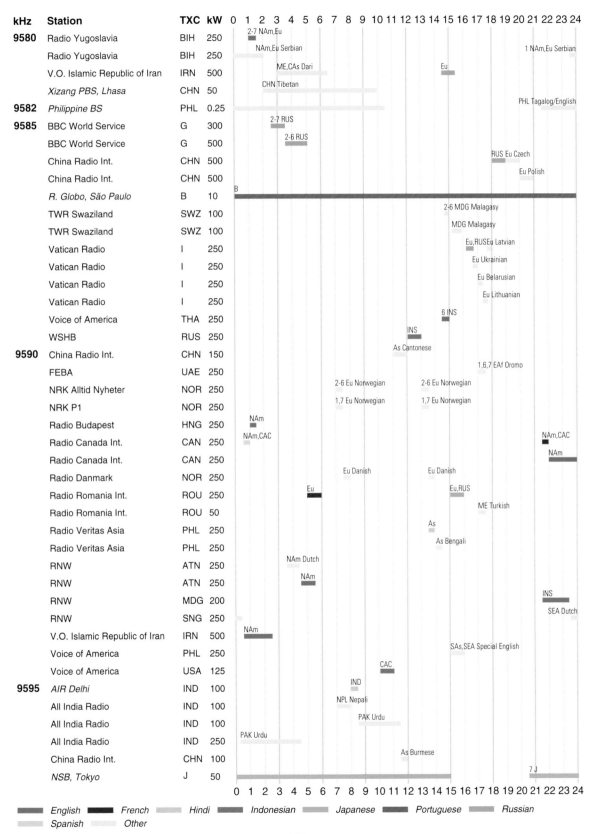

kHz	Station	TXC	kW	Schedule
9580	Radio Yugoslavia	BIH	250	2-7 NAm,Eu
	Radio Yugoslavia	BIH	250	NAm,Eu Serbian / 1 NAm,Eu Serbian
	V.O. Islamic Republic of Iran	IRN	500	ME,CAs Dari / Eu
	Xizang PBS, Lhasa	CHN	50	CHN Tibetan
9582	Philippine BS	PHL	0.25	PHL Tagalog/English
9585	BBC World Service	G	300	2-7 RUS
	BBC World Service	G	500	2-6 RUS
	China Radio Int.	CHN	500	RUS Eu Czech
	China Radio Int.	CHN	500	Eu Polish
	R. Globo, São Paulo	B	10	B
	TWR Swaziland	SWZ	100	2-6 MDG Malagasy
	TWR Swaziland	SWZ	100	MDG Malagasy
	Vatican Radio	I	250	Eu,RUSEu Latvian
	Vatican Radio	I	250	Eu Ukrainian
	Vatican Radio	I	250	Eu Belarusian
	Vatican Radio	I	250	Eu Lithuanian
	Voice of America	THA	250	6 INS
	WSHB	RUS	250	INS
9590	China Radio Int.	CHN	150	As Cantonese
	FEBA	UAE	250	1,6,7 EAf Oromo
	NRK Alltid Nyheter	NOR	250	2-6 Eu Norwegian / 2-6 Eu Norwegian
	NRK P1	NOR	250	1,7 Eu Norwegian / 1,7 Eu Norwegian
	Radio Budapest	HNG	250	NAm
	Radio Canada Int.	CAN	250	NAm,CAC / NAm,CAC
	Radio Canada Int.	CAN	250	NAm
	Radio Danmark	NOR	250	Eu Danish / Eu Danish
	Radio Romania Int.	ROU	250	Eu / Eu,RUS
	Radio Romania Int.	ROU	50	ME Turkish
	Radio Veritas Asia	PHL	250	As
	Radio Veritas Asia	PHL	250	As Bengali
	RNW	ATN	250	NAm Dutch
	RNW	ATN	250	NAm
	RNW	MDG	200	INS
	RNW	SNG	250	SEA Dutch
	V.O. Islamic Republic of Iran	IRN	500	NAm
	Voice of America	PHL	250	SAs,SEA Special English
	Voice of America	USA	125	CAC
9595	AIR Delhi	IND	100	IND
	All India Radio	IND	100	NPL Nepali
	All India Radio	IND	100	PAK Urdu
	All India Radio	IND	250	PAK Urdu
	China Radio Int.	CHN	100	As Burmese
	NSB, Tokyo	J	50	7 J

English   French   Hindi   Indonesian   Japanese   Portuguese   Russian
Spanish   Other

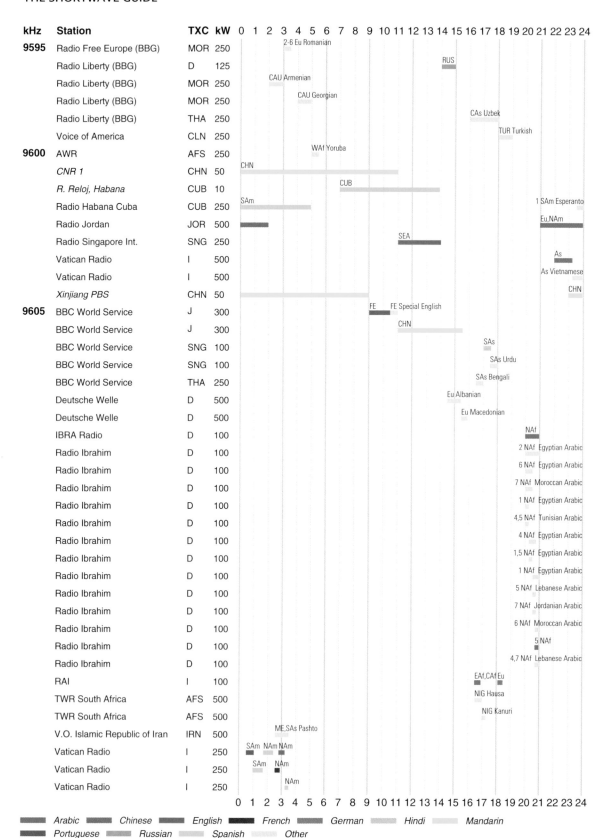

kHz	Station	TXC	kW
9595	Radio Free Europe (BBG)	MOR	250
	Radio Liberty (BBG)	D	125
	Radio Liberty (BBG)	MOR	250
	Radio Liberty (BBG)	MOR	250
	Radio Liberty (BBG)	THA	250
	Voice of America	CLN	250
9600	AWR	AFS	250
	CNR 1	CHN	50
	R. Reloj, Habana	CUB	10
	Radio Habana Cuba	CUB	250
	Radio Jordan	JOR	500
	Radio Singapore Int.	SNG	250
	Vatican Radio	I	500
	Vatican Radio	I	500
	Xinjiang PBS	CHN	50
9605	BBC World Service	J	300
	BBC World Service	J	300
	BBC World Service	SNG	100
	BBC World Service	SNG	100
	BBC World Service	THA	250
	Deutsche Welle	D	500
	Deutsche Welle	D	500
	IBRA Radio	D	100
	Radio Ibrahim	D	100
	Radio Ibrahim	D	100
	Radio Ibrahim	D	100
	Radio Ibrahim	D	100
	Radio Ibrahim	D	100
	Radio Ibrahim	D	100
	Radio Ibrahim	D	100
	Radio Ibrahim	D	100
	Radio Ibrahim	D	100
	Radio Ibrahim	D	100
	Radio Ibrahim	D	100
	Radio Ibrahim	D	100
	Radio Ibrahim	D	100
	RAI	I	100
	TWR South Africa	AFS	500
	TWR South Africa	AFS	500
	V.O. Islamic Republic of Iran	IRN	500
	Vatican Radio	I	250
	Vatican Radio	I	250
	Vatican Radio	I	250

Legend: Arabic Chinese English French German Hindi Mandarin Portuguese Russian Spanish Other

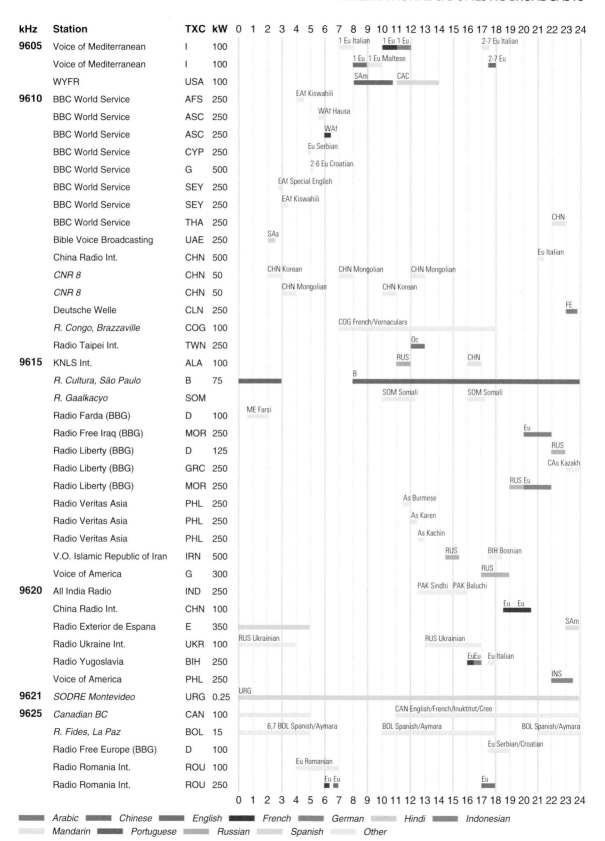

kHz	Station	TXC	kW	0	1	2	3	4	5	6	7	8	9	10	11	12	13	14	15	16	17	18	19	20	21	22	23	24
9605	Voice of Mediterranean	I	100									1 Eu Italian		1 Eu 1 Eu							2-7 Eu Italian							
	Voice of Mediterranean	I	100									1 Eu 1 Eu Maltese									2-7 Eu							
	WYFR	USA	100									SAm			CAC													
9610	BBC World Service	AFS	250				EAf Kiswahili																				
	BBC World Service	ASC	250					WAf Hausa																			
	BBC World Service	ASC	250						WAf																		
	BBC World Service	CYP	250				Eu Serbian																				
	BBC World Service	G	500				2-6 Eu Croatian																				
	BBC World Service	SEY	250			EAf Special English																					
	BBC World Service	SEY	250			EAf Kiswahili																					
	BBC World Service	THA	250																						CHN		
	Bible Voice Broadcasting	UAE	250			SAs																					
	China Radio Int.	CHN	500																					Eu Italian			
	CNR 8	CHN	50			CHN Korean					CHN Mongolian				CHN Mongolian											
	CNR 8	CHN	50				CHN Mongolian							CHN Korean												
	Deutsche Welle	CLN	250																							FE	
	R. Congo, Brazzaville	COG	100								COG French/Vernaculars																
	Radio Taipei Int.	TWN	250													Oc											
9615	KNLS Int.	ALA	100									RUS			CHN												
	R. Cultura, São Paulo	B	75								B																
	R. Gaalkacyo	SOM										SOM Somali				SOM Somali											
	Radio Farda (BBG)	D	100		ME Farsi																						
	Radio Free Iraq (BBG)	MOR	250																		Eu						
	Radio Liberty (BBG)	D	125																				RUS				
	Radio Liberty (BBG)	GRC	250																					CAs Kazakh			
	Radio Liberty (BBG)	MOR	250																		RUS Eu						
	Radio Veritas Asia	PHL	250									As Burmese															
	Radio Veritas Asia	PHL	250									As Karen															
	Radio Veritas Asia	PHL	250									As Kachin															
	V.O. Islamic Republic of Iran	IRN	500													RUS			BIH Bosnian								
	Voice of America	G	300																RUS								
9620	All India Radio	IND	250											PAK Sindhi PAK Baluchi													
	China Radio Int.	CHN	100																	Eu Eu							
	Radio Exterior de Espana	E	350																					SAm			
	Radio Ukraine Int.	UKR	100	RUS Ukrainian											RUS Ukrainian												
	Radio Yugoslavia	BIH	250																EuEu Eu Italian								
	Voice of America	PHL	250																					INS			
9621	SODRE Montevideo	URG	0.25	URG																							
9625	Canadian BC	CAN	100							CAN English/French/Inuktitut/Cree																	
	R. Fides, La Paz	BOL	15			6,7 BOL Spanish/Aymara							BOL Spanish/Aymara									BOL Spanish/Aymara				
	Radio Free Europe (BBG)	D	100																	Eu Serbian/Croatian							
	Radio Romania Int.	ROU	100			Eu Romanian																					
	Radio Romania Int.	ROU	250						Eu Eu											Eu							

| | | 0 | 1 | 2 | 3 | 4 | 5 | 6 | 7 | 8 | 9 | 10 | 11 | 12 | 13 | 14 | 15 | 16 | 17 | 18 | 19 | 20 | 21 | 22 | 23 | 24 |

Arabic   Chinese   English   French   German   Hindi   Indonesian
Mandarin   Portuguese   Russian   Spanish   Other

81

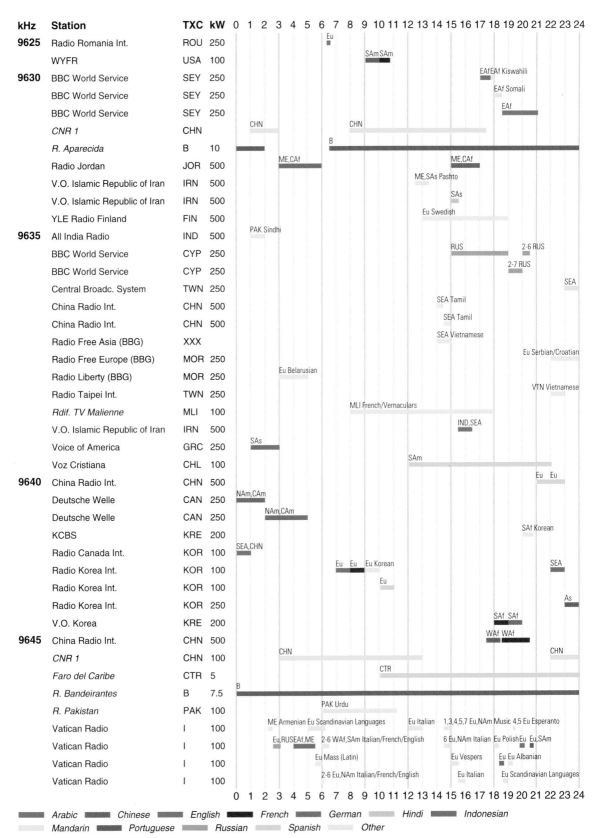

kHz	Station	TXC	kW	0-24 transmission schedule
9625	Radio Romania Int.	ROU	250	Eu
	WYFR	USA	100	SAm SAm
9630	BBC World Service	SEY	250	EAf EAf Kiswahili
	BBC World Service	SEY	250	EAf Somali
	BBC World Service	SEY	250	EAf
	CNR 1	CHN		CHN CHN
	R. Aparecida	B	10	B
	Radio Jordan	JOR	500	ME,CAf ME,CAf
	V.O. Islamic Republic of Iran	IRN	500	ME,SAs Pashto
	V.O. Islamic Republic of Iran	IRN	500	SAs
	YLE Radio Finland	FIN	500	Eu Swedish
9635	All India Radio	IND	500	PAK Sindhi
	BBC World Service	CYP	250	RUS 2-6 RUS
	BBC World Service	CYP	250	2-7 RUS
	Central Broadc. System	TWN	250	SEA
	China Radio Int.	CHN	500	SEA Tamil
	China Radio Int.	CHN	500	SEA Tamil
	Radio Free Asia (BBG)	XXX		SEA Vietnamese
	Radio Free Europe (BBG)	MOR	250	Eu Serbian/Croatian
	Radio Liberty (BBG)	MOR	250	Eu Belarusian
	Radio Taipei Int.	TWN	250	VTN Vietnamese
	Rdif. TV Malienne	MLI	100	MLI French/Vernaculars
	V.O. Islamic Republic of Iran	IRN	500	IND,SEA
	Voice of America	GRC	250	SAs
	Voz Cristiana	CHL	100	SAm
9640	China Radio Int.	CHN	500	Eu Eu
	Deutsche Welle	CAN	250	NAm,CAm
	Deutsche Welle	CAN	250	NAm,CAm
	KCBS	KRE	200	SAf Korean
	Radio Canada Int.	KOR	100	SEA,CHN
	Radio Korea Int.	KOR	100	Eu Eu Eu Korean SEA
	Radio Korea Int.	KOR	100	Eu
	Radio Korea Int.	KOR	250	As
	V.O. Korea	KRE	200	SAf SAf
9645	China Radio Int.	CHN	500	WAf WAf
	CNR 1	CHN	100	CHN CHN
	Faro del Caribe	CTR	5	CTR
	R. Bandeirantes	B	7.5	B
	R. Pakistan	PAK	100	PAK Urdu
	Vatican Radio	I	100	ME Armenian Eu Scandinavian Languages Eu Italian 1,3,4,5,7 Eu,NAm Music 4,5 Eu Esperanto
	Vatican Radio	I	100	Eu,RUSEAf,ME 2-6 WAf,SAm Italian/French/English 6 Eu,NAm Italian Eu PolishEu Eu,SAm
	Vatican Radio	I	100	Eu Mass (Latin) Eu Vespers Eu Eu Albanian
	Vatican Radio	I	100	2-6 Eu,NAm Italian/French/English Eu Italian Eu Scandinavian Languages

Legend:

Arabic	Chinese	English	French
German	Hindi	Indonesian	
Mandarin	Portuguese	Russian	Spanish
Other			

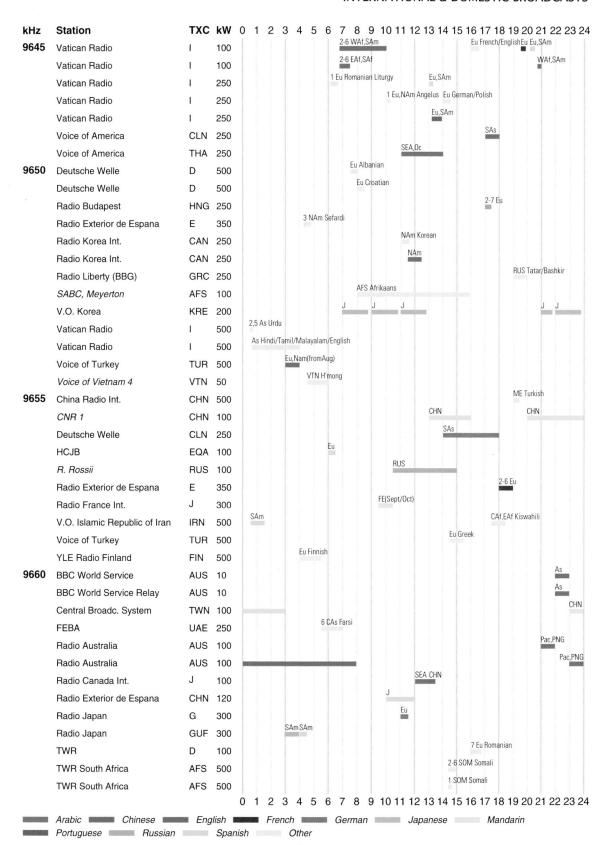

kHz	Station	TXC	kW	Broadcast schedule (0–24)
9645	Vatican Radio	I	100	2-6 WAf,SAm; Eu French/English Eu Eu,SAm
	Vatican Radio	I	100	2-6 EAf,SAf; WAf,SAm
	Vatican Radio	I	250	1 Eu Romanian Liturgy; Eu,SAm
	Vatican Radio	I	250	1 Eu,NAm Angelus Eu German/Polish
	Vatican Radio	I	250	Eu,SAm
	Voice of America	CLN	250	SAs
	Voice of America	THA	250	SEA,Oc
9650	Deutsche Welle	D	500	Eu Albanian
	Deutsche Welle	D	500	Eu Croatian
	Radio Budapest	HNG	250	2-7 Eu
	Radio Exterior de Espana	E	350	3 NAm Sefardi
	Radio Korea Int.	CAN	250	NAm Korean
	Radio Korea Int.	CAN	250	NAm
	Radio Liberty (BBG)	GRC	250	RUS Tatar/Bashkir
	SABC, Meyerton	AFS	100	AFS Afrikaans
	V.O. Korea	KRE	200	J J J J J
	Vatican Radio	I	500	2,5 As Urdu
	Vatican Radio	I	500	As Hindi/Tamil/Malayalam/English
	Voice of Turkey	TUR	500	Eu,Nam(fromAug)
	Voice of Vietnam 4	VTN	50	VTN H'mong
9655	China Radio Int.	CHN	500	ME Turkish
	CNR 1	CHN	100	CHN CHN
	Deutsche Welle	CLN	250	SAs
	HCJB	EQA	100	Eu
	R. Rossii	RUS	100	RUS
	Radio Exterior de Espana	E	350	2-6 Eu
	Radio France Int.	J	300	FE(Sept/Oct)
	V.O. Islamic Republic of Iran	IRN	500	SAm CAf,EAf Kiswahili
	Voice of Turkey	TUR	500	Eu Greek
	YLE Radio Finland	FIN	500	Eu Finnish
9660	BBC World Service	AUS	10	As
	BBC World Service Relay	AUS	10	As
	Central Broadc. System	TWN	100	CHN
	FEBA	UAE	250	6 CAs Farsi
	Radio Australia	AUS	100	Pac,PNG
	Radio Australia	AUS	100	Pac,PNG
	Radio Canada Int.	J	100	SEA CHN
	Radio Exterior de Espana	CHN	120	J
	Radio Japan	G	300	Eu
	Radio Japan	GUF	300	SAm SAm
	TWR	D	100	7 Eu Romanian
	TWR South Africa	AFS	500	2-6 SOM Somali
	TWR South Africa	AFS	500	1 SOM Somali

Legend: Arabic · Chinese · English · French · German · Japanese · Mandarin · Portuguese · Russian · Spanish · Other

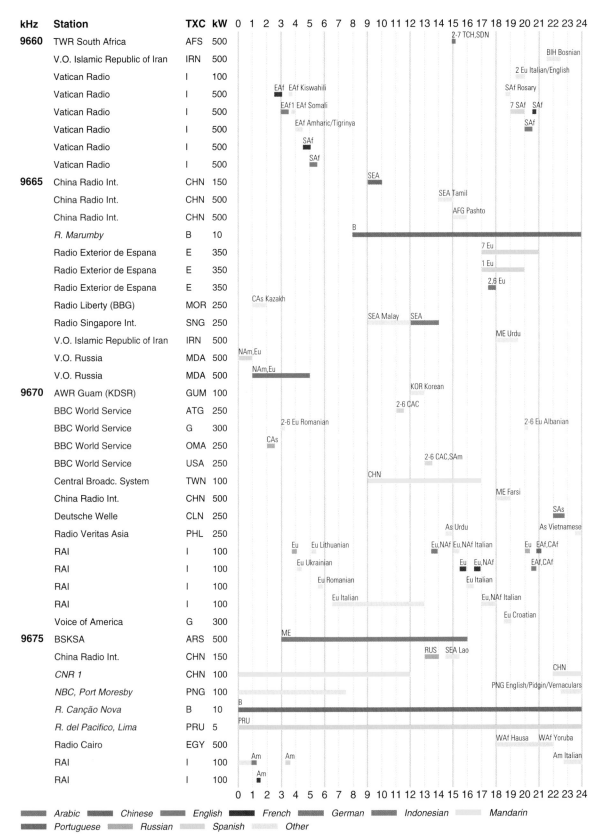

kHz	Station	TXC	kW
9660	TWR South Africa	AFS	500
	V.O. Islamic Republic of Iran	IRN	500
	Vatican Radio	I	100
	Vatican Radio	I	500
	Vatican Radio	I	500
	Vatican Radio	I	500
	Vatican Radio	I	500
	Vatican Radio	I	500
9665	China Radio Int.	CHN	150
	China Radio Int.	CHN	500
	China Radio Int.	CHN	500
	R. Marumby	B	10
	Radio Exterior de Espana	E	350
	Radio Exterior de Espana	E	350
	Radio Exterior de Espana	E	350
	Radio Liberty (BBG)	MOR	250
	Radio Singapore Int.	SNG	250
	V.O. Islamic Republic of Iran	IRN	500
	V.O. Russia	MDA	500
	V.O. Russia	MDA	500
9670	AWR Guam (KDSR)	GUM	100
	BBC World Service	ATG	250
	BBC World Service	G	300
	BBC World Service	OMA	250
	BBC World Service	USA	250
	Central Broadc. System	TWN	100
	China Radio Int.	CHN	500
	Deutsche Welle	CLN	250
	Radio Veritas Asia	PHL	250
	RAI	I	100
	RAI	I	100
	RAI	I	100
	RAI	I	100
	Voice of America	G	300
9675	BSKSA	ARS	500
	China Radio Int.	CHN	150
	CNR 1	CHN	100
	NBC, Port Moresby	PNG	100
	R. Canção Nova	B	10
	R. del Pacifico, Lima	PRU	5
	Radio Cairo	EGY	500
	RAI	I	100
	RAI	I	100

Legend: Arabic, Chinese, English, French, German, Indonesian, Mandarin, Portuguese, Russian, Spanish, Other

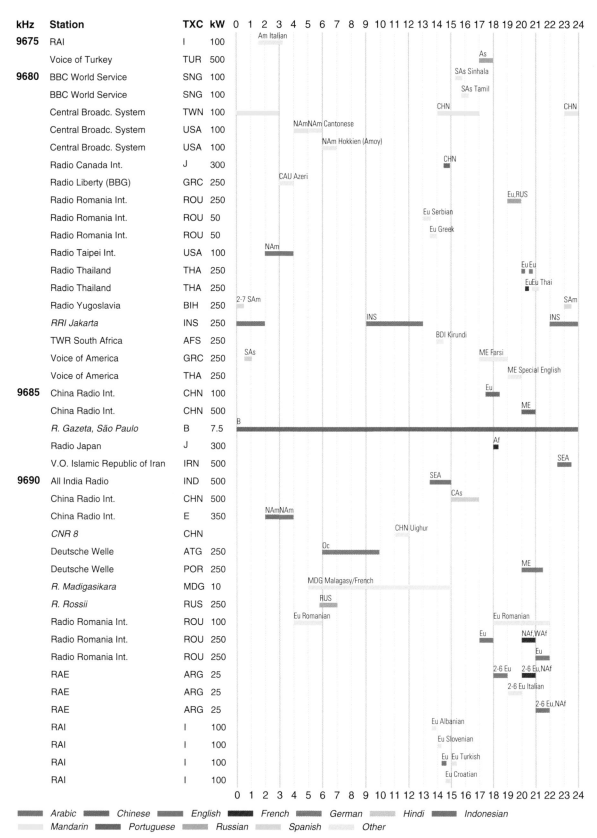

kHz	Station	TXC	kW																						
				0	1	2	3	4	5	6	7	8	9	10	11	12	13	14	15	16	17	18	19	20	21	22	23	24
9675	RAI	I	100	Am Italian
	Voice of Turkey	TUR	500	As
9680	BBC World Service	SNG	100	SAs Sinhala
	BBC World Service	SNG	100	SAs Tamil
	Central Broadc. System	TWN	100	CHN	CHN
	Central Broadc. System	USA	100	NAmNAm Cantonese
	Central Broadc. System	USA	100	NAm Hokkien (Amoy)
	Radio Canada Int.	J	300	CHN
	Radio Liberty (BBG)	GRC	250	CAU Azeri
	Radio Romania Int.	ROU	250	Eu,RUS
	Radio Romania Int.	ROU	50	Eu Serbian
	Radio Romania Int.	ROU	50	Eu Greek
	Radio Taipei Int.	USA	100	NAm
	Radio Thailand	THA	250	Eu Eu
	Radio Thailand	THA	250	EuEu Thai
	Radio Yugoslavia	BIH	250	2-7 SAm	SAm
	RRI Jakarta	INS	250	INS	INS
	TWR South Africa	AFS	250	BDI Kirundi
	Voice of America	GRC	250	SAs	ME Farsi
	Voice of America	THA	250	ME Special English
9685	China Radio Int.	CHN	100	Eu
	China Radio Int.	CHN	500	ME
	R. Gazeta, São Paulo	B	7.5	B
	Radio Japan	J	300	Af
	V.O. Islamic Republic of Iran	IRN	500	SEA
9690	All India Radio	IND	500	SEA
	China Radio Int.	CHN	500	CAs
	China Radio Int.	E	350	NAmNAm
	CNR 8	CHN		CHN Uighur
	Deutsche Welle	ATG	250	Oc
	Deutsche Welle	POR	250	ME
	R. Madigasikara	MDG	10	MDG Malagasy/French
	R. Rossii	RUS	250	RUS
	Radio Romania Int.	ROU	100	Eu Romanian	Eu Romanian
	Radio Romania Int.	ROU	250	Eu	NAf,WAf
	Radio Romania Int.	ROU	250	Eu
	RAE	ARG	25	2-6 Eu	2-6 Eu,NAf
	RAE	ARG	25	2-6 Eu Italian
	RAE	ARG	25	2-6 Eu,NAf
	RAI	I	100	Eu Albanian
	RAI	I	100	Eu Slovenian
	RAI	I	100	Eu Eu Turkish
	RAI	I	100	Eu Croatian

0 1 2 3 4 5 6 7 8 9 10 11 12 13 14 15 16 17 18 19 20 21 22 23 24

▬▬▬ *Arabic* ▬▬▬ *Chinese* ▬▬▬ *English* ▬▬▬ *French* ▬▬▬ *German* ▬▬▬ *Hindi* ▬▬▬ *Indonesian*
▬▬▬ *Mandarin* ▬▬▬ *Portuguese* ▬▬▬ *Russian* ▬▬▬ *Spanish* ▬▬▬ *Other*

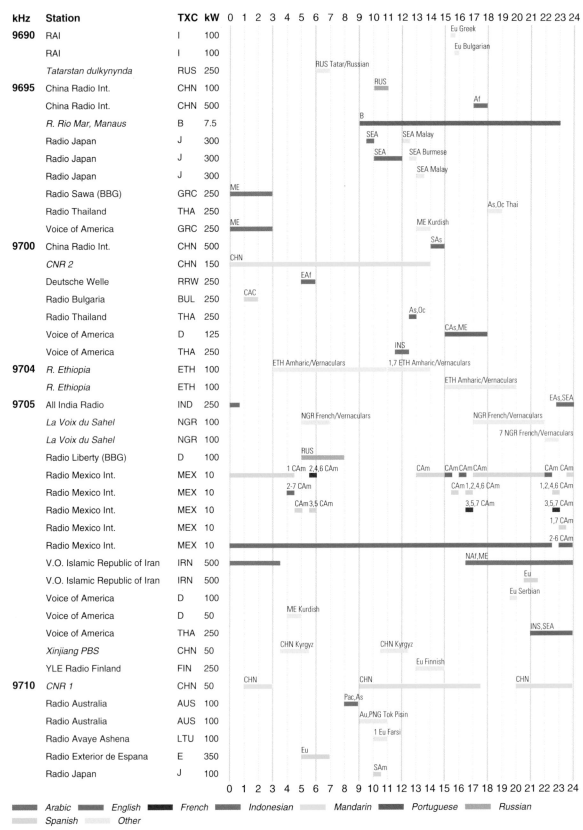

kHz	Station	TXC	kW
9690	RAI	I	100
	RAI	I	100
	Tatarstan dulkynynda	RUS	250
9695	China Radio Int.	CHN	100
	China Radio Int.	CHN	500
	R. Rio Mar, Manaus	B	7.5
	Radio Japan	J	300
	Radio Japan	J	300
	Radio Japan	J	300
	Radio Sawa (BBG)	GRC	250
	Radio Thailand	THA	250
	Voice of America	GRC	250
9700	China Radio Int.	CHN	500
	CNR 2	CHN	150
	Deutsche Welle	RRW	250
	Radio Bulgaria	BUL	250
	Radio Thailand	THA	250
	Voice of America	D	125
	Voice of America	THA	250
9704	*R. Ethiopia*	ETH	100
	R. Ethiopia	ETH	100
9705	All India Radio	IND	250
	La Voix du Sahel	NGR	100
	La Voix du Sahel	NGR	100
	Radio Liberty (BBG)	D	100
	Radio Mexico Int.	MEX	10
	Radio Mexico Int.	MEX	10
	Radio Mexico Int.	MEX	10
	Radio Mexico Int.	MEX	10
	Radio Mexico Int.	MEX	10
	V.O. Islamic Republic of Iran	IRN	500
	V.O. Islamic Republic of Iran	IRN	500
	Voice of America	D	100
	Voice of America	D	50
	Voice of America	THA	250
	Xinjiang PBS	CHN	50
	YLE Radio Finland	FIN	250
9710	*CNR 1*	CHN	50
	Radio Australia	AUS	100
	Radio Australia	AUS	100
	Radio Avaye Ashena	LTU	100
	Radio Exterior de Espana	E	350
	Radio Japan	J	100

Legend: Arabic · English · French · Indonesian · Mandarin · Portuguese · Russian · Spanish · Other

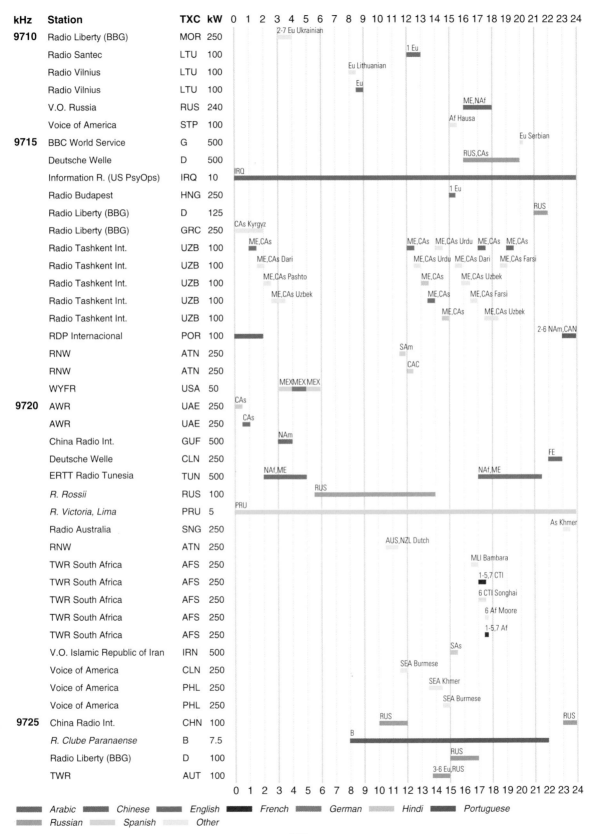

kHz	Station	TXC	kW	Schedule
9710	Radio Liberty (BBG)	MOR	250	2-7 Eu Ukrainian
	Radio Santec	LTU	100	1 Eu
	Radio Vilnius	LTU	100	Eu Lithuanian
	Radio Vilnius	LTU	100	Eu
	V.O. Russia	RUS	240	ME,NAf
	Voice of America	STP	100	Af Hausa
9715	BBC World Service	G	500	Eu Serbian
	Deutsche Welle	D	500	RUS,CAs
	Information R. (US PsyOps)	IRQ	10	IRQ
	Radio Budapest	HNG	250	1 Eu
	Radio Liberty (BBG)	D	125	RUS
	Radio Liberty (BBG)	GRC	250	CAs Kyrgyz
	Radio Tashkent Int.	UZB	100	ME,CAs / ME,CAs / ME,CAs Urdu / ME,CAs / ME,CAs
	Radio Tashkent Int.	UZB	100	ME,CAs Dari / ME,CAs Urdu / ME,CAs Dari / ME,CAs Farsi
	Radio Tashkent Int.	UZB	100	ME,CAs Pashto / ME,CAs / ME,CAs Uzbek
	Radio Tashkent Int.	UZB	100	ME,CAs Uzbek / ME,CAs / ME,CAs Farsi
	Radio Tashkent Int.	UZB	100	ME,CAs / ME,CAs Uzbek
	RDP Internacional	POR	100	2-6 NAm,CAN
	RNW	ATN	250	SAm
	RNW	ATN	250	CAC
	WYFR	USA	50	MEXMEX MEX
9720	AWR	UAE	250	CAs
	AWR	UAE	250	CAs
	China Radio Int.	GUF	500	NAm
	Deutsche Welle	CLN	250	FE
	ERTT Radio Tunesia	TUN	500	NAf,ME / NAf,ME
	R. Rossii	RUS	100	RUS
	R. Victoria, Lima	PRU	5	PRU
	Radio Australia	SNG	250	As Khmer
	RNW	ATN	250	AUS,NZL Dutch
	TWR South Africa	AFS	250	MLI Bambara
	TWR South Africa	AFS	250	1-5,7 CTI
	TWR South Africa	AFS	250	6 CTI Songhai
	TWR South Africa	AFS	250	6 Af Moore
	TWR South Africa	AFS	250	1-5,7 Af
	V.O. Islamic Republic of Iran	IRN	500	SAs
	Voice of America	CLN	250	SEA Burmese
	Voice of America	PHL	250	SEA Khmer
	Voice of America	PHL	250	SEA Burmese
9725	China Radio Int.	CHN	100	RUS / RUS
	R. Clube Paranaense	B	7.5	B
	Radio Liberty (BBG)	D	100	RUS
	TWR	AUT	100	3-6 Eu,RUS

Arabic Chinese English French German Hindi Portuguese
Russian Spanish Other

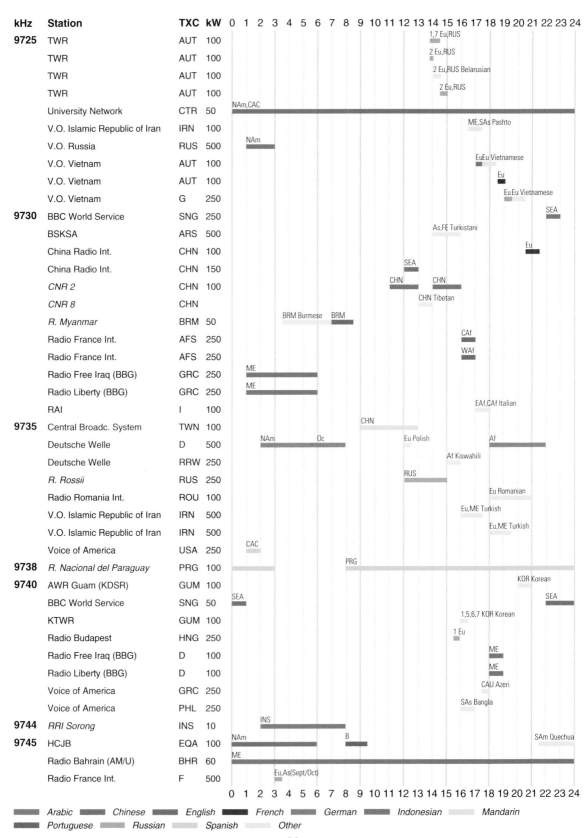

kHz	Station	TXC	kW
9725	TWR	AUT	100	1,7 Eu,RUS
	TWR	AUT	100	2 Eu,RUS
	TWR	AUT	100	2 Eu,RUS Belarusian
	TWR	AUT	100	2 Eu,RUS
	University Network	CTR	50	NAm,CAC
	V.O. Islamic Republic of Iran	IRN	100	ME,SAs Pashto
	V.O. Russia	RUS	500	NAm
	V.O. Vietnam	AUT	100	EuEu Vietnamese
	V.O. Vietnam	AUT	100	Eu
	V.O. Vietnam	G	250	Eu Eu Vietnamese
9730	BBC World Service	SNG	250	SEA
	BSKSA	ARS	500	As,FE Turkistani
	China Radio Int.	CHN	100	Eu
	China Radio Int.	CHN	150	SEA
	CNR 2	CHN	100	CHN / CHN
	CNR 8	CHN		CHN Tibetan
	R. Myanmar	BRM	50	BRM Burmese / BRM
	Radio France Int.	AFS	250	CAf
	Radio France Int.	AFS	250	WAf
	Radio Free Iraq (BBG)	GRC	250	ME
	Radio Liberty (BBG)	GRC	250	ME
	RAI	I	100	EAf,CAf Italian
9735	Central Broadc. System	TWN	100	CHN
	Deutsche Welle	D	500	NAm / Oc / Eu Polish / Af
	Deutsche Welle	RRW	250	Af Kiswahili
	R. Rossii	RUS	250	RUS
	Radio Romania Int.	ROU	100	Eu Romanian
	V.O. Islamic Republic of Iran	IRN	500	Eu,ME Turkish
	V.O. Islamic Republic of Iran	IRN	500	Eu,ME Turkish
	Voice of America	USA	250	CAC
9738	R. Nacional del Paraguay	PRG	100	PRG
9740	AWR Guam (KDSR)	GUM	100	KOR Korean
	BBC World Service	SNG	50	SEA / SEA
	KTWR	GUM	100	1,5,6,7 KOR Korean
	Radio Budapest	HNG	250	1 Eu
	Radio Free Iraq (BBG)	D	100	ME
	Radio Liberty (BBG)	D	100	ME
	Voice of America	GRC	250	CAU Azeri
	Voice of America	PHL	250	SAs Bangla
9744	RRI Sorong	INS	10	INS
9745	HCJB	EQA	100	NAm / B / SAm Quechua
	Radio Bahrain (AM/U)	BHR	60	ME
	Radio France Int.	F	500	Eu,As(Sept/Oct)

Arabic — Chinese — English — French — German — Indonesian — Mandarin
Portuguese — Russian — Spanish — Other

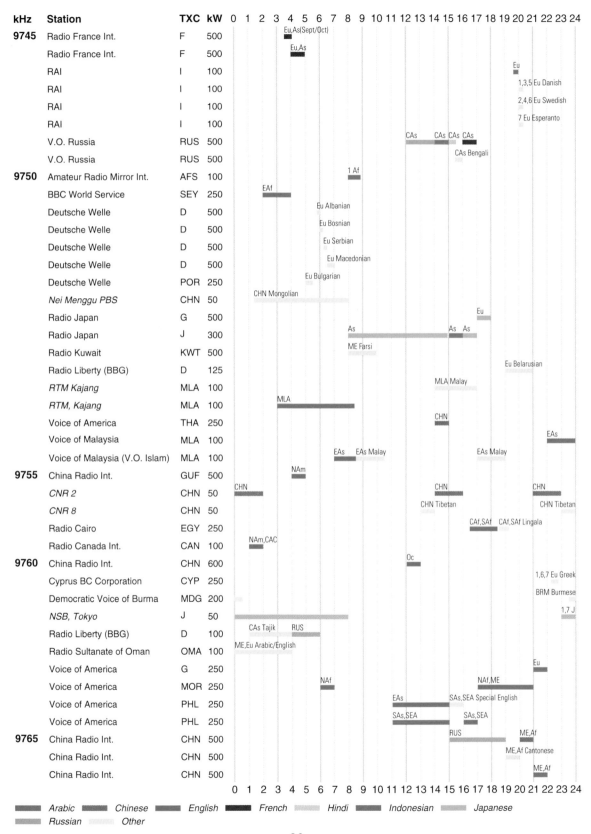

kHz	Station	TXC	kW																							
				0	1	2	3	4	5	6	7	8	9	10	11	12	13	14	15	16	17	18	19	20	21	22	23	24

9745 Radio France Int. — F — 500 — Eu,As(Sept/Oct)
Radio France Int. — F — 500 — Eu,As
RAI — I — 100 — Eu
RAI — I — 100 — 1,3,5 Eu Danish
RAI — I — 100 — 2,4,6 Eu Swedish
RAI — I — 100 — 7 Eu Esperanto
V.O. Russia — RUS — 500 — CAs CAs CAs CAs
V.O. Russia — RUS — 500 — CAs Bengali

9750 Amateur Radio Mirror Int. — AFS — 100 — 1 Af
BBC World Service — SEY — 250 — EAf
Deutsche Welle — D — 500 — Eu Albanian
Deutsche Welle — D — 500 — Eu Bosnian
Deutsche Welle — D — 500 — Eu Serbian
Deutsche Welle — D — 500 — Eu Macedonian
Deutsche Welle — POR — 250 — Eu Bulgarian
Nei Menggu PBS — CHN — 50 — CHN Mongolian
Radio Japan — G — 500 — Eu
Radio Japan — J — 300 — As — As As
Radio Kuwait — KWT — 500 — ME Farsi
Radio Liberty (BBG) — D — 125 — Eu Belarusian
RTM Kajang — MLA — 100 — MLA Malay
RTM, Kajang — MLA — 100 — MLA
Voice of America — THA — 250 — CHN
Voice of Malaysia — MLA — 100 — EAs
Voice of Malaysia (V.O. Islam) — MLA — 100 — EAs EAs Malay — EAs Malay

9755 China Radio Int. — GUF — 500 — NAm
CNR 2 — CHN — 50 — CHN — CHN — CHN
CNR 8 — CHN — 50 — CHN Tibetan — CHN Tibetan
Radio Cairo — EGY — 250 — CAf,SAf CAf,SAf Lingala
Radio Canada Int. — CAN — 100 — NAm,CAC

9760 China Radio Int. — CHN — 600 — Oc
Cyprus BC Corporation — CYP — 250 — 1,6,7 Eu Greek
Democratic Voice of Burma — MDG — 200 — BRM Burmese
NSB, Tokyo — J — 50 — 1,7 J
Radio Liberty (BBG) — D — 100 — CAs Tajik RUS
Radio Sultanate of Oman — OMA — 100 — ME,Eu Arabic/English
Voice of America — G — 250 — Eu
Voice of America — MOR — 250 — NAf — NAf,ME
Voice of America — PHL — 250 — EAs SAs,SEA Special English
Voice of America — PHL — 250 — SAs,SEA SAs,SEA

9765 China Radio Int. — CHN — 500 — RUS ME,Af
China Radio Int. — CHN — 500 — ME,Af Cantonese
China Radio Int. — CHN — 500 — ME,Af

				0	1	2	3	4	5	6	7	8	9	10	11	12	13	14	15	16	17	18	19	20	21	22	23	24

Legend: Arabic — Chinese — *English* — French — Hindi — Indonesian — Japanese — Russian — Other

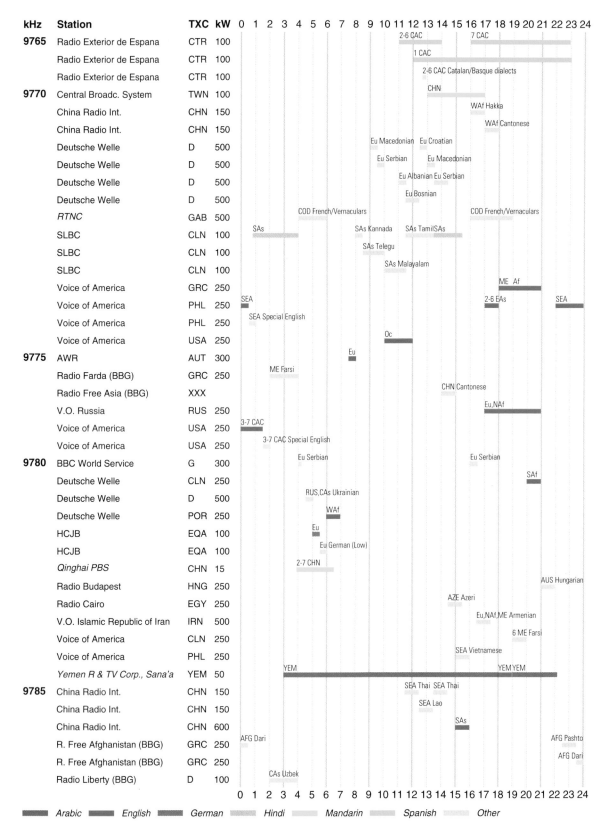

kHz	Station	TXC	kW
9765	Radio Exterior de Espana	CTR	100
	Radio Exterior de Espana	CTR	100
	Radio Exterior de Espana	CTR	100
9770	Central Broadc. System	TWN	100
	China Radio Int.	CHN	150
	China Radio Int.	CHN	150
	Deutsche Welle	D	500
	Deutsche Welle	D	500
	Deutsche Welle	D	500
	Deutsche Welle	D	500
	RTNC	GAB	500
	SLBC	CLN	100
	SLBC	CLN	100
	SLBC	CLN	100
	Voice of America	GRC	250
	Voice of America	PHL	250
	Voice of America	PHL	250
	Voice of America	USA	250
9775	AWR	AUT	300
	Radio Farda (BBG)	GRC	250
	Radio Free Asia (BBG)	XXX	
	V.O. Russia	RUS	250
	Voice of America	USA	250
	Voice of America	USA	250
9780	BBC World Service	G	300
	Deutsche Welle	CLN	250
	Deutsche Welle	D	500
	Deutsche Welle	POR	250
	HCJB	EQA	100
	HCJB	EQA	100
	Qinghai PBS	CHN	15
	Radio Budapest	HNG	250
	Radio Cairo	EGY	250
	V.O. Islamic Republic of Iran	IRN	500
	Voice of America	CLN	250
	Voice of America	PHL	250
	Yemen R & TV Corp., Sana'a	YEM	50
9785	China Radio Int.	CHN	150
	China Radio Int.	CHN	150
	China Radio Int.	CHN	600
	R. Free Afghanistan (BBG)	GRC	250
	R. Free Afghanistan (BBG)	GRC	250
	Radio Liberty (BBG)	D	100

Legend: Arabic · English · German · Hindi · Mandarin · Spanish · Other

90

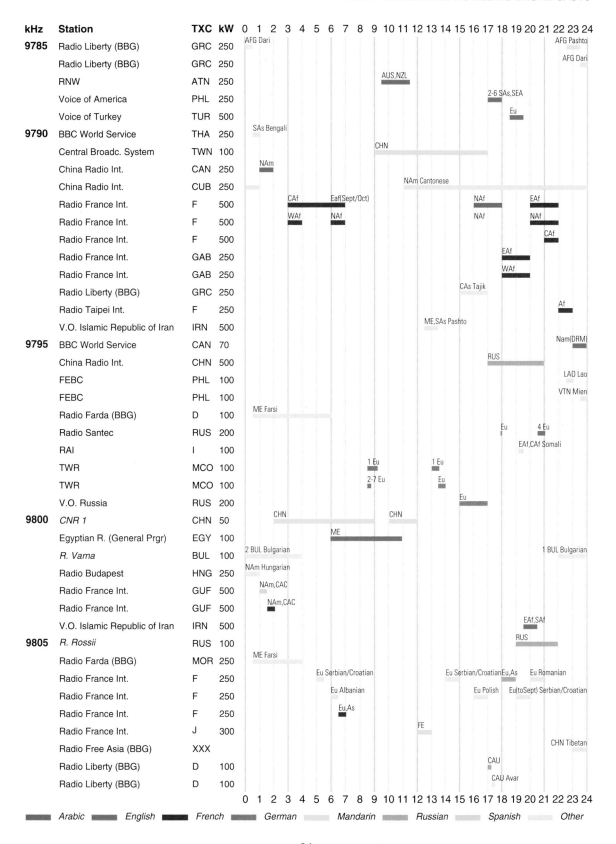

kHz	Station	TXC	kW	Time / Target (0–24)
9785	Radio Liberty (BBG)	GRC	250	AFG Dari / AFG Pashto
	Radio Liberty (BBG)	GRC	250	AFG Dari
	RNW	ATN	250	AUS,NZL
	Voice of America	PHL	250	2-6 SAs,SEA
	Voice of Turkey	TUR	500	Eu
9790	BBC World Service	THA	250	SAs Bengali
	Central Broadc. System	TWN	100	CHN
	China Radio Int.	CAN	250	NAm
	China Radio Int.	CUB	250	NAm Cantonese
	Radio France Int.	F	500	CAf / Eaf(Sept/Oct) / NAf / EAf
	Radio France Int.	F	500	WAf / NAf / NAf / NAf
	Radio France Int.	F	500	CAf
	Radio France Int.	GAB	250	EAf
	Radio France Int.	GAB	250	WAf
	Radio Liberty (BBG)	GRC	250	CAs Tajik
	Radio Taipei Int.	F	250	Af
	V.O. Islamic Republic of Iran	IRN	500	ME,SAs Pashto
9795	BBC World Service	CAN	70	Nam(DRM)
	China Radio Int.	CHN	500	RUS
	FEBC	PHL	100	LAO Lao
	FEBC	PHL	100	VTN Mien
	Radio Farda (BBG)	D	100	ME Farsi
	Radio Santec	RUS	200	Eu / 4 Eu
	RAI	I	100	EAf,CAf Somali
	TWR	MCO	100	1 Eu / 1 Eu
	TWR	MCO	100	2-7 Eu / Eu
	V.O. Russia	RUS	200	Eu
9800	CNR 1	CHN	50	CHN / CHN
	Egyptian R. (General Prgr)	EGY	100	ME
	R. Varna	BUL	100	2 BUL Bulgarian / 1 BUL Bulgarian
	Radio Budapest	HNG	250	NAm Hungarian
	Radio France Int.	GUF	500	NAm,CAC
	Radio France Int.	GUF	500	NAm,CAC
	V.O. Islamic Republic of Iran	IRN	500	EAf,SAf
9805	R. Rossii	RUS	100	RUS
	Radio Farda (BBG)	MOR	250	ME Farsi
	Radio France Int.	F	250	Eu Serbian/Croatian / Eu Serbian/Croatian Eu,As / Eu Romanian
	Radio France Int.	F	250	Eu Albanian / Eu Polish / Eu(toSept) Serbian/Croatian
	Radio France Int.	F	250	Eu,As
	Radio France Int.	J	300	FE
	Radio Free Asia (BBG)	XXX		CHN Tibetan
	Radio Liberty (BBG)	D	100	CAU
	Radio Liberty (BBG)	D	100	CAU Avar

Legend: ▮ Arabic ▮ English ▮ French ▮ German ▮ Mandarin ▮ Russian ▮ Spanish ▮ Other

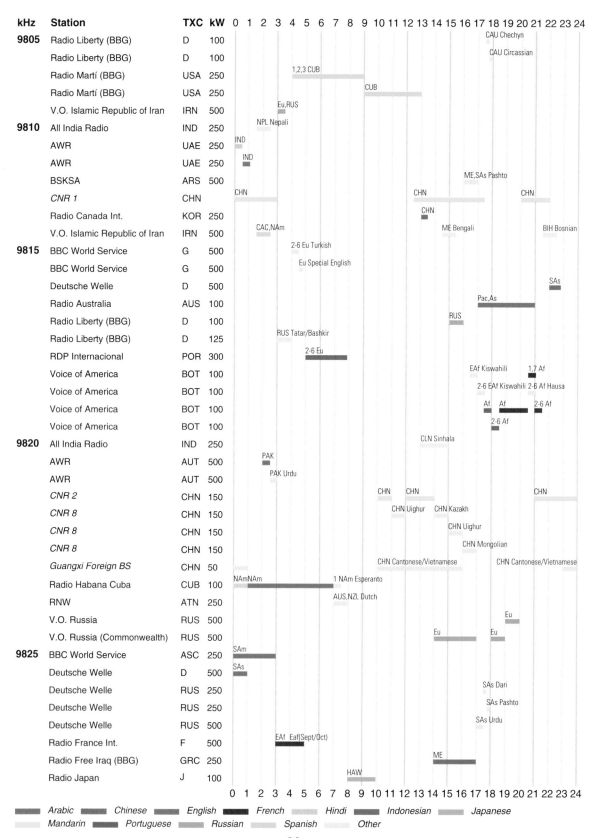

kHz	Station	TXC	kW
9805	Radio Liberty (BBG)	D	100
	Radio Liberty (BBG)	D	100
	Radio Martí (BBG)	USA	250
	Radio Martí (BBG)	USA	250
	V.O. Islamic Republic of Iran	IRN	500
9810	All India Radio	IND	250
	AWR	UAE	250
	AWR	UAE	250
	BSKSA	ARS	500
	CNR 1	CHN	
	Radio Canada Int.	KOR	250
	V.O. Islamic Republic of Iran	IRN	500
9815	BBC World Service	G	500
	BBC World Service	G	500
	Deutsche Welle	D	500
	Radio Australia	AUS	100
	Radio Liberty (BBG)	D	100
	Radio Liberty (BBG)	D	125
	RDP Internacional	POR	300
	Voice of America	BOT	100
	Voice of America	BOT	100
	Voice of America	BOT	100
	Voice of America	BOT	100
9820	All India Radio	IND	250
	AWR	AUT	500
	AWR	AUT	500
	CNR 2	CHN	150
	CNR 8	CHN	150
	CNR 8	CHN	150
	CNR 8	CHN	150
	Guangxi Foreign BS	CHN	50
	Radio Habana Cuba	CUB	100
	RNW	ATN	250
	V.O. Russia	RUS	500
	V.O. Russia (Commonwealth)	RUS	500
9825	BBC World Service	ASC	250
	Deutsche Welle	D	500
	Deutsche Welle	RUS	250
	Deutsche Welle	RUS	250
	Deutsche Welle	RUS	500
	Radio France Int.	F	500
	Radio Free Iraq (BBG)	GRC	250
	Radio Japan	J	100

Schedule annotations (left to right across 0–24 h):

- Radio Liberty (BBG) D 100 — CAU Chechyn
- Radio Liberty (BBG) D 100 — CAU Circassian
- Radio Martí (BBG) USA 250 — 1,2,3 CUB
- Radio Martí (BBG) USA 250 — CUB
- V.O. Islamic Republic of Iran IRN 500 — Eu,RUS
- All India Radio IND 250 — NPL Nepali
- AWR UAE 250 — IND
- AWR UAE 250 — IND
- BSKSA ARS 500 — ME,SAs Pashto
- CNR 1 CHN — CHN ... CHN ... CHN
- Radio Canada Int. KOR 250 — CHN
- V.O. Islamic Republic of Iran IRN 500 — CAC,NAm ... ME Bengali ... BIH Bosnian
- BBC World Service G 500 — 2-6 Eu Turkish
- BBC World Service G 500 — Eu Special English
- Deutsche Welle D 500 — SAs
- Radio Australia AUS 100 — Pac,As
- Radio Liberty (BBG) D 100 — RUS
- Radio Liberty (BBG) D 125 — RUS Tatar/Bashkir
- RDP Internacional POR 300 — 2-6 Eu
- Voice of America BOT 100 — EAf Kiswahili ... 1,7 Af
- Voice of America BOT 100 — 2-6 EAf Kiswahili ... 2-6 Af Hausa
- Voice of America BOT 100 — Af ... Af ... 2-6 Af
- Voice of America BOT 100 — 2-6 Af
- All India Radio IND 250 — CLN Sinhala
- AWR AUT 500 — PAK
- AWR AUT 500 — PAK Urdu
- CNR 2 CHN 150 — CHN ... CHN ... CHN
- CNR 8 CHN 150 — CHN Uighur ... CHN Kazakh
- CNR 8 CHN 150 — CHN Uighur
- CNR 8 CHN 150 — CHN Mongolian
- Guangxi Foreign BS CHN 50 — CHN Cantonese/Vietnamese ... CHN Cantonese/Vietnamese
- Radio Habana Cuba CUB 100 — NAmNAm ... 1 NAm Esperanto
- RNW ATN 250 — AUS,NZL Dutch
- V.O. Russia RUS 500 — Eu
- V.O. Russia (Commonwealth) RUS 500 — Eu ... Eu
- BBC World Service ASC 250 — SAm
- Deutsche Welle D 500 — SAs
- Deutsche Welle RUS 250 — SAs Dari
- Deutsche Welle RUS 250 — SAs Pashto
- Deutsche Welle RUS 500 — SAs Urdu
- Radio France Int. F 500 — EAf Eaf(Sept/Oct)
- Radio Free Iraq (BBG) GRC 250 — ME
- Radio Japan J 100 — HAW

Legend: Arabic · Chinese · English · French · Hindi · Indonesian · Japanese · Mandarin · Portuguese · Russian · Spanish · Other

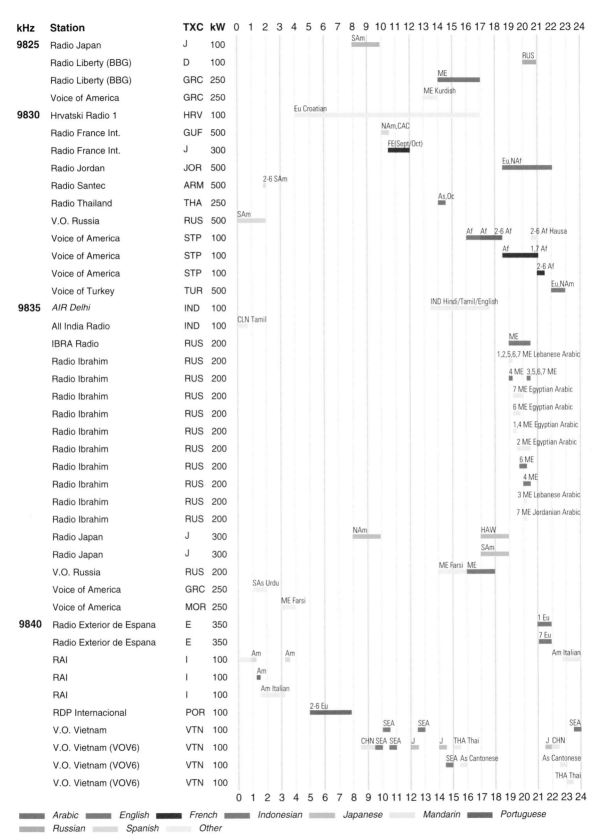

kHz	Station	TXC	kW
9825	Radio Japan	J	100
	Radio Liberty (BBG)	D	100
	Radio Liberty (BBG)	GRC	250
	Voice of America	GRC	250
9830	Hrvatski Radio 1	HRV	100
	Radio France Int.	GUF	500
	Radio France Int.	J	300
	Radio Jordan	JOR	500
	Radio Santec	ARM	500
	Radio Thailand	THA	250
	V.O. Russia	RUS	500
	Voice of America	STP	100
	Voice of America	STP	100
	Voice of America	STP	100
	Voice of Turkey	TUR	500
9835	AIR Delhi	IND	100
	All India Radio	IND	100
	IBRA Radio	RUS	200
	Radio Ibrahim	RUS	200
	Radio Ibrahim	RUS	200
	Radio Ibrahim	RUS	200
	Radio Ibrahim	RUS	200
	Radio Ibrahim	RUS	200
	Radio Ibrahim	RUS	200
	Radio Ibrahim	RUS	200
	Radio Ibrahim	RUS	200
	Radio Ibrahim	RUS	200
	Radio Ibrahim	RUS	200
	Radio Japan	J	300
	Radio Japan	J	300
	V.O. Russia	RUS	200
	Voice of America	GRC	250
	Voice of America	MOR	250
9840	Radio Exterior de Espana	E	350
	Radio Exterior de Espana	E	350
	RAI	I	100
	RAI	I	100
	RAI	I	100
	RDP Internacional	POR	100
	V.O. Vietnam	VTN	100
	V.O. Vietnam (VOV6)	VTN	100
	V.O. Vietnam (VOV6)	VTN	100
	V.O. Vietnam (VOV6)	VTN	100

Legend: Arabic, English, French, Indonesian, Japanese, Mandarin, Portuguese, Russian, Spanish, Other

kHz	Station	TXC	kW	0 1 2 3 4 5 6 7 8 9 10 11 12 13 14 15 16 17 18 19 20 21 22 23 24
9840	Voice of America	D	100	Eu Albanian
	Voice of America	THA	250	2-6 SAs,SEA
	WSHB	USA	500	SAf
9845	All India Radio	IND	100	PAK,AFG Pashto
	All India Radio	IND	100	AFG Dari
	AWR	AFS	250	WAf
	CNR 1	CHN	100	CHN CHN
	R. Free Afghanistan (BBG)	THA	250	AFG Pashto
	R. Free Afghanistan (BBG)	THA	250	AFG Dari
	R. Rossii	RUS	250	RUS
	Radio Liberty (BBG)	THA	250	AFG Pashto
	Radio Liberty (BBG)	THA	250	AFG Dari
	RAI	I	100	Eu
	RNW	ATN	250	MEX,CAC NAm
	Voice of America	PHL	250	CHN EAs Special English
	WSHB	USA	500	1-5,7 NZL
	WSHB	USA	500	6 NZL
9850	Radio Budapest	HNG	250	SAm Hungarian
	Radio Budapest	HNG	250	2 SAm Hungarian
	Radio Free Asia (BBG)	XXX		CHN
	Radio Free Asia (BBG)	XXX		KOR Korean
	Radio Liberty (BBG)	MOR	250	CAU
	Radio Liberty (BBG)	MOR	250	CAU Avar
	Radio Liberty (BBG)	MOR	250	CAU Chechyn
	Radio Liberty (BBG)	MOR	250	CAU Circassian
	TWR	ALB	100	Eu Polish
	Vatican Radio	I	250	1 Eu Ukrainian Liturgy
	Voice of America	STP	100	Af 2-6 Af
	Voice of Vietnam 4	VTN	50	VTN H'mong
	WHRI World Harvest R.	USA	100	NAm,Eu,As
9855	Egyptian R. (General Prgr)	EGY	250	ME
	FEBC	PHL	100	CBG Khmer
	FEBC Saipan (KFBS)	MRA	100	2 GEO Ossetic
	FEBC Saipan (KFBS)	MRA	100	3,4 KAZ Kazakh
	FEBC Saipan (KFBS)	MRA	100	5,6 KAZ Kyrgyz
	FEBC Saipan (KFBS)	MRA	100	1,7 UZB Uzbek
	FEBC Saipan (KFBS)	MRA	100	6 RUS Udmurt
	FEBC Saipan (KFBS)	MRA	100	7 RUS Tatar
	FEBC Saipan (KFBS)	MRA	100	2 RUS Mari-Low
	FEBC Saipan (KFBS)	MRA	100	4 RUS Chuvash
	FEBC Saipan (KFBS)	MRA	100	5 RUS Udmurt
	FEBC Saipan (KFBS)	MRA	100	1 RUS Tatar
	FEBC Saipan (KFBS)	MRA	100	3 UKR Ukrainian

0 1 2 3 4 5 6 7 8 9 10 11 12 13 14 15 16 17 18 19 20 21 22 23 24

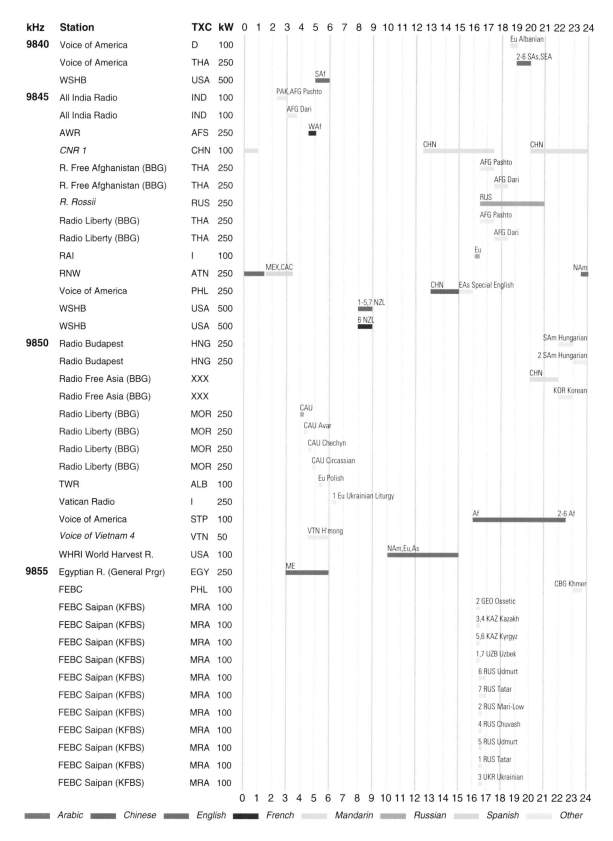

Arabic   Chinese   English   French   Mandarin   Russian   Spanish   Other

kHz	Station	TXC	kW																								
				0 1 2 3 4 5 6 7 8 9 10 11 12 13 14 15 16 17 18 19 20 21 22 23 24																								
9855	FEBC Saipan (KFBS)	MRA	100	3,4,5 RUS,KAZ
	FEBC Saipan (KFBS)	MRA	100	2 RUS Komu
	FEBC Saipan (KFBS)	MRA	100	1 RUS Bashkir
	Radio Kuwait	KWT	500	Eu,NAm
	Radio Liberty (BBG)	D	100	Eu Ukrainian
	Radio Prague	SVK	250	Eu
	Radio Sawa (BBG)	GRC	250	NAf
	Voice of America	GRC	250	NAf
9860	China Radio Int.	CHN	150	Eu Hungarian As Esperanto
	China Radio Int.	CHN	150	Eu Bulgarian
	HCJB	EQA	100	Eu
	Radio Santec	RUS	500	2-6 B 2-6 SAm
	RNW	D	125	Eu
	Voice of America	PHL	250	SEA Vietnamese
	WSHB	USA	500	1,7 Eu
	WSHB	USA	500	2,4 Eu
	WSHB	USA	500	3,5 Eu
	WSHB	USA	500	6 Eu 3,5 Eu
	WSHB	USA	500	2 Eu
	WSHB	USA	500	4,6 Eu
	WSHB	USA	500	1,7 Eu
9865	BBC World Service	UZB	200	SAs Bengali
	BBC World Service	UZB	200	SAs
	Christian Voice	ZMB	100	CAf,SAf
	Radio Free Iraq (BBG)	D	100	ME
	Radio Free Iraq (BBG)	MOR	250	ME
	Radio Liberty (BBG)	CLN	250	CAs Kazakh
	Radio Liberty (BBG)	D	100	ME
	Radio Liberty (BBG)	MOR	250	ME
	Radio Prague	CZE	100	Eu
	Radio Vlaanderen Int.	RUS	250	EAs
	Radio Vlaanderen Int.	RUS	250	EAs Dutch
	V.O. Russia	RUS	500	Eu
9870	BBC World Service	ASC	250	SAm Brazilian
	BSKSA	ARS	500	Eu,NAf
	China Radio Int.	CHN	150	RUS RUS MNG Mongolian
	Radio Free Europe (BBG)	D	100	Eu Romanian
	Radio Free Europe (BBG)	D	100	2-6 Eu Romanian
	Radio Korea Int.	KOR	100	ME,Af ME,Af Korean
	Radio Korea Int.	KOR	100	ME,Af
	Radio Österreich 1 Int.	AUT	300	SAm NAm SAm
	Radio Prague	CZE	100	NAm Czech
	Radio Prague	CZE	100	NAm

0 1 2 3 4 5 6 7 8 9 10 11 12 13 14 15 16 17 18 19 20 21 22 23 24

Arabic   English   French   German   Hindi   Portuguese   Russian

Spanish   Other

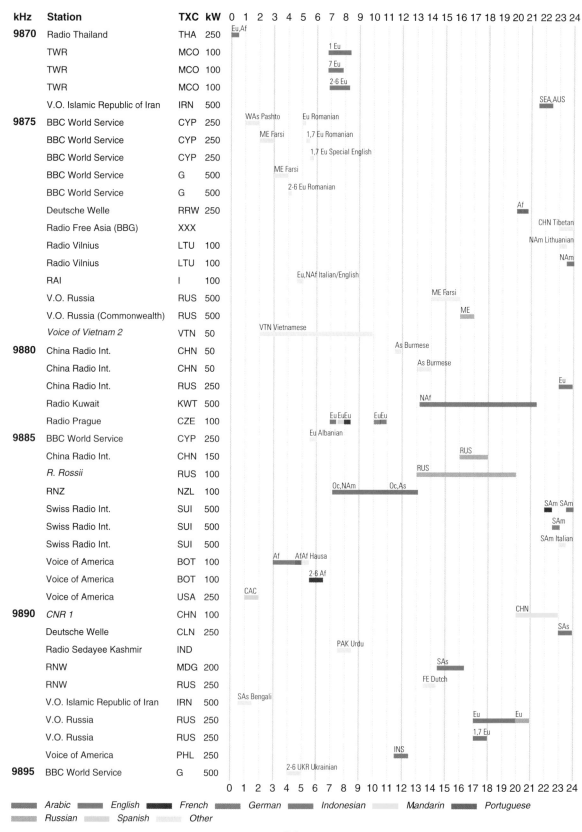

kHz	Station	TXC	kW
9870	Radio Thailand	THA	250
	TWR	MCO	100
	TWR	MCO	100
	TWR	MCO	100
	V.O. Islamic Republic of Iran	IRN	500
9875	BBC World Service	CYP	250
	BBC World Service	CYP	250
	BBC World Service	CYP	250
	BBC World Service	G	500
	BBC World Service	G	500
	Deutsche Welle	RRW	250
	Radio Free Asia (BBG)	XXX	
	Radio Vilnius	LTU	100
	Radio Vilnius	LTU	100
	RAI	I	100
	V.O. Russia	RUS	500
	V.O. Russia (Commonwealth)	RUS	500
	Voice of Vietnam 2	VTN	50
9880	China Radio Int.	CHN	50
	China Radio Int.	CHN	50
	China Radio Int.	RUS	250
	Radio Kuwait	KWT	500
	Radio Prague	CZE	100
9885	BBC World Service	CYP	250
	China Radio Int.	CHN	150
	R. Rossii	RUS	100
	RNZ	NZL	100
	Swiss Radio Int.	SUI	500
	Swiss Radio Int.	SUI	500
	Swiss Radio Int.	SUI	500
	Voice of America	BOT	100
	Voice of America	BOT	100
	Voice of America	USA	250
9890	*CNR 1*	CHN	100
	Deutsche Welle	CLN	250
	Radio Sedayee Kashmir	IND	
	RNW	MDG	200
	RNW	RUS	250
	V.O. Islamic Republic of Iran	IRN	500
	V.O. Russia	RUS	250
	V.O. Russia	RUS	250
	Voice of America	PHL	250
9895	BBC World Service	G	500

Legend:
Arabic, English, French, German, Indonesian, Mandarin, Portuguese, Russian, Spanish, Other

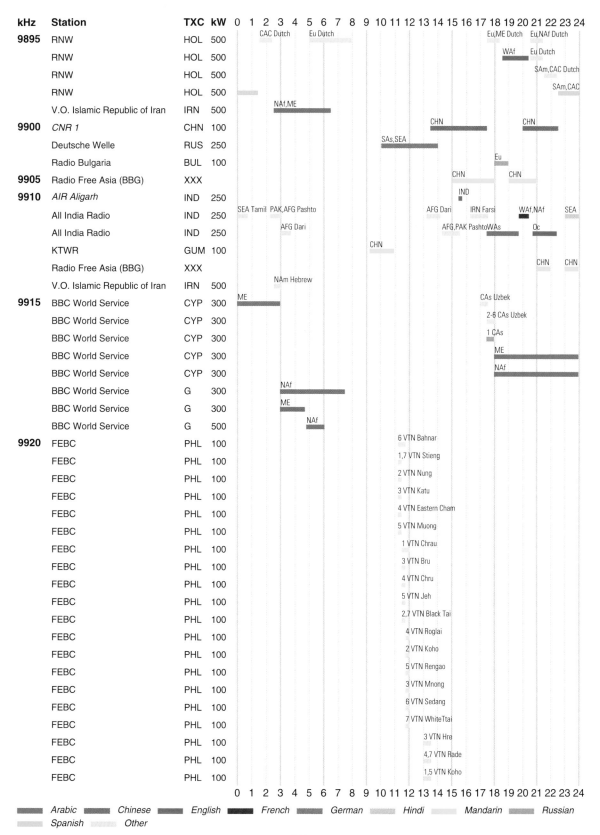

kHz	Station	TXC	kW	0 1 2 3 4 5 6 7 8 9 10 11 12 13 14 15 16 17 18 19 20 21 22 23 24
9895	RNW	HOL	500	CAC Dutch — Eu Dutch — Eu,ME Dutch — Eu,NAf Dutch
	RNW	HOL	500	WAf — Eu Dutch
	RNW	HOL	500	SAm,CAC Dutch
	RNW	HOL	500	SAm,CAC
	V.O. Islamic Republic of Iran	IRN	500	NAf,ME
9900	*CNR 1*	CHN	100	CHN — CHN
	Deutsche Welle	RUS	250	SAs,SEA
	Radio Bulgaria	BUL	100	Eu
9905	Radio Free Asia (BBG)	XXX		CHN — CHN
9910	*AIR Aligarh*	IND	250	IND
	All India Radio	IND	250	SEA Tamil PAK,AFG Pashto — AFG Dari — IRN Farsi — WAf,NAf — SEA
	All India Radio	IND	250	AFG Dari — AFG,PAK Pashto WAs — Oc
	KTWR	GUM	100	CHN
	Radio Free Asia (BBG)	XXX		CHN — CHN
	V.O. Islamic Republic of Iran	IRN	500	NAm Hebrew
9915	BBC World Service	CYP	300	ME — CAs Uzbek
	BBC World Service	CYP	300	2-6 CAs Uzbek
	BBC World Service	CYP	300	1 CAs
	BBC World Service	CYP	300	ME
	BBC World Service	CYP	300	NAf
	BBC World Service	G	300	NAf
	BBC World Service	G	300	ME
	BBC World Service	G	500	NAf
9920	FEBC	PHL	100	6 VTN Bahnar
	FEBC	PHL	100	1,7 VTN Stieng
	FEBC	PHL	100	2 VTN Nung
	FEBC	PHL	100	3 VTN Katu
	FEBC	PHL	100	4 VTN Eastern Cham
	FEBC	PHL	100	5 VTN Muong
	FEBC	PHL	100	1 VTN Chrau
	FEBC	PHL	100	3 VTN Bru
	FEBC	PHL	100	4 VTN Chru
	FEBC	PHL	100	5 VTN Jeh
	FEBC	PHL	100	2,7 VTN Black Tai
	FEBC	PHL	100	4 VTN Roglai
	FEBC	PHL	100	2 VTN Koho
	FEBC	PHL	100	5 VTN Rengao
	FEBC	PHL	100	3 VTN Mnong
	FEBC	PHL	100	6 VTN Sedang
	FEBC	PHL	100	7 VTN WhiteTtai
	FEBC	PHL	100	3 VTN Hre
	FEBC	PHL	100	4,7 VTN Rade
	FEBC	PHL	100	1,5 VTN Koho

0 1 2 3 4 5 6 7 8 9 10 11 12 13 14 15 16 17 18 19 20 21 22 23 24

Arabic Chinese English French German Hindi Mandarin Russian
Spanish Other

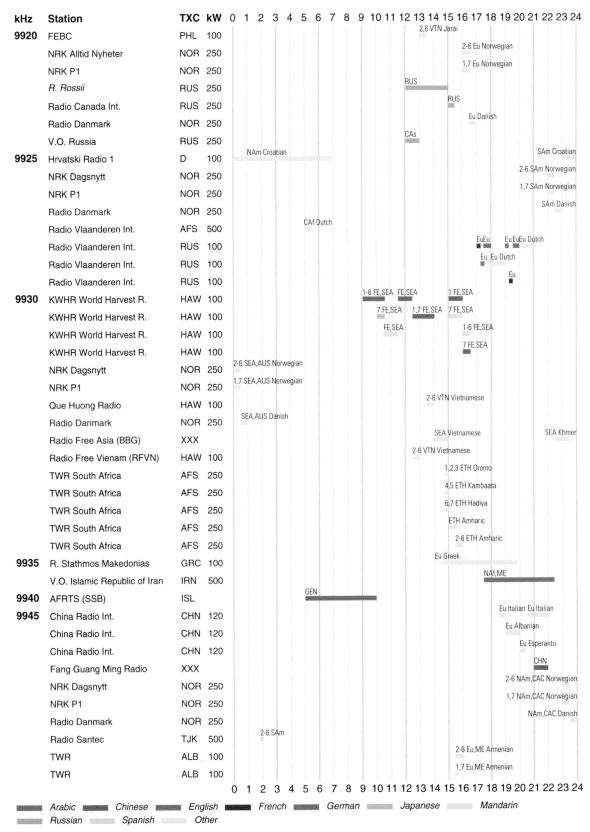

kHz	Station	TXC	kW
9920	FEBC	PHL	100
	NRK Alltid Nyheter	NOR	250
	NRK P1	NOR	250
	R. Rossii	RUS	250
	Radio Canada Int.	RUS	250
	Radio Danmark	NOR	250
	V.O. Russia	RUS	250
9925	Hrvatski Radio 1	D	100
	NRK Dagsnytt	NOR	250
	NRK P1	NOR	250
	Radio Danmark	NOR	250
	Radio Vlaanderen Int.	AFS	500
	Radio Vlaanderen Int.	RUS	100
	Radio Vlaanderen Int.	RUS	100
	Radio Vlaanderen Int.	RUS	100
9930	KWHR World Harvest R.	HAW	100
	KWHR World Harvest R.	HAW	100
	KWHR World Harvest R.	HAW	100
	KWHR World Harvest R.	HAW	100
	NRK Dagsnytt	NOR	250
	NRK P1	NOR	250
	Que Huong Radio	HAW	100
	Radio Danmark	NOR	250
	Radio Free Asia (BBG)	XXX	
	Radio Free Vienam (RFVN)	HAW	100
	TWR South Africa	AFS	250
	TWR South Africa	AFS	250
	TWR South Africa	AFS	250
	TWR South Africa	AFS	250
	TWR South Africa	AFS	250
9935	R. Stathmos Makedonias	GRC	100
	V.O. Islamic Republic of Iran	IRN	500
9940	AFRTS (SSB)	ISL	
9945	China Radio Int.	CHN	120
	China Radio Int.	CHN	120
	China Radio Int.	CHN	120
	Fang Guang Ming Radio	XXX	
	NRK Dagsnytt	NOR	250
	NRK P1	NOR	250
	Radio Danmark	NOR	250
	Radio Santec	TJK	500
	TWR	ALB	100
	TWR	ALB	100

Legend: Arabic · Chinese · English · French · German · Japanese · Mandarin · Russian · Spanish · Other

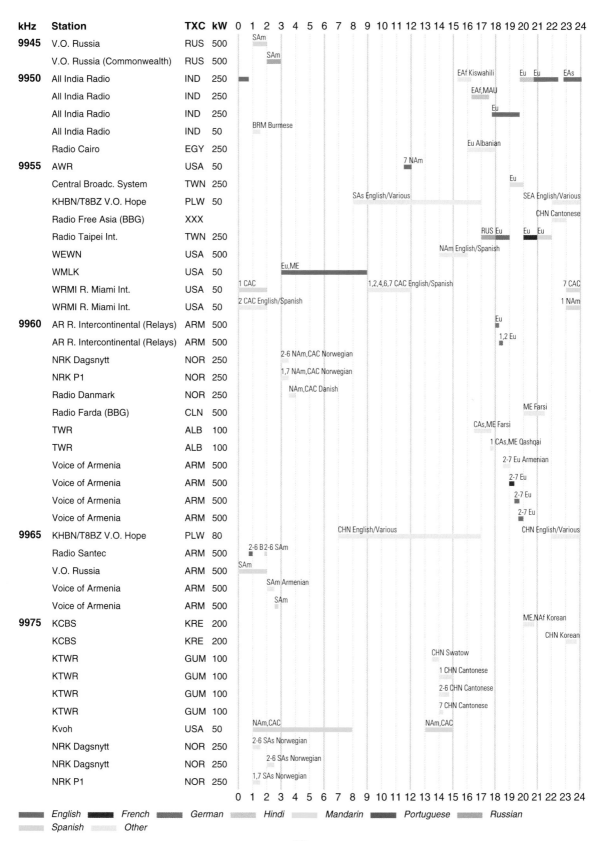

kHz	Station	TXC	kW	Broadcast Schedule
9945	V.O. Russia	RUS	500	SAm
	V.O. Russia (Commonwealth)	RUS	500	SAm
9950	All India Radio	IND	250	EAf Kiswahili · Eu · Eu · EAs
	All India Radio	IND	250	EAf,MAU
	All India Radio	IND	250	Eu
	All India Radio	IND	50	BRM Burmese
	Radio Cairo	EGY	250	Eu Albanian
9955	AWR	USA	50	7 NAm
	Central Broadc. System	TWN	250	Eu
	KHBN/T8BZ V.O. Hope	PLW	50	SAs English/Various · SEA English/Various
	Radio Free Asia (BBG)	XXX		CHN Cantonese
	Radio Taipei Int.	TWN	250	RUS Eu · Eu Eu
	WEWN	USA	500	NAm English/Spanish
	WMLK	USA	50	Eu,ME
	WRMI R. Miami Int.	USA	50	1 CAC · 1,2,4,6,7 CAC English/Spanish · 7 CAC
	WRMI R. Miami Int.	USA	50	2 CAC English/Spanish · 1 NAm
9960	AR R. Intercontinental (Relays)	ARM	500	Eu
	AR R. Intercontinental (Relays)	ARM	500	1,2 Eu
	NRK Dagsnytt	NOR	250	2-6 NAm,CAC Norwegian
	NRK P1	NOR	250	1,7 NAm,CAC Norwegian
	Radio Danmark	NOR	250	NAm,CAC Danish
	Radio Farda (BBG)	CLN	500	ME Farsi
	TWR	ALB	100	CAs,ME Farsi
	TWR	ALB	100	1 CAs,ME Qashqai
	Voice of Armenia	ARM	500	2-7 Eu Armenian
	Voice of Armenia	ARM	500	2-7 Eu
	Voice of Armenia	ARM	500	2-7 Eu
	Voice of Armenia	ARM	500	2-7 Eu
9965	KHBN/T8BZ V.O. Hope	PLW	80	CHN English/Various · CHN English/Various
	Radio Santec	ARM	500	2-6 B 2-6 SAm
	V.O. Russia	ARM	500	SAm
	Voice of Armenia	ARM	500	SAm Armenian
	Voice of Armenia	ARM	500	SAm
9975	KCBS	KRE	200	ME,NAf Korean
	KCBS	KRE	200	CHN Korean
	KTWR	GUM	100	CHN Swatow
	KTWR	GUM	100	1 CHN Cantonese
	KTWR	GUM	100	2-6 CHN Cantonese
	KTWR	GUM	100	7 CHN Cantonese
	Kvoh	USA	50	NAm,CAC · NAm,CAC
	NRK Dagsnytt	NOR	250	2-6 SAs Norwegian
	NRK Dagsnytt	NOR	250	2-6 SAs Norwegian
	NRK P1	NOR	250	1,7 SAs Norwegian

Legend: ▬ English ▬ French ▬ German ▬ Hindi ▬ Mandarin ▬ Portuguese ▬ Russian ▬ Spanish ▬ Other

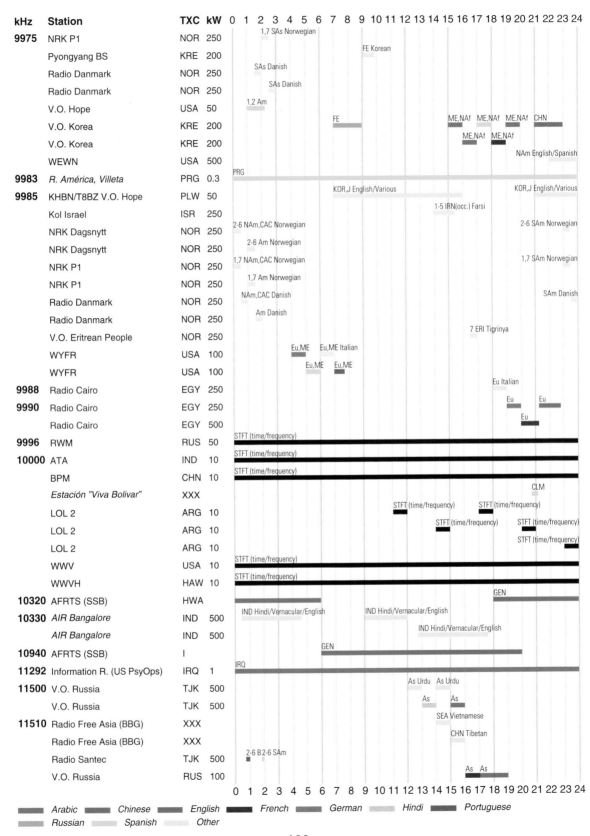

kHz	Station	TXC	kW																						

kHz	Station	TXC	kW
9975	NRK P1	NOR	250
	Pyongyang BS	KRE	200
	Radio Danmark	NOR	250
	Radio Danmark	NOR	250
	V.O. Hope	USA	50
	V.O. Korea	KRE	200
	V.O. Korea	KRE	200
	WEWN	USA	500
9983	*R. América, Villeta*	PRG	0.3
9985	KHBN/T8BZ V.O. Hope	PLW	50
	Kol Israel	ISR	250
	NRK Dagsnytt	NOR	250
	NRK Dagsnytt	NOR	250
	NRK P1	NOR	250
	NRK P1	NOR	250
	Radio Danmark	NOR	250
	Radio Danmark	NOR	250
	V.O. Eritrean People	NOR	250
	WYFR	USA	100
	WYFR	USA	100
9988	Radio Cairo	EGY	250
9990	Radio Cairo	EGY	250
	Radio Cairo	EGY	500
9996	RWM	RUS	50
10000	ATA	IND	10
	BPM	CHN	10
	Estación "Viva Bolivar"	XXX	
	LOL 2	ARG	10
	LOL 2	ARG	10
	LOL 2	ARG	10
	WWV	USA	10
	WWVH	HAW	10
10320	AFRTS (SSB)	HWA	
10330	*AIR Bangalore*	IND	500
	AIR Bangalore	IND	500
10940	AFRTS (SSB)	I	
11292	Information R. (US PsyOps)	IRQ	1
11500	V.O. Russia	TJK	500
	V.O. Russia	TJK	500
11510	Radio Free Asia (BBG)	XXX	
	Radio Free Asia (BBG)	XXX	
	Radio Santec	TJK	500
	V.O. Russia	RUS	100

Legend: Arabic · Chinese · English · French · German · Hindi · Portuguese · Russian · Spanish · Other

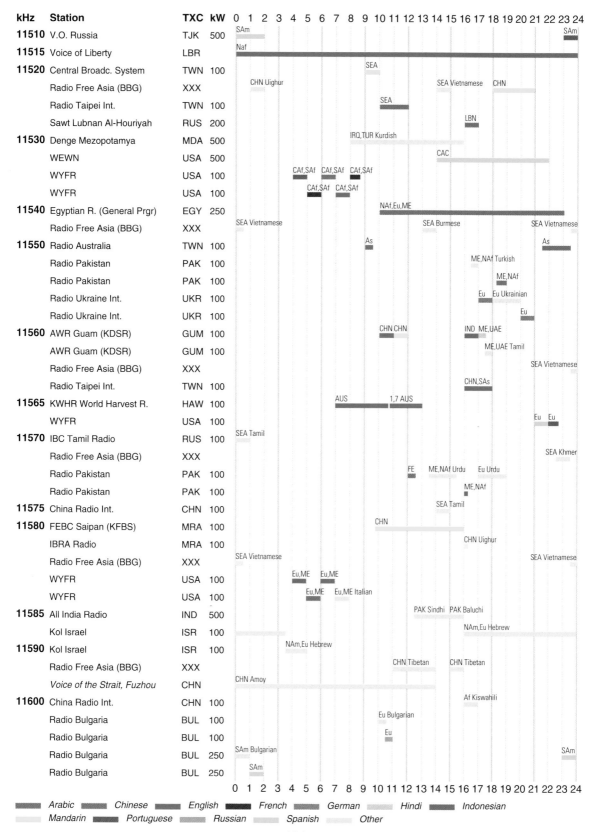

kHz	Station	TXC	kW
11510	V.O. Russia	TJK	500
11515	Voice of Liberty	LBR	
11520	Central Broadc. System	TWN	100
	Radio Free Asia (BBG)	XXX	
	Radio Taipei Int.	TWN	100
	Sawt Lubnan Al-Houriyah	RUS	200
11530	Denge Mezopotamya	MDA	500
	WEWN	USA	500
	WYFR	USA	100
	WYFR	USA	100
11540	Egyptian R. (General Prgr)	EGY	250
	Radio Free Asia (BBG)	XXX	
11550	Radio Australia	TWN	100
	Radio Pakistan	PAK	100
	Radio Pakistan	PAK	100
	Radio Ukraine Int.	UKR	100
	Radio Ukraine Int.	UKR	100
11560	AWR Guam (KDSR)	GUM	100
	AWR Guam (KDSR)	GUM	100
	Radio Free Asia (BBG)	XXX	
	Radio Taipei Int.	TWN	100
11565	KWHR World Harvest R.	HAW	100
	WYFR	USA	100
11570	IBC Tamil Radio	RUS	100
	Radio Free Asia (BBG)	XXX	
	Radio Pakistan	PAK	100
	Radio Pakistan	PAK	100
11575	China Radio Int.	CHN	100
11580	FEBC Saipan (KFBS)	MRA	100
	IBRA Radio	MRA	100
	Radio Free Asia (BBG)	XXX	
	WYFR	USA	100
	WYFR	USA	100
11585	All India Radio	IND	500
	Kol Israel	ISR	100
11590	Kol Israel	ISR	100
	Radio Free Asia (BBG)	XXX	
	Voice of the Strait, Fuzhou	CHN	
11600	China Radio Int.	CHN	100
	Radio Bulgaria	BUL	100
	Radio Bulgaria	BUL	100
	Radio Bulgaria	BUL	250
	Radio Bulgaria	BUL	250

Legend: Arabic · Chinese · English · French · German · Hindi · Indonesian · Mandarin · Portuguese · Russian · Spanish · Other

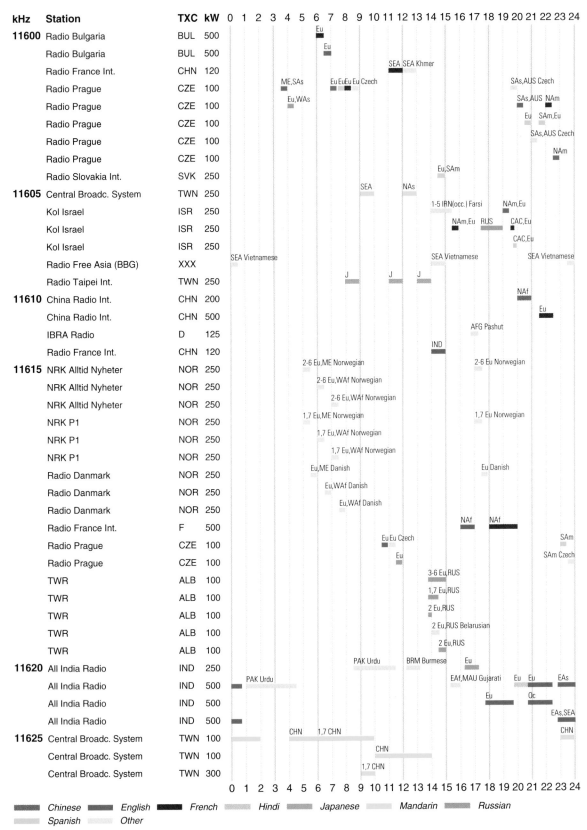

kHz	Station	TXC	kW
11600	Radio Bulgaria	BUL	500
	Radio Bulgaria	BUL	500
	Radio France Int.	CHN	120
	Radio Prague	CZE	100
	Radio Prague	CZE	100
	Radio Prague	CZE	100
	Radio Prague	CZE	100
	Radio Prague	CZE	100
	Radio Slovakia Int.	SVK	250
11605	Central Broadc. System	TWN	250
	Kol Israel	ISR	250
	Kol Israel	ISR	250
	Kol Israel	ISR	250
	Radio Free Asia (BBG)	XXX	
	Radio Taipei Int.	TWN	250
11610	China Radio Int.	CHN	200
	China Radio Int.	CHN	500
	IBRA Radio	D	125
	Radio France Int.	CHN	120
11615	NRK Alltid Nyheter	NOR	250
	NRK Alltid Nyheter	NOR	250
	NRK Alltid Nyheter	NOR	250
	NRK P1	NOR	250
	NRK P1	NOR	250
	NRK P1	NOR	250
	Radio Danmark	NOR	250
	Radio Danmark	NOR	250
	Radio Danmark	NOR	250
	Radio France Int.	F	500
	Radio Prague	CZE	100
	Radio Prague	CZE	100
	TWR	ALB	100
	TWR	ALB	100
	TWR	ALB	100
	TWR	ALB	100
	TWR	ALB	100
11620	All India Radio	IND	250
	All India Radio	IND	500
	All India Radio	IND	500
	All India Radio	IND	500
11625	Central Broadc. System	TWN	100
	Central Broadc. System	TWN	100
	Central Broadc. System	TWN	300

Legend: Chinese, English, French, Hindi, Japanese, Mandarin, Russian, Spanish, Other

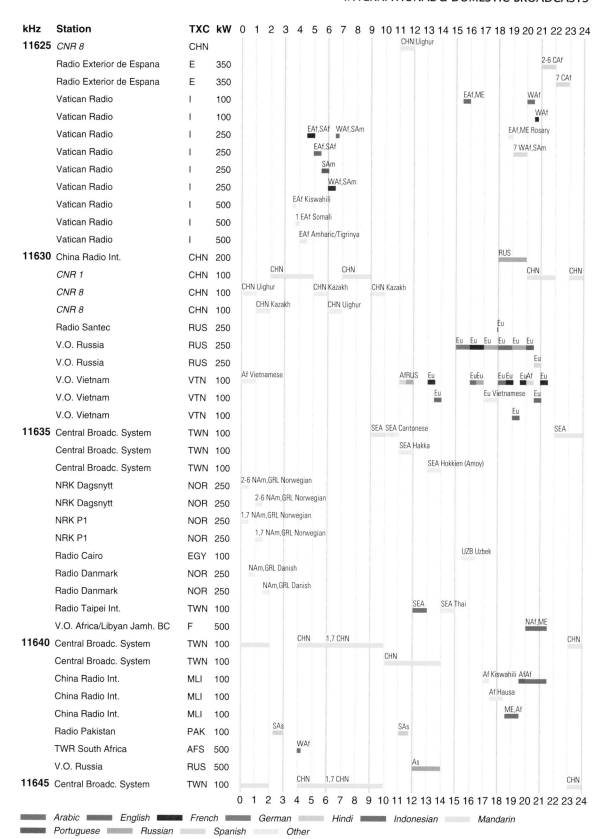

kHz	Station	TXC	kW	Broadcast schedule / target
11625	CNR 8	CHN		CHN Uighur
	Radio Exterior de Espana	E	350	2-6 CAf
	Radio Exterior de Espana	E	350	7 CAf
	Vatican Radio	I	100	EAf,ME
	Vatican Radio	I	100	WAf
	Vatican Radio	I	250	EAf,SAf WAf,SAm / WAf EAf,ME Rosary
	Vatican Radio	I	250	EAf,SAf / 7 WAf,SAm
	Vatican Radio	I	250	SAm
	Vatican Radio	I	250	WAf,SAm
	Vatican Radio	I	500	EAf Kiswahili
	Vatican Radio	I	500	1 EAf Somali
	Vatican Radio	I	500	EAf Amharic/Tigrinya
11630	China Radio Int.	CHN	200	RUS
	CNR 1	CHN	100	CHN CHN CHN CHN
	CNR 8	CHN	100	CHN Uighur CHN Kazakh CHN Kazakh
	CNR 8	CHN	100	CHN Kazakh CHN Uighur
	Radio Santec	RUS	250	Eu
	V.O. Russia	RUS	250	Eu Eu Eu Eu Eu Eu Eu
	V.O. Russia	RUS	250	Eu
	V.O. Vietnam	VTN	100	Af Vietnamese AfRUS Eu EuEu EuEu EuAf Eu
	V.O. Vietnam	VTN	100	Eu Eu Vietnamese Eu
	V.O. Vietnam	VTN	100	Eu
11635	Central Broadc. System	TWN	100	SEA SEA Cantonese SEA
	Central Broadc. System	TWN	100	SEA Hakka
	Central Broadc. System	TWN	100	SEA Hokkien (Amoy)
	NRK Dagsnytt	NOR	250	2-6 NAm,GRL Norwegian
	NRK Dagsnytt	NOR	250	2-6 NAm,GRL Norwegian
	NRK P1	NOR	250	1,7 NAm,GRL Norwegian
	NRK P1	NOR	250	1,7 NAm,GRL Norwegian
	Radio Cairo	EGY	100	UZB Uzbek
	Radio Danmark	NOR	250	NAm,GRL Danish
	Radio Danmark	NOR	250	NAm,GRL Danish
	Radio Taipei Int.	TWN	100	SEA SEA Thai
	V.O. Africa/Libyan Jamh. BC	F	500	NAf,ME
11640	Central Broadc. System	TWN	100	CHN 1,7 CHN CHN
	Central Broadc. System	TWN	100	CHN
	China Radio Int.	MLI	100	Af Kiswahili AfAf
	China Radio Int.	MLI	100	Af Hausa
	China Radio Int.	MLI	100	ME,Af
	Radio Pakistan	PAK	100	SAs SAs
	TWR South Africa	AFS	500	WAf
	V.O. Russia	RUS	500	As
11645	Central Broadc. System	TWN	100	CHN 1,7 CHN CHN

Legend: Arabic — English — French — German — Hindi — Indonesian — Mandarin — Portuguese — Russian — Spanish — Other

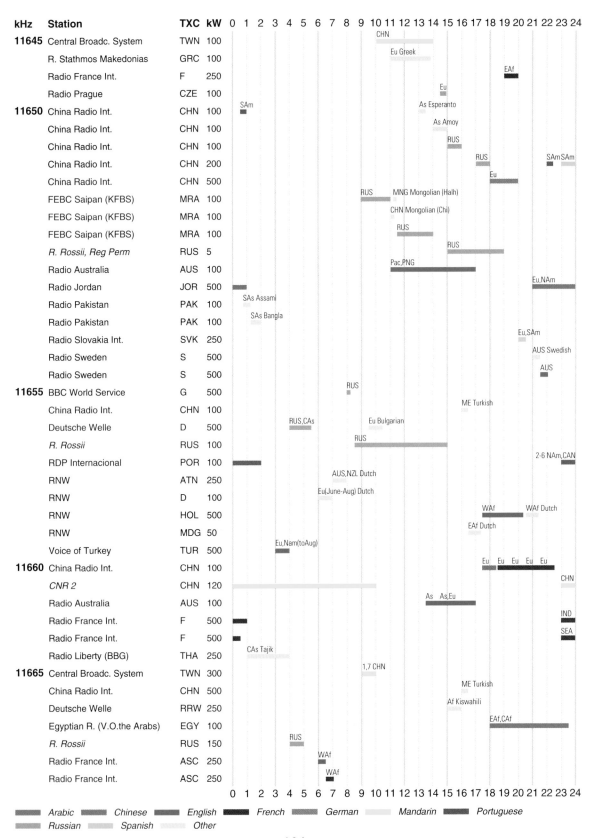

kHz	Station	TXC	kW
11645	Central Broadc. System	TWN	100
	R. Stathmos Makedonias	GRC	100
	Radio France Int.	F	250
	Radio Prague	CZE	100
11650	China Radio Int.	CHN	100
	China Radio Int.	CHN	100
	China Radio Int.	CHN	100
	China Radio Int.	CHN	200
	China Radio Int.	CHN	500
	FEBC Saipan (KFBS)	MRA	100
	FEBC Saipan (KFBS)	MRA	100
	FEBC Saipan (KFBS)	MRA	100
	R. Rossii, Reg Perm	RUS	5
	Radio Australia	AUS	100
	Radio Jordan	JOR	500
	Radio Pakistan	PAK	100
	Radio Pakistan	PAK	100
	Radio Slovakia Int.	SVK	250
	Radio Sweden	S	500
	Radio Sweden	S	500
11655	BBC World Service	G	500
	China Radio Int.	CHN	100
	Deutsche Welle	D	500
	R. Rossii	RUS	100
	RDP Internacional	POR	100
	RNW	ATN	250
	RNW	D	100
	RNW	HOL	500
	RNW	MDG	50
	Voice of Turkey	TUR	500
11660	China Radio Int.	CHN	100
	CNR 2	CHN	120
	Radio Australia	AUS	100
	Radio France Int.	F	500
	Radio France Int.	F	500
	Radio Liberty (BBG)	THA	250
11665	Central Broadc. System	TWN	300
	China Radio Int.	CHN	500
	Deutsche Welle	RRW	250
	Egyptian R. (V.O.the Arabs)	EGY	100
	R. Rossii	RUS	150
	Radio France Int.	ASC	250
	Radio France Int.	ASC	250

Legend: Arabic · Chinese · English · French · German · Mandarin · Portuguese · Russian · Spanish · Other

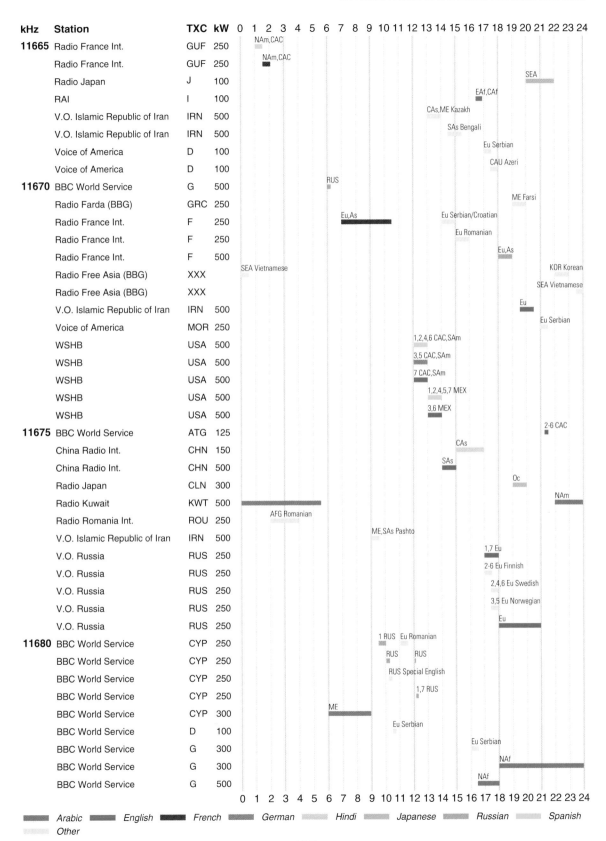

kHz	Station	TXC	kW
11665	Radio France Int.	GUF	250	NAm,CAC
	Radio France Int.	GUF	250	NAm,CAC
	Radio Japan	J	100	SEA
	RAI	I	100	EAf,CAf
	V.O. Islamic Republic of Iran	IRN	500	CAs,ME Kazakh
	V.O. Islamic Republic of Iran	IRN	500	SAs Bengali
	Voice of America	D	100	Eu Serbian
	Voice of America	D	100	CAU Azeri
11670	BBC World Service	G	500	RUS
	Radio Farda (BBG)	GRC	250	ME Farsi
	Radio France Int.	F	250	Eu,As / Eu Serbian/Croatian
	Radio France Int.	F	250	Eu Romanian
	Radio France Int.	F	500	Eu,As
	Radio Free Asia (BBG)	XXX		SEA Vietnamese / KOR Korean
	Radio Free Asia (BBG)	XXX		SEA Vietnamese
	V.O. Islamic Republic of Iran	IRN	500	Eu
	Voice of America	MOR	250	Eu Serbian
	WSHB	USA	500	1,2,4,6 CAC,SAm
	WSHB	USA	500	3,5 CAC,SAm
	WSHB	USA	500	7 CAC,SAm
	WSHB	USA	500	1,2,4,5,7 MEX
	WSHB	USA	500	3,6 MEX
11675	BBC World Service	ATG	125	2-6 CAC
	China Radio Int.	CHN	150	CAs
	China Radio Int.	CHN	500	SAs
	Radio Japan	CLN	300	Oc
	Radio Kuwait	KWT	500	NAm
	Radio Romania Int.	ROU	250	AFG Romanian
	V.O. Islamic Republic of Iran	IRN	500	ME,SAs Pashto
	V.O. Russia	RUS	250	1,7 Eu
	V.O. Russia	RUS	250	2-6 Eu Finnish
	V.O. Russia	RUS	250	2,4,6 Eu Swedish
	V.O. Russia	RUS	250	3,5 Eu Norwegian
	V.O. Russia	RUS	250	Eu
11680	BBC World Service	CYP	250	1 RUS Eu Romanian
	BBC World Service	CYP	250	RUS RUS
	BBC World Service	CYP	250	RUS Special English
	BBC World Service	CYP	250	1,7 RUS
	BBC World Service	CYP	300	ME
	BBC World Service	D	100	Eu Serbian
	BBC World Service	G	300	Eu Serbian
	BBC World Service	G	300	NAf
	BBC World Service	G	500	NAf

Legend: Arabic · English · French · German · Hindi · Japanese · Russian · Spanish · Other

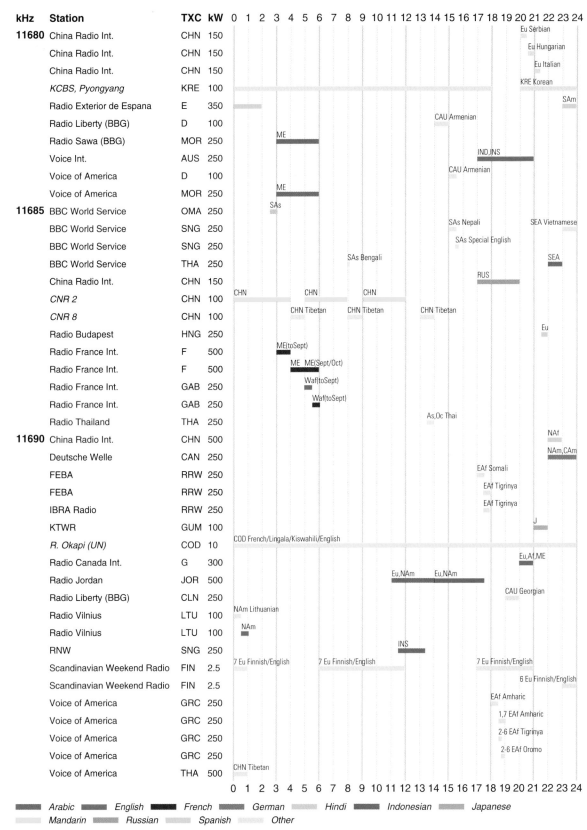

kHz	Station	TXC	kW																								
				0	1	2	3	4	5	6	7	8	9	10	11	12	13	14	15	16	17	18	19	20	21	22	23 24
11680	China Radio Int.	CHN	150																						Eu Serbian
	China Radio Int.	CHN	150																					Eu Hungarian
	China Radio Int.	CHN	150																					Eu Italian
	KCBS, Pyongyang	KRE	100																			KRE Korean
	Radio Exterior de Espana	E	350	SAm																					
	Radio Liberty (BBG)	D	100														CAU Armenian
	Radio Sawa (BBG)	MOR	250				ME										
	Voice Int.	AUS	250																IND,INS
	Voice of America	D	100														CAU Armenian
	Voice of America	MOR	250				ME										
11685	BBC World Service	OMA	250			SAs																		
	BBC World Service	SNG	250														SAs Nepali					SEA Vietnamese
	BBC World Service	SNG	250														SAs Special English
	BBC World Service	THA	250										SAs Bengali								SEA
	China Radio Int.	CHN	150														RUS
	CNR 2	CHN	100	CHN					CHN			CHN										
	CNR 8	CHN	100					CHN Tibetan			CHN Tibetan		CHN Tibetan		
	Radio Budapest	HNG	250																			Eu
	Radio France Int.	F	500				ME(toSept)											
	Radio France Int.	F	500					ME ME(Sept/Oct)									
	Radio France Int.	GAB	250					Waf(toSept)									
	Radio France Int.	GAB	250						Waf(toSept)								
	Radio Thailand	THA	250														As,Oc Thai
11690	China Radio Int.	CHN	500																				NAf
	Deutsche Welle	CAN	250																				NAm,CAm
	FEBA	RRW	250																EAf Somali
	FEBA	RRW	250																EAf Tigrinya
	IBRA Radio	RRW	250																EAf Tigrinya
	KTWR	GUM	100																				J
	R. Okapi (UN)	COD	10	COD French/Lingala/Kiswahili/English														
	Radio Canada Int.	G	300																			Eu,Af,ME
	Radio Jordan	JOR	500													Eu,NAm		Eu,NAm
	Radio Liberty (BBG)	CLN	250																				CAU Georgian
	Radio Vilnius	LTU	100	NAm Lithuanian														
	Radio Vilnius	LTU	100		NAm												
	RNW	SNG	250														INS
	Scandinavian Weekend Radio	FIN	2.5	7 Eu Finnish/English							7 Eu Finnish/English										7 Eu Finnish/English
	Scandinavian Weekend Radio	FIN	2.5																				6 Eu Finnish/English
	Voice of America	GRC	250																		EAf Amharic
	Voice of America	GRC	250																		1,7 EAf Amharic
	Voice of America	GRC	250																		2-6 EAf Tigrinya
	Voice of America	GRC	250																		2-6 EAf Oromo
	Voice of America	THA	500	CHN Tibetan													
				0	1	2	3	4	5	6	7	8	9	10	11	12	13	14	15	16	17	18	19	20	21	22	23 24

■ *Arabic*	■ *English*	■ *French*	■ *German*	▤ *Hindi*	■ *Indonesian*	▤ *Japanese*
▤ *Mandarin*	■ *Russian*	▤ *Spanish*	▤ *Other*			

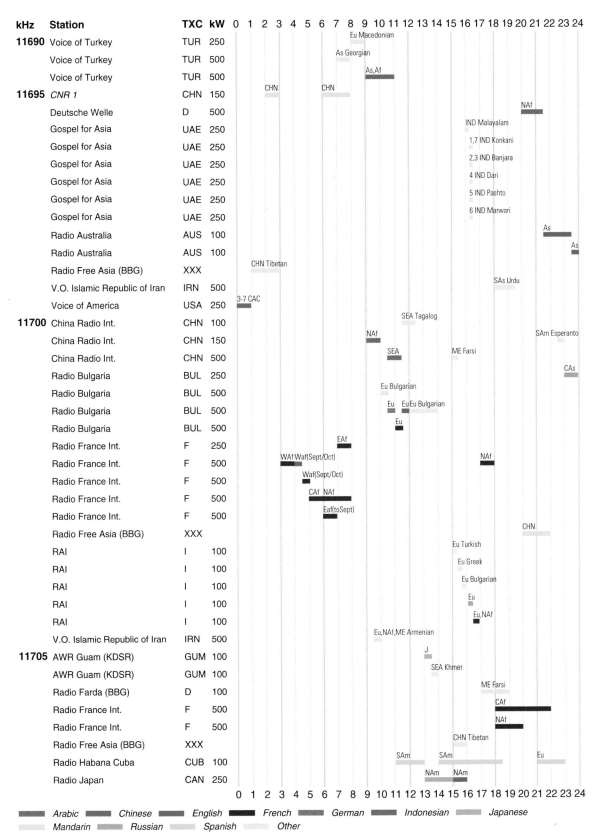

kHz	Station	TXC	kW																							
11690	Voice of Turkey	TUR	250										Eu Macedonian													
	Voice of Turkey	TUR	500								As Georgian															
	Voice of Turkey	TUR	500									As,Af													
11695	*CNR 1*	CHN	150			CHN				CHN																
	Deutsche Welle	D	500																			NAf				
	Gospel for Asia	UAE	250													IND Malayalam										
	Gospel for Asia	UAE	250													1,7 IND Konkani									
	Gospel for Asia	UAE	250													2,3 IND Banjara									
	Gospel for Asia	UAE	250													4 IND Dari									
	Gospel for Asia	UAE	250													5 IND Pashto									
	Gospel for Asia	UAE	250													6 IND Marwari									
	Radio Australia	AUS	100																				As				
	Radio Australia	AUS	100																							As	
	Radio Free Asia (BBG)	XXX				CHN Tibetan																				
	V.O. Islamic Republic of Iran	IRN	500															SAs Urdu								
	Voice of America	USA	250	3-7 CAC																						
11700	China Radio Int.	CHN	100										SEA Tagalog													
	China Radio Int.	CHN	150								NAf												SAm Esperanto			
	China Radio Int.	CHN	500									SEA				ME Farsi								CAs		
	Radio Bulgaria	BUL	250																						CAs		
	Radio Bulgaria	BUL	500								Eu Bulgarian																
	Radio Bulgaria	BUL	500									Eu		Eu Eu Bulgarian												
	Radio Bulgaria	BUL	500									Eu															
	Radio France Int.	F	250							EAf																	
	Radio France Int.	F	500				WAf Waf(Sept/Oct)											NAf									
	Radio France Int.	F	500					Waf(Sept/Oct)																			
	Radio France Int.	F	500				CAf NAf																			
	Radio France Int.	F	500					Eaf(toSept)																			
	Radio Free Asia (BBG)	XXX																			CHN					
	RAI	I	100													Eu Turkish										
	RAI	I	100												Eu Greek										
	RAI	I	100													Eu Bulgarian									
	RAI	I	100													Eu										
	RAI	I	100													Eu,NAf									
	V.O. Islamic Republic of Iran	IRN	500								Eu,NAf,ME Armenian															
11705	AWR Guam (KDSR)	GUM	100												J											
	AWR Guam (KDSR)	GUM	100											SEA Khmer												
	Radio Farda (BBG)	D	100															ME Farsi								
	Radio France Int.	F	500														CAf									
	Radio France Int.	F	500														NAf									
	Radio Free Asia (BBG)	XXX														CHN Tibetan									
	Radio Habana Cuba	CUB	100										SAm			SAm							Eu			
	Radio Japan	CAN	250												NAm		NAm									

Arabic   Chinese   English   French   German   Indonesian   Japanese
Mandarin   Russian   Spanish   Other

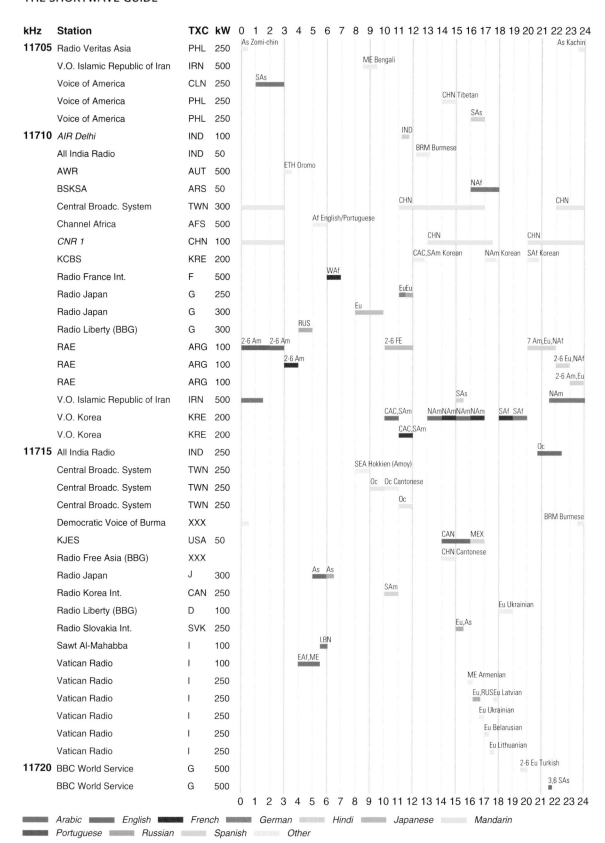

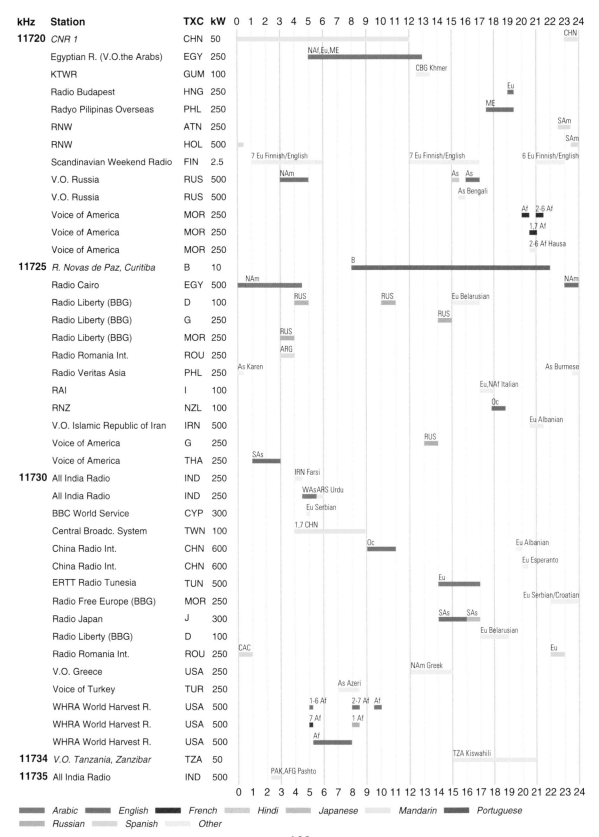

kHz	Station	TXC	kW
11720	*CNR 1*	CHN	50
	Egyptian R. (V.O.the Arabs)	EGY	250
	KTWR	GUM	100
	Radio Budapest	HNG	250
	Radyo Pilipinas Overseas	PHL	250
	RNW	ATN	250
	RNW	HOL	500
	Scandinavian Weekend Radio	FIN	2.5
	V.O. Russia	RUS	500
	V.O. Russia	RUS	500
	Voice of America	MOR	250
	Voice of America	MOR	250
	Voice of America	MOR	250
11725	*R. Novas de Paz, Curitiba*	B	10
	Radio Cairo	EGY	500
	Radio Liberty (BBG)	D	100
	Radio Liberty (BBG)	G	250
	Radio Liberty (BBG)	MOR	250
	Radio Romania Int.	ROU	250
	Radio Veritas Asia	PHL	250
	RAI	I	100
	RNZ	NZL	100
	V.O. Islamic Republic of Iran	IRN	500
	Voice of America	G	250
	Voice of America	THA	250
11730	All India Radio	IND	250
	All India Radio	IND	250
	BBC World Service	CYP	300
	Central Broadc. System	TWN	100
	China Radio Int.	CHN	600
	China Radio Int.	CHN	600
	ERTT Radio Tunesia	TUN	500
	Radio Free Europe (BBG)	MOR	250
	Radio Japan	J	300
	Radio Liberty (BBG)	D	100
	Radio Romania Int.	ROU	250
	V.O. Greece	USA	250
	Voice of Turkey	TUR	250
	WHRA World Harvest R.	USA	500
	WHRA World Harvest R.	USA	500
	WHRA World Harvest R.	USA	500
11734	*V.O. Tanzania, Zanzibar*	TZA	50
11735	All India Radio	IND	500

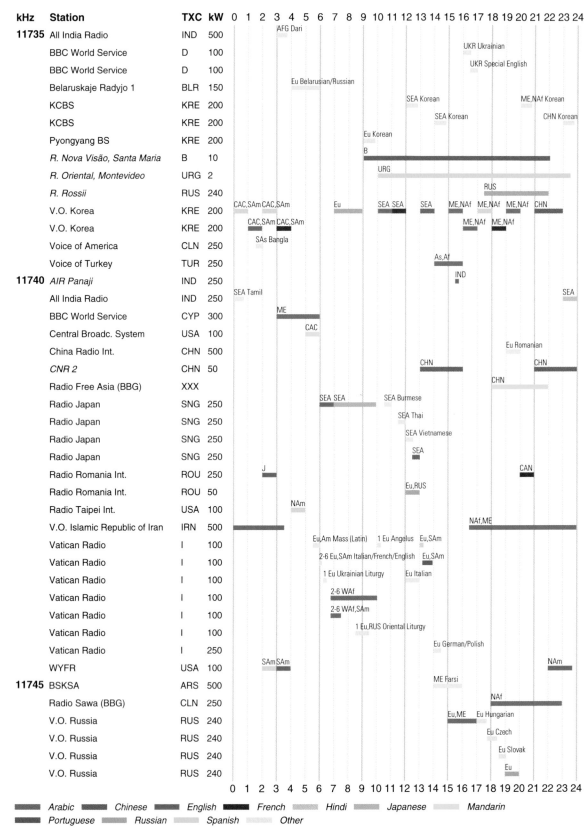

kHz	Station	TXC	kW	0 1 2 3 4 5 6 7 8 9 10 11 12 13 14 15 16 17 18 19 20 21 22 23 24
11735	All India Radio	IND	500	AFG Dari
	BBC World Service	D	100	UKR Ukrainian
	BBC World Service	D	100	UKR Special English
	Belaruskaje Radyjo 1	BLR	150	Eu Belarusian/Russian
	KCBS	KRE	200	SEA Korean ME,NAf Korean
	KCBS	KRE	200	SEA Korean CHN Korean
	Pyongyang BS	KRE	200	Eu Korean
	R. Nova Visão, Santa Maria	B	10	B
	R. Oriental, Montevideo	URG	2	URG
	R. Rossii	RUS	240	RUS
	V.O. Korea	KRE	200	CAC,SAm CAC,SAm Eu SEA SEA SEA ME,NAf ME,NAf ME,NAf CHN
	V.O. Korea	KRE	200	CAC,SAm CAC,SAm ME,NAf ME,NAf
	Voice of America	CLN	250	SAs Bangla
	Voice of Turkey	TUR	250	As,Af
11740	AIR Panaji	IND	250	IND
	All India Radio	IND	250	SEA Tamil SEA
	BBC World Service	CYP	300	ME
	Central Broadc. System	USA	100	CAC
	China Radio Int.	CHN	500	Eu Romanian
	CNR 2	CHN	50	CHN CHN
	Radio Free Asia (BBG)	XXX		CHN
	Radio Japan	SNG	250	SEA SEA SEA Burmese
	Radio Japan	SNG	250	SEA Thai
	Radio Japan	SNG	250	SEA Vietnamese
	Radio Japan	SNG	250	SEA
	Radio Romania Int.	ROU	250	J CAN
	Radio Romania Int.	ROU	50	Eu,RUS
	Radio Taipei Int.	USA	100	NAm
	V.O. Islamic Republic of Iran	IRN	500	NAf,ME
	Vatican Radio	I	100	Eu,Am Mass (Latin) 1 Eu Angelus Eu,SAm
	Vatican Radio	I	100	2-6 Eu,SAm Italian/French/English Eu,SAm
	Vatican Radio	I	100	1 Eu Ukrainian Liturgy Eu Italian
	Vatican Radio	I	100	2-6 WAf
	Vatican Radio	I	100	2-6 WAf,SAm
	Vatican Radio	I	100	1 Eu,RUS Oriental Liturgy
	Vatican Radio	I	250	Eu German/Polish
	WYFR	USA	100	SAm SAm NAm
11745	BSKSA	ARS	500	ME Farsi
	Radio Sawa (BBG)	CLN	250	NAf
	V.O. Russia	RUS	240	Eu,ME Eu Hungarian
	V.O. Russia	RUS	240	Eu Czech
	V.O. Russia	RUS	240	Eu Slovak
	V.O. Russia	RUS	240	Eu

Arabic	Chinese	English	French	Hindi	Japanese	Mandarin
Portuguese	Russian	Spanish	Other			

110

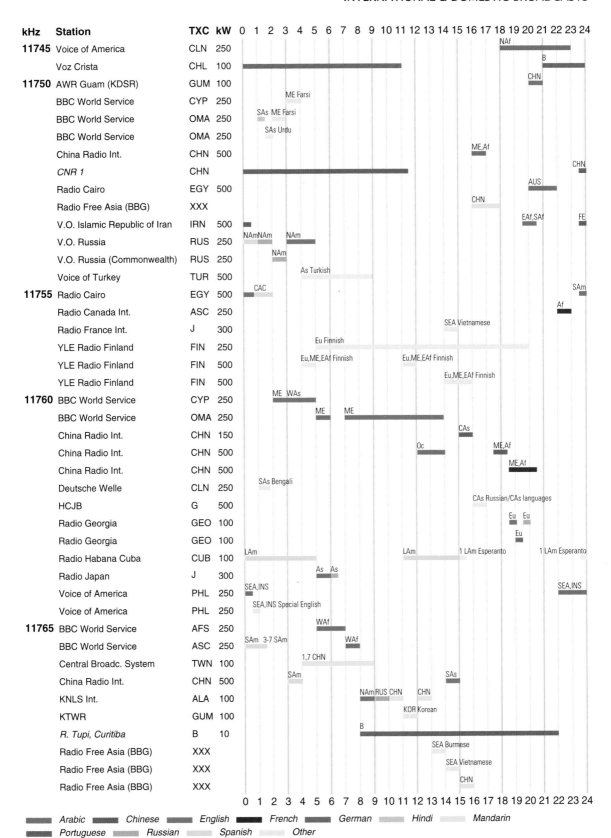

kHz	Station	TXC	kW
11745	Voice of America	CLN	250
	Voz Crista	CHL	100
11750	AWR Guam (KDSR)	GUM	100
	BBC World Service	CYP	250
	BBC World Service	OMA	250
	BBC World Service	OMA	250
	China Radio Int.	CHN	500
	CNR 1	CHN	
	Radio Cairo	EGY	500
	Radio Free Asia (BBG)	XXX	
	V.O. Islamic Republic of Iran	IRN	500
	V.O. Russia	RUS	250
	V.O. Russia (Commonwealth)	RUS	250
	Voice of Turkey	TUR	500
11755	Radio Cairo	EGY	500
	Radio Canada Int.	ASC	250
	Radio France Int.	J	300
	YLE Radio Finland	FIN	250
	YLE Radio Finland	FIN	500
	YLE Radio Finland	FIN	500
11760	BBC World Service	CYP	250
	BBC World Service	OMA	250
	China Radio Int.	CHN	150
	China Radio Int.	CHN	500
	China Radio Int.	CHN	500
	Deutsche Welle	CLN	250
	HCJB	G	500
	Radio Georgia	GEO	100
	Radio Georgia	GEO	100
	Radio Habana Cuba	CUB	100
	Radio Japan	J	300
	Voice of America	PHL	250
	Voice of America	PHL	250
11765	BBC World Service	AFS	250
	BBC World Service	ASC	250
	Central Broadc. System	TWN	100
	China Radio Int.	CHN	500
	KNLS Int.	ALA	100
	KTWR	GUM	100
	R. Tupi, Curitiba	B	10
	Radio Free Asia (BBG)	XXX	
	Radio Free Asia (BBG)	XXX	
	Radio Free Asia (BBG)	XXX	

Arabic   Chinese   English   French   German   Hindi   Mandarin
Portuguese   Russian   Spanish   Other

111

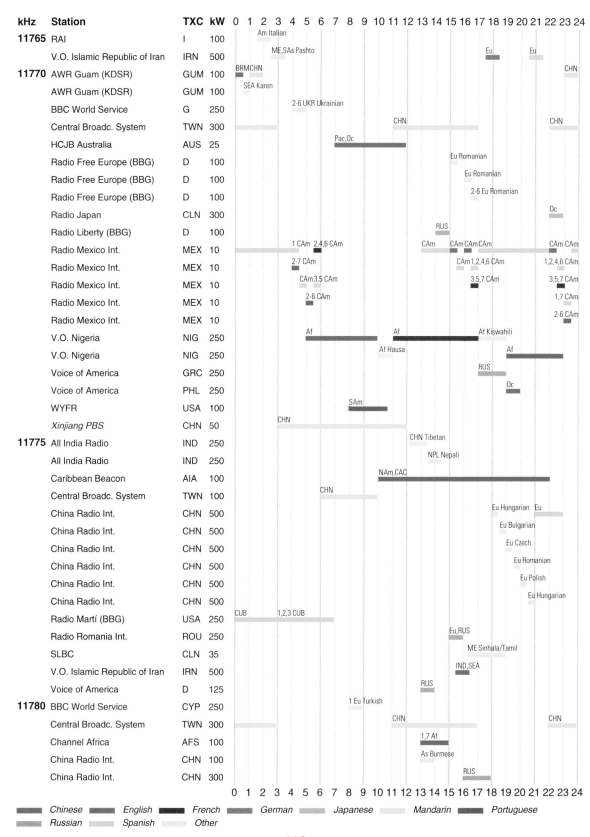

kHz	Station	TXC	kW	Schedule (0–24)
11765	RAI	I	100	Am Italian
	V.O. Islamic Republic of Iran	IRN	500	ME,SAs Pashto; Eu; Eu
11770	AWR Guam (KDSR)	GUM	100	BRM CHN; CHN
	AWR Guam (KDSR)	GUM	100	SEA Karen
	BBC World Service	G	250	2-6 UKR Ukrainian
	Central Broadc. System	TWN	300	CHN; CHN
	HCJB Australia	AUS	25	Pac,Oc
	Radio Free Europe (BBG)	D	100	Eu Romanian
	Radio Free Europe (BBG)	D	100	Eu Romanian
	Radio Free Europe (BBG)	D	100	2-6 Eu Romanian
	Radio Japan	CLN	300	Oc
	Radio Liberty (BBG)	D	100	RUS
	Radio Mexico Int.	MEX	10	1 CAm; 2,4,6 CAm; CAm; CAm CAm CAm; CAm CAm
	Radio Mexico Int.	MEX	10	2-7 CAm; CAm 1,2,4,6 CAm; 1,2,4,6 CAm
	Radio Mexico Int.	MEX	10	CAm 3,5 CAm; 3,5,7 CAm; 3,5,7 CAm
	Radio Mexico Int.	MEX	10	2-6 CAm; 1,7 CAm
	Radio Mexico Int.	MEX	10	2-6 CAm
	V.O. Nigeria	NIG	250	Af; Af; Af Kiswahili
	V.O. Nigeria	NIG	250	Af Hausa; Af
	Voice of America	GRC	250	RUS
	Voice of America	PHL	250	Oc
	WYFR	USA	100	SAm
	Xinjiang PBS	CHN	50	CHN
11775	All India Radio	IND	250	CHN Tibetan
	All India Radio	IND	250	NPL Nepali
	Caribbean Beacon	AIA	100	NAm,CAC
	Central Broadc. System	TWN	100	CHN
	China Radio Int.	CHN	500	Eu Hungarian Eu
	China Radio Int.	CHN	500	Eu Bulgarian
	China Radio Int.	CHN	500	Eu Czech
	China Radio Int.	CHN	500	Eu Romanian
	China Radio Int.	CHN	500	Eu Polish
	China Radio Int.	CHN	500	Eu Hungarian
	Radio Martí (BBG)	USA	250	CUB; 1,2,3 CUB
	Radio Romania Int.	ROU	250	Eu,RUS
	SLBC	CLN	35	ME Sinhala/Tamil
	V.O. Islamic Republic of Iran	IRN	500	IND,SEA
	Voice of America	D	125	RUS
11780	BBC World Service	CYP	250	1 Eu Turkish
	Central Broadc. System	TWN	300	CHN; CHN
	Channel Africa	AFS	100	1,7 Af
	China Radio Int.	CHN	100	As Burmese
	China Radio Int.	CHN	300	RUS

Legend: Chinese · English · French · German · Japanese · Mandarin · Portuguese · Russian · Spanish · Other

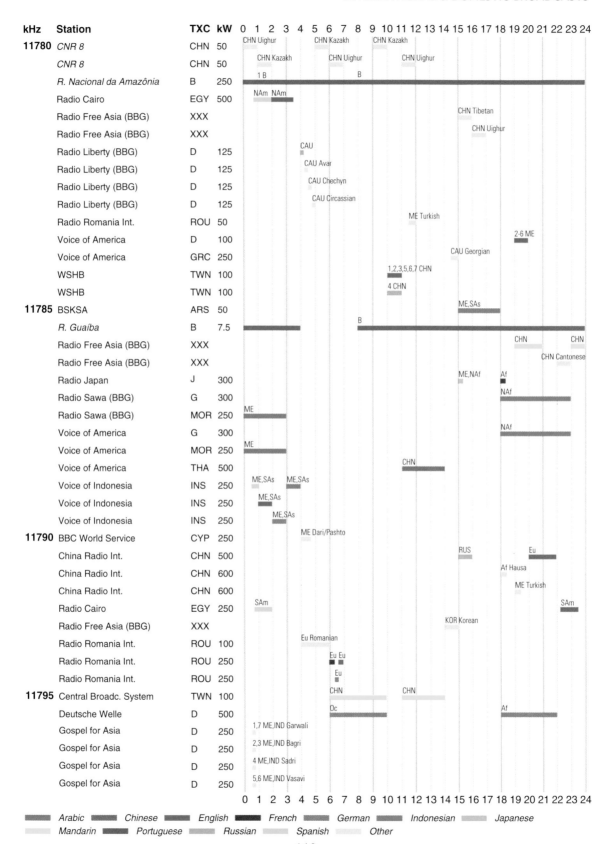

kHz	Station	TXC	kW
11780	CNR 8	CHN	50
	CNR 8	CHN	50
	R. Nacional da Amazônia	B	250
	Radio Cairo	EGY	500
	Radio Free Asia (BBG)	XXX	
	Radio Free Asia (BBG)	XXX	
	Radio Liberty (BBG)	D	125
	Radio Liberty (BBG)	D	125
	Radio Liberty (BBG)	D	125
	Radio Liberty (BBG)	D	125
	Radio Romania Int.	ROU	50
	Voice of America	D	100
	Voice of America	GRC	250
	WSHB	TWN	100
	WSHB	TWN	100
11785	BSKSA	ARS	50
	R. Guaíba	B	7.5
	Radio Free Asia (BBG)	XXX	
	Radio Free Asia (BBG)	XXX	
	Radio Japan	J	300
	Radio Sawa (BBG)	G	300
	Radio Sawa (BBG)	MOR	250
	Voice of America	G	300
	Voice of America	MOR	250
	Voice of America	THA	500
	Voice of Indonesia	INS	250
	Voice of Indonesia	INS	250
	Voice of Indonesia	INS	250
11790	BBC World Service	CYP	250
	China Radio Int.	CHN	500
	China Radio Int.	CHN	600
	China Radio Int.	CHN	600
	Radio Cairo	EGY	250
	Radio Free Asia (BBG)	XXX	
	Radio Romania Int.	ROU	100
	Radio Romania Int.	ROU	250
	Radio Romania Int.	ROU	250
11795	Central Broadc. System	TWN	100
	Deutsche Welle	D	500
	Gospel for Asia	D	250
	Gospel for Asia	D	250
	Gospel for Asia	D	250
	Gospel for Asia	D	250

Arabic   Chinese   English   French   German   Indonesian   Japanese
Mandarin   Portuguese   Russian   Spanish   Other

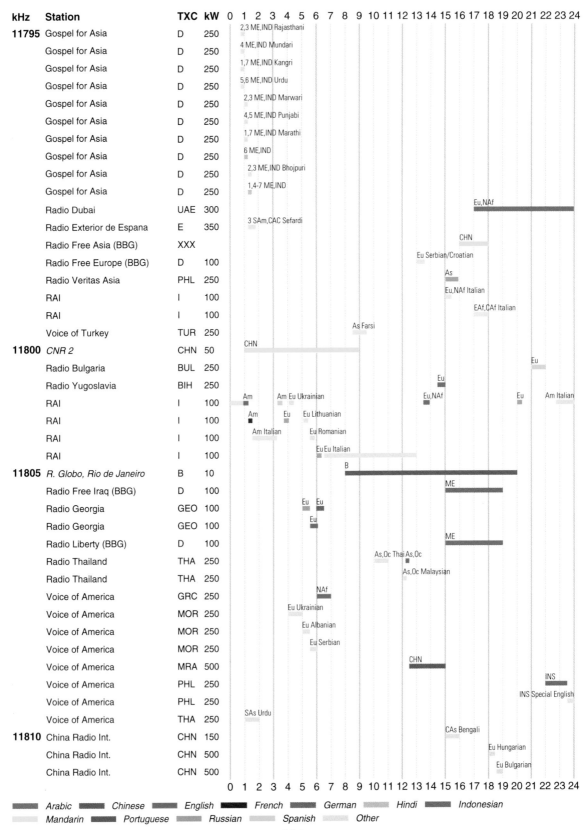

kHz	Station	TXC	kW	Schedule (0–24h)
11795	Gospel for Asia	D	250	2,3 ME,IND Rajasthani
	Gospel for Asia	D	250	4 ME,IND Mundari
	Gospel for Asia	D	250	1,7 ME,IND Kangri
	Gospel for Asia	D	250	5,6 ME,IND Urdu
	Gospel for Asia	D	250	2,3 ME,IND Marwari
	Gospel for Asia	D	250	4,5 ME,IND Punjabi
	Gospel for Asia	D	250	1,7 ME,IND Marathi
	Gospel for Asia	D	250	6 ME,IND
	Gospel for Asia	D	250	2,3 ME,IND Bhojpuri
	Gospel for Asia	D	250	1,4-7 ME,IND
	Radio Dubai	UAE	300	Eu,NAf
	Radio Exterior de Espana	E	350	3 SAm,CAC Sefardi
	Radio Free Asia (BBG)	XXX		CHN
	Radio Free Europe (BBG)	D	100	Eu Serbian/Croatian
	Radio Veritas Asia	PHL	250	As
	RAI	I	100	Eu,NAf Italian
	RAI	I	100	EAf,CAf Italian
	Voice of Turkey	TUR	250	As Farsi
11800	CNR 2	CHN	50	CHN
	Radio Bulgaria	BUL	250	Eu
	Radio Yugoslavia	BIH	250	Eu
	RAI	I	100	Am / Am Eu Ukrainian / Eu,NAf / Eu / Am Italian
	RAI	I	100	Am / Eu / Eu Lithuanian
	RAI	I	100	Am Italian / Eu Romanian
	RAI	I	100	Eu Eu Italian
11805	R. Globo, Rio de Janeiro	B	10	B
	Radio Free Iraq (BBG)	D	100	ME
	Radio Georgia	GEO	100	Eu Eu
	Radio Georgia	GEO	100	Eu
	Radio Liberty (BBG)	D	100	ME
	Radio Thailand	THA	250	As,Oc Thai As,Oc
	Radio Thailand	THA	250	As,Oc Malaysian
	Voice of America	GRC	250	NAf
	Voice of America	MOR	250	Eu Ukrainian
	Voice of America	MOR	250	Eu Albanian
	Voice of America	MOR	250	Eu Serbian
	Voice of America	MRA	500	CHN
	Voice of America	PHL	250	INS
	Voice of America	PHL	250	INS Special English
	Voice of America	THA	250	SAs Urdu
11810	China Radio Int.	CHN	150	CAs Bengali
	China Radio Int.	CHN	500	Eu Hungarian
	China Radio Int.	CHN	500	Eu Bulgarian

Arabic Chinese English French German Hindi Indonesian
Mandarin Portuguese Russian Spanish Other

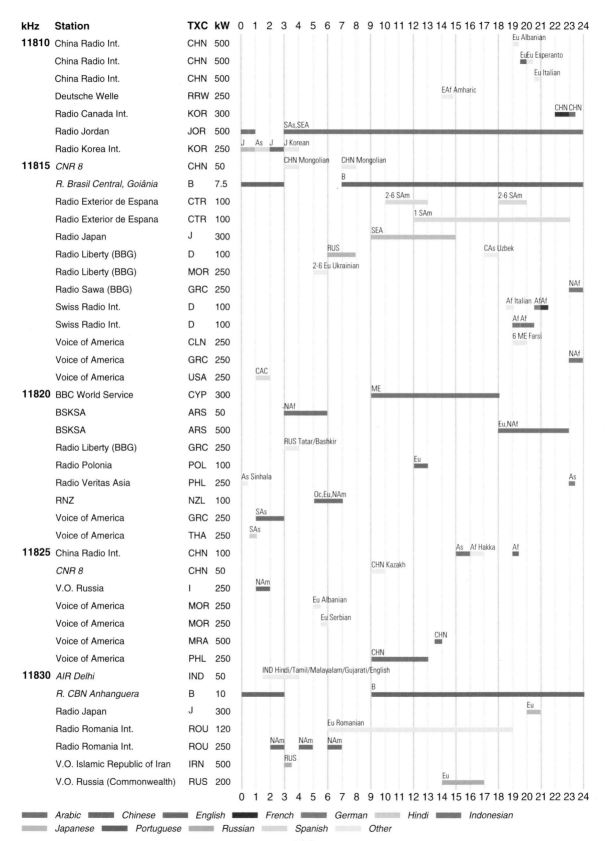

kHz	Station	TXC	kW																								

The chart spans 0–24 hours. Station details:

- **11810** China Radio Int. — CHN 500 — Eu Albanian
- China Radio Int. — CHN 500 — EuEu Esperanto
- China Radio Int. — CHN 500 — Eu Italian
- Deutsche Welle — RRW 250 — EAf Amharic
- Radio Canada Int. — KOR 300 — CHN CHN
- Radio Jordan — JOR 500 — SAs,SEA
- Radio Korea Int. — KOR 250 — J As J — J Korean
- **11815** *CNR 8* — CHN 50 — CHN Mongolian CHN Mongolian
- *R. Brasil Central, Goiânia* — B 7.5 — B
- Radio Exterior de Espana — CTR 100 — 2-6 SAm / 2-6 SAm
- Radio Exterior de Espana — CTR 100 — 1 SAm
- Radio Japan — J 300 — SEA
- Radio Liberty (BBG) — D 100 — RUS / CAs Uzbek
- Radio Liberty (BBG) — MOR 250 — 2-6 Eu Ukrainian
- Radio Sawa (BBG) — GRC 250 — NAf
- Swiss Radio Int. — D 100 — Af Italian AfAf
- Swiss Radio Int. — D 100 — Af Af
- Voice of America — CLN 250 — 6 ME Farsi
- Voice of America — GRC 250 — NAf
- Voice of America — USA 250 — CAC
- **11820** BBC World Service — CYP 300 — ME
- BSKSA — ARS 50 — NAf
- BSKSA — ARS 500 — Eu,NAf
- Radio Liberty (BBG) — GRC 250 — RUS Tatar/Bashkir
- Radio Polonia — POL 100 — Eu
- Radio Veritas Asia — PHL 250 — As Sinhala / As
- RNZ — NZL 100 — Oc,Eu,NAm
- Voice of America — GRC 250 — SAs
- Voice of America — THA 250 — SAs
- **11825** China Radio Int. — CHN 100 — As Af Hakka Af
- *CNR 8* — CHN 50 — CHN Kazakh
- V.O. Russia — I 250 — NAm
- Voice of America — MOR 250 — Eu Albanian
- Voice of America — MOR 250 — Eu Serbian
- Voice of America — MRA 500 — CHN
- Voice of America — PHL 250 — CHN
- **11830** *AIR Delhi* — IND 50 — IND Hindi/Tamil/Malayalam/Gujarati/English
- *R. CBN Anhanguera* — B 10 — B
- Radio Japan — J 300 — Eu
- Radio Romania Int. — ROU 120 — Eu Romanian
- Radio Romania Int. — ROU 250 — NAm NAm NAm
- V.O. Islamic Republic of Iran — IRN 500 — RUS
- V.O. Russia (Commonwealth) — RUS 200 — Eu

Legend: Arabic, Chinese, English, French, German, Hindi, Indonesian, Japanese, Portuguese, Russian, Spanish, Other

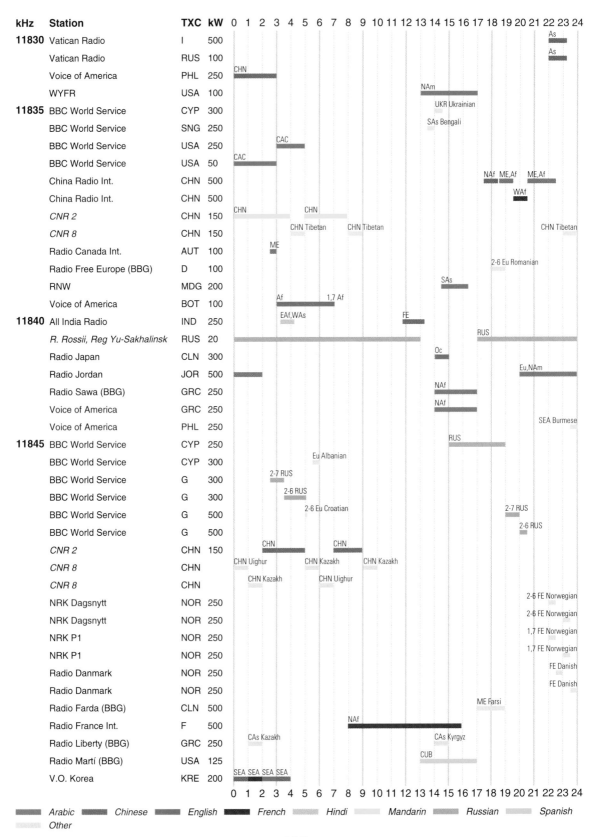

kHz	Station	TXC	kW
11830	Vatican Radio	I	500
	Vatican Radio	RUS	100
	Voice of America	PHL	250
	WYFR	USA	100
11835	BBC World Service	CYP	300
	BBC World Service	SNG	250
	BBC World Service	USA	250
	BBC World Service	USA	50
	China Radio Int.	CHN	500
	China Radio Int.	CHN	500
	CNR 2	CHN	150
	CNR 8	CHN	150
	Radio Canada Int.	AUT	100
	Radio Free Europe (BBG)	D	100
	RNW	MDG	200
	Voice of America	BOT	100
11840	All India Radio	IND	250
	R. Rossii, Reg Yu-Sakhalinsk	RUS	20
	Radio Japan	CLN	300
	Radio Jordan	JOR	500
	Radio Sawa (BBG)	GRC	250
	Voice of America	GRC	250
	Voice of America	PHL	250
11845	BBC World Service	CYP	250
	BBC World Service	CYP	300
	BBC World Service	G	300
	BBC World Service	G	300
	BBC World Service	G	500
	BBC World Service	G	500
	CNR 2	CHN	150
	CNR 8	CHN	
	CNR 8	CHN	
	NRK Dagsnytt	NOR	250
	NRK Dagsnytt	NOR	250
	NRK P1	NOR	250
	NRK P1	NOR	250
	Radio Danmark	NOR	250
	Radio Danmark	NOR	250
	Radio Farda (BBG)	CLN	500
	Radio France Int.	F	500
	Radio Liberty (BBG)	GRC	250
	Radio Martí (BBG)	USA	125
	V.O. Korea	KRE	200

Legend: Arabic, Chinese, English, French, Hindi, Mandarin, Russian, Spanish, Other

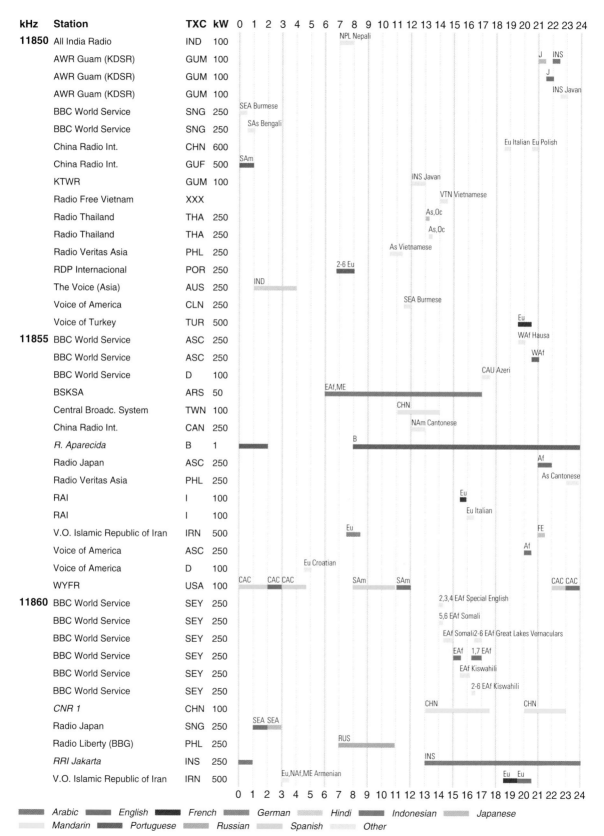

kHz	Station	TXC	kW	Schedule
11850	All India Radio	IND	100	NPL Nepali
	AWR Guam (KDSR)	GUM	100	J INS
	AWR Guam (KDSR)	GUM	100	J
	AWR Guam (KDSR)	GUM	100	INS Javan
	BBC World Service	SNG	250	SEA Burmese
	BBC World Service	SNG	250	SAs Bengali
	China Radio Int.	CHN	600	Eu Italian Eu Polish
	China Radio Int.	GUF	500	SAm
	KTWR	GUM	100	INS Javan
	Radio Free Vietnam	XXX		VTN Vietnamese
	Radio Thailand	THA	250	As,Oc
	Radio Thailand	THA	250	As,Oc
	Radio Veritas Asia	PHL	250	As Vietnamese
	RDP Internacional	POR	250	2-6 Eu
	The Voice (Asia)	AUS	250	IND
	Voice of America	CLN	250	SEA Burmese
	Voice of Turkey	TUR	500	Eu
11855	BBC World Service	ASC	250	WAf Hausa
	BBC World Service	ASC	250	WAf
	BBC World Service	D	100	CAU Azeri
	BSKSA	ARS	50	EAf,ME
	Central Broadc. System	TWN	100	CHN
	China Radio Int.	CAN	250	NAm Cantonese
	R. Aparecida	B	1	B
	Radio Japan	ASC	250	Af
	Radio Veritas Asia	PHL	250	As Cantonese
	RAI	I	100	Eu
	RAI	I	100	Eu Italian
	V.O. Islamic Republic of Iran	IRN	500	Eu FE
	Voice of America	ASC	250	Af
	Voice of America	D	100	Eu Croatian
	WYFR	USA	100	CAC CAC CAC SAm SAm CAC CAC
11860	BBC World Service	SEY	250	2,3,4 EAf Special English
	BBC World Service	SEY	250	5,6 EAf Somali
	BBC World Service	SEY	250	EAf Somali2-6 EAf Great Lakes Vernaculars
	BBC World Service	SEY	250	EAf 1,7 EAf
	BBC World Service	SEY	250	EAf Kiswahili
	BBC World Service	SEY	250	2-6 EAf Kiswahili
	CNR 1	CHN	100	CHN CHN
	Radio Japan	SNG	250	SEA SEA
	Radio Liberty (BBG)	PHL	250	RUS
	RRI Jakarta	INS	250	INS
	V.O. Islamic Republic of Iran	IRN	500	Eu,NAf,ME Armenian Eu Eu

Legend: Arabic | English | French | German | Hindi | Indonesian | Japanese | Mandarin | Portuguese | Russian | Spanish | Other

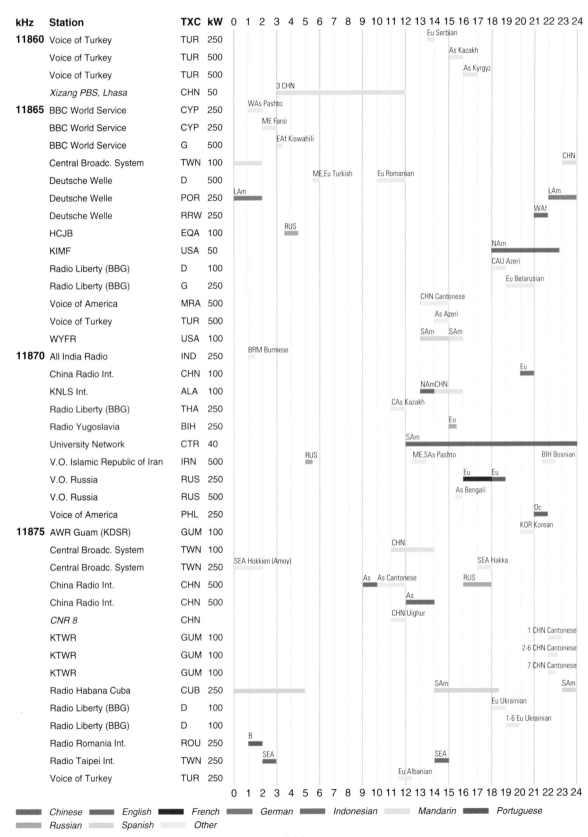

kHz	Station	TXC	kW
11860	Voice of Turkey	TUR	250
	Voice of Turkey	TUR	500
	Voice of Turkey	TUR	500
	Xizang PBS, Lhasa	CHN	50
11865	BBC World Service	CYP	250
	BBC World Service	CYP	250
	BBC World Service	G	500
	Central Broadc. System	TWN	100
	Deutsche Welle	D	500
	Deutsche Welle	POR	250
	Deutsche Welle	RRW	250
	HCJB	EQA	100
	KIMF	USA	50
	Radio Liberty (BBG)	D	100
	Radio Liberty (BBG)	G	250
	Voice of America	MRA	500
	Voice of Turkey	TUR	500
	WYFR	USA	100
11870	All India Radio	IND	250
	China Radio Int.	CHN	100
	KNLS Int.	ALA	100
	Radio Liberty (BBG)	THA	250
	Radio Yugoslavia	BIH	250
	University Network	CTR	40
	V.O. Islamic Republic of Iran	IRN	500
	V.O. Russia	RUS	250
	V.O. Russia	RUS	500
	Voice of America	PHL	250
11875	AWR Guam (KDSR)	GUM	100
	Central Broadc. System	TWN	100
	Central Broadc. System	TWN	250
	China Radio Int.	CHN	500
	China Radio Int.	CHN	500
	CNR 8	CHN	
	KTWR	GUM	100
	KTWR	GUM	100
	KTWR	GUM	100
	Radio Habana Cuba	CUB	250
	Radio Liberty (BBG)	D	100
	Radio Liberty (BBG)	D	100
	Radio Romania Int.	ROU	250
	Radio Taipei Int.	TWN	250
	Voice of Turkey	TUR	250

Legend: Chinese · English · French · German · Indonesian · Mandarin · Portuguese · Russian · Spanish · Other

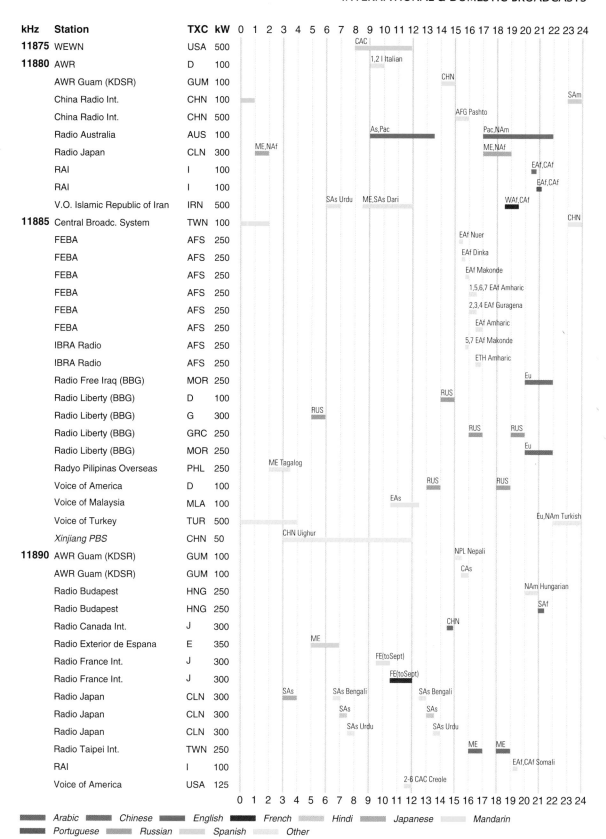

kHz	Station	TXC	kW
11875	WEWN	USA	500
11880	AWR	D	100
	AWR Guam (KDSR)	GUM	100
	China Radio Int.	CHN	100
	China Radio Int.	CHN	500
	Radio Australia	AUS	100
	Radio Japan	CLN	300
	RAI	I	100
	RAI	I	100
	V.O. Islamic Republic of Iran	IRN	500
11885	Central Broadc. System	TWN	100
	FEBA	AFS	250
	FEBA	AFS	250
	FEBA	AFS	250
	FEBA	AFS	250
	FEBA	AFS	250
	FEBA	AFS	250
	IBRA Radio	AFS	250
	IBRA Radio	AFS	250
	Radio Free Iraq (BBG)	MOR	250
	Radio Liberty (BBG)	D	100
	Radio Liberty (BBG)	G	300
	Radio Liberty (BBG)	GRC	250
	Radio Liberty (BBG)	MOR	250
	Radyo Pilipinas Overseas	PHL	250
	Voice of America	D	100
	Voice of Malaysia	MLA	100
	Voice of Turkey	TUR	500
	Xinjiang PBS	CHN	50
11890	AWR Guam (KDSR)	GUM	100
	AWR Guam (KDSR)	GUM	100
	Radio Budapest	HNG	250
	Radio Budapest	HNG	250
	Radio Canada Int.	J	300
	Radio Exterior de Espana	E	350
	Radio France Int.	J	300
	Radio France Int.	J	300
	Radio Japan	CLN	300
	Radio Japan	CLN	300
	Radio Japan	CLN	300
	Radio Taipei Int.	TWN	250
	RAI	I	100
	Voice of America	USA	125

Legend: Arabic — Chinese — English — French — Hindi — Japanese — Mandarin — Portuguese — Russian — Spanish — Other

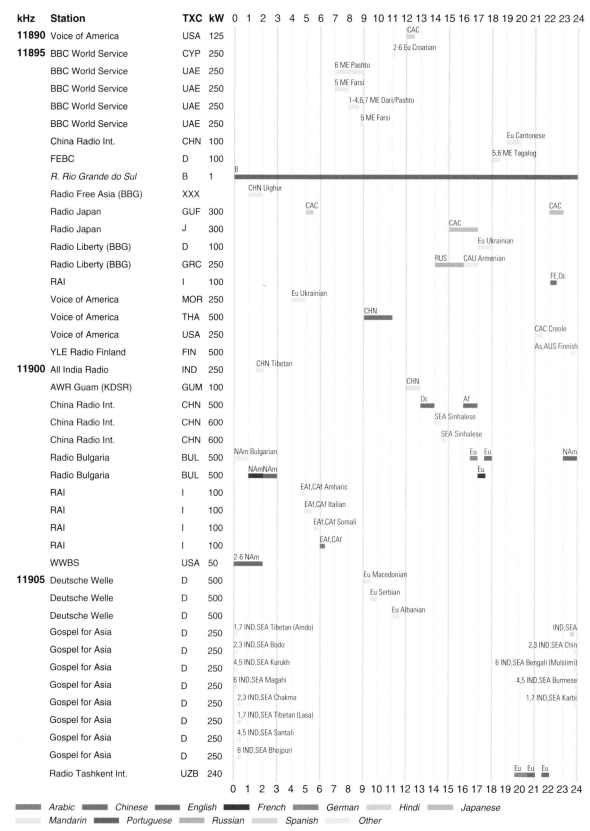

kHz	Station	TXC	kW	
11890	Voice of America	USA	125	CAC
11895	BBC World Service	CYP	250	2-6 Eu Croatian
	BBC World Service	UAE	250	6 ME Pashto
	BBC World Service	UAE	250	5 ME Farsi
	BBC World Service	UAE	250	1-4,6,7 ME Dari/Pashto
	BBC World Service	UAE	250	5 ME Farsi
	China Radio Int.	CHN	100	Eu Cantonese
	FEBC	D	100	5,6 ME Tagalog
	R. Rio Grande do Sul	B	1	B
	Radio Free Asia (BBG)	XXX		CHN Uighur
	Radio Japan	GUF	300	CAC / CAC
	Radio Japan	J	300	CAC
	Radio Liberty (BBG)	D	100	Eu Ukrainian
	Radio Liberty (BBG)	GRC	250	RUS CAU Armenian
	RAI	I	100	FE,Oc
	Voice of America	MOR	250	Eu Ukrainian
	Voice of America	THA	500	CHN
	Voice of America	USA	250	CAC Creole
	YLE Radio Finland	FIN	500	As,AUS Finnish
11900	All India Radio	IND	250	CHN Tibetan
	AWR Guam (KDSR)	GUM	100	CHN
	China Radio Int.	CHN	500	Oc Af
	China Radio Int.	CHN	600	SEA Sinhalese
	China Radio Int.	CHN	600	SEA Sinhalese
	Radio Bulgaria	BUL	500	NAm Bulgarian Eu Eu NAm
	Radio Bulgaria	BUL	500	NAmNAm Eu
	RAI	I	100	EAf,CAf Amharic
	RAI	I	100	EAf,CAf Italian
	RAI	I	100	EAf,CAf Somali
	RAI	I	100	EAf,CAf
	WWBS	USA	50	2-6 NAm
11905	Deutsche Welle	D	500	Eu Macedonian
	Deutsche Welle	D	500	Eu Serbian
	Deutsche Welle	D	500	Eu Albanian
	Gospel for Asia	D	250	1,7 IND,SEA Tibetan (Amdo) IND,SEA
	Gospel for Asia	D	250	2,3 IND,SEA Bodo 2,3 IND,SEA Chin
	Gospel for Asia	D	250	4,5 IND,SEA Kurukh 6 IND,SEA Bengali (Mulslimi)
	Gospel for Asia	D	250	6 IND,SEA Magahi 4,5 IND,SEA Burmese
	Gospel for Asia	D	250	2,3 IND,SEA Chakma 1,7 IND,SEA Karbi
	Gospel for Asia	D	250	1,7 IND,SEA Tibetan (Lasa)
	Gospel for Asia	D	250	4,5 IND,SEA Santali
	Gospel for Asia	D	250	6 IND,SEA Bhojpuri
	Radio Tashkent Int.	UZB	240	Eu Eu Eu

Arabic   Chinese   English   French   German   Hindi   Japanese

Mandarin   Portuguese   Russian   Spanish   Other

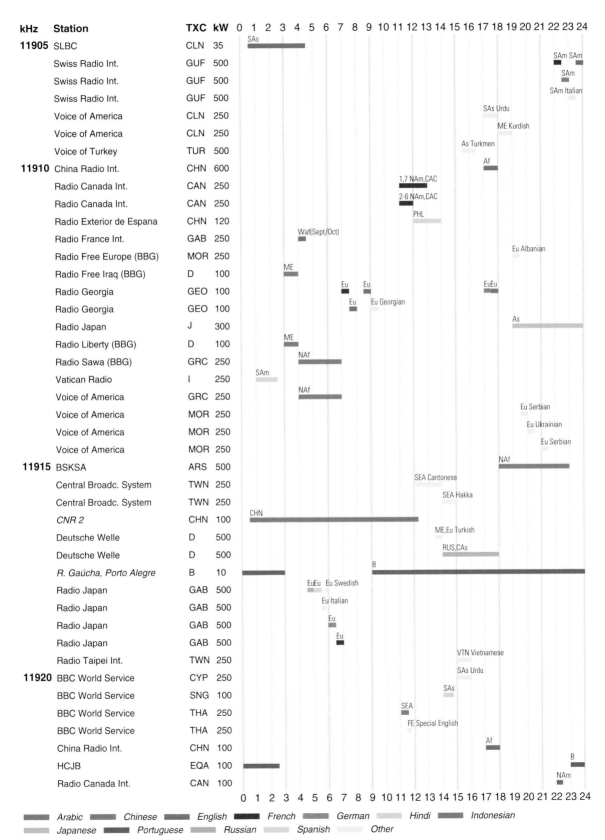

kHz	Station	TXC	kW	Schedule (0–24)
11905	SLBC	CLN	35	SAs
	Swiss Radio Int.	GUF	500	SAm SAm
	Swiss Radio Int.	GUF	500	SAm
	Swiss Radio Int.	GUF	500	SAm Italian
	Voice of America	CLN	250	SAs Urdu
	Voice of America	CLN	250	ME Kurdish
	Voice of Turkey	TUR	500	As Turkmen
11910	China Radio Int.	CHN	600	Af
	Radio Canada Int.	CAN	250	1,7 NAm,CAC
	Radio Canada Int.	CAN	250	2-6 NAm,CAC
	Radio Exterior de Espana	CHN	120	PHL
	Radio France Int.	GAB	250	Waf(Sept/Oct)
	Radio Free Europe (BBG)	MOR	250	Eu Albanian
	Radio Free Iraq (BBG)	D	100	ME
	Radio Georgia	GEO	100	Eu Eu EuEu
	Radio Georgia	GEO	100	Eu Eu Georgian
	Radio Japan	J	300	As
	Radio Liberty (BBG)	D	100	ME
	Radio Sawa (BBG)	GRC	250	NAf
	Vatican Radio	I	250	SAm
	Voice of America	GRC	250	NAf
	Voice of America	MOR	250	Eu Serbian
	Voice of America	MOR	250	Eu Ukrainian
	Voice of America	MOR	250	Eu Serbian
11915	BSKSA	ARS	500	NAf
	Central Broadc. System	TWN	250	SEA Cantonese
	Central Broadc. System	TWN	250	SEA Hakka
	CNR 2	CHN	100	CHN
	Deutsche Welle	D	500	ME,Eu Turkish
	Deutsche Welle	D	500	RUS,CAs
	R. Gaúcha, Porto Alegre	B	10	B
	Radio Japan	GAB	500	EuEu Eu Swedish
	Radio Japan	GAB	500	Eu Italian
	Radio Japan	GAB	500	Eu
	Radio Japan	GAB	500	Eu
	Radio Taipei Int.	TWN	250	VTN Vietnamese
11920	BBC World Service	CYP	250	SAs Urdu
	BBC World Service	SNG	100	SAs
	BBC World Service	THA	250	SEA
	BBC World Service	THA	250	FE Special English
	China Radio Int.	CHN	100	Af
	HCJB	EQA	100	B
	Radio Canada Int.	CAN	100	NAm

Legend: Arabic, Chinese, English, French, German, Hindi, Indonesian, Japanese, Portuguese, Russian, Spanish, Other

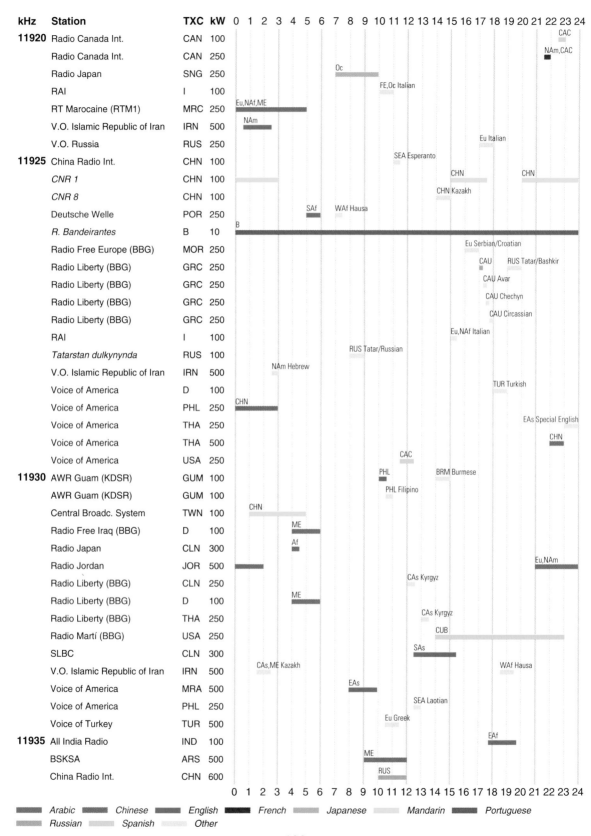

kHz	Station	TXC	kW
11920	Radio Canada Int.	CAN	100
	Radio Canada Int.	CAN	250
	Radio Japan	SNG	250
	RAI	I	100
	RT Marocaine (RTM1)	MRC	250
	V.O. Islamic Republic of Iran	IRN	500
	V.O. Russia	RUS	250
11925	China Radio Int.	CHN	100
	CNR 1	CHN	100
	CNR 8	CHN	100
	Deutsche Welle	POR	250
	R. Bandeirantes	B	10
	Radio Free Europe (BBG)	MOR	250
	Radio Liberty (BBG)	GRC	250
	Radio Liberty (BBG)	GRC	250
	Radio Liberty (BBG)	GRC	250
	Radio Liberty (BBG)	GRC	250
	RAI	I	100
	Tatarstan dulkynynda	RUS	100
	V.O. Islamic Republic of Iran	IRN	500
	Voice of America	D	100
	Voice of America	PHL	250
	Voice of America	THA	250
	Voice of America	THA	500
	Voice of America	USA	250
11930	AWR Guam (KDSR)	GUM	100
	AWR Guam (KDSR)	GUM	100
	Central Broadc. System	TWN	100
	Radio Free Iraq (BBG)	D	100
	Radio Japan	CLN	300
	Radio Jordan	JOR	500
	Radio Liberty (BBG)	CLN	250
	Radio Liberty (BBG)	D	100
	Radio Liberty (BBG)	THA	250
	Radio Martí (BBG)	USA	250
	SLBC	CLN	300
	V.O. Islamic Republic of Iran	IRN	500
	Voice of America	MRA	500
	Voice of America	PHL	250
	Voice of Turkey	TUR	500
11935	All India Radio	IND	100
	BSKSA	ARS	500
	China Radio Int.	CHN	600

Legend:
Arabic   *Chinese*   *English*   *French*   *Japanese*   *Mandarin*   *Portuguese*
Russian   *Spanish*   *Other*

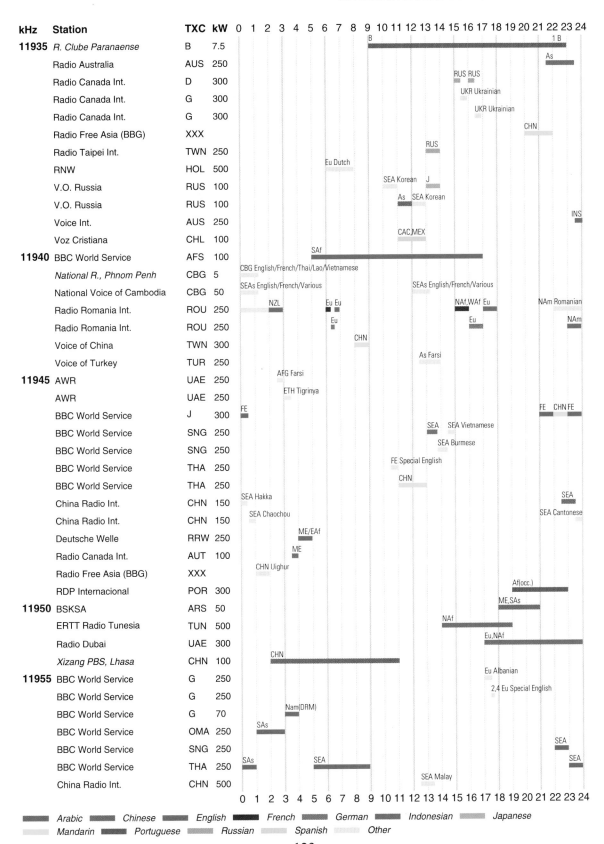

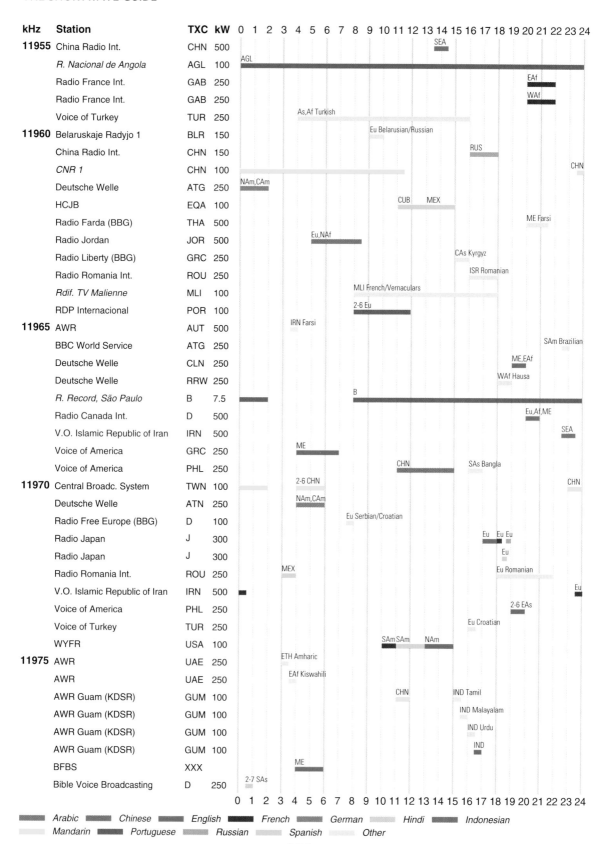

kHz	Station	TXC	kW	
11955	China Radio Int.	CHN	500	SEA
	R. Nacional de Angola	AGL	100	AGL
	Radio France Int.	GAB	250	EAf
	Radio France Int.	GAB	250	WAf
	Voice of Turkey	TUR	250	As,Af Turkish
11960	Belaruskaje Radyjo 1	BLR	150	Eu Belarusian/Russian
	China Radio Int.	CHN	150	RUS
	CNR 1	CHN	100	CHN
	Deutsche Welle	ATG	250	NAm,CAm
	HCJB	EQA	100	CUB MEX
	Radio Farda (BBG)	THA	500	ME Farsi
	Radio Jordan	JOR	500	Eu,NAf
	Radio Liberty (BBG)	GRC	250	CAs Kyrgyz
	Radio Romania Int.	ROU	250	ISR Romanian
	Rdif. TV Malienne	MLI	100	MLI French/Vernaculars
	RDP Internacional	POR	100	2-6 Eu
11965	AWR	AUT	500	IRN Farsi
	BBC World Service	ATG	250	SAm Brazilian
	Deutsche Welle	CLN	250	ME,EAf
	Deutsche Welle	RRW	250	WAf Hausa
	R. Record, São Paulo	B	7.5	B
	Radio Canada Int.	D	500	Eu,Af,ME
	V.O. Islamic Republic of Iran	IRN	500	SEA
	Voice of America	GRC	250	ME
	Voice of America	PHL	250	CHN SAs Bangla
11970	Central Broadc. System	TWN	100	2-6 CHN CHN
	Deutsche Welle	ATN	250	NAm,CAm
	Radio Free Europe (BBG)	D	100	Eu Serbian/Croatian
	Radio Japan	J	300	Eu Eu Eu
	Radio Japan	J	300	Eu
	Radio Romania Int.	ROU	250	MEX Eu Romanian
	V.O. Islamic Republic of Iran	IRN	500	Eu
	Voice of America	PHL	250	2-6 EAs
	Voice of Turkey	TUR	250	Eu Croatian
	WYFR	USA	100	SAm SAm NAm
11975	AWR	UAE	250	ETH Amharic
	AWR	UAE	250	EAf Kiswahili
	AWR Guam (KDSR)	GUM	100	CHN IND Tamil
	AWR Guam (KDSR)	GUM	100	IND Malayalam
	AWR Guam (KDSR)	GUM	100	IND Urdu
	AWR Guam (KDSR)	GUM	100	IND
	BFBS	XXX		ME
	Bible Voice Broadcasting	D	250	2-7 SAs

Legend: Arabic, Chinese, English, French, German, Hindi, Indonesian, Mandarin, Portuguese, Russian, Spanish, Other

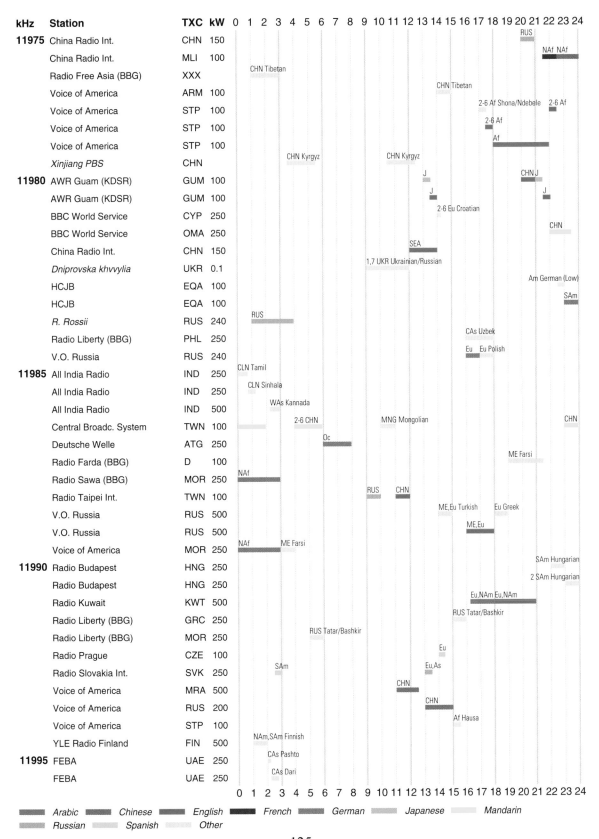

kHz	Station	TXC	kW	Schedule
11975	China Radio Int.	CHN	150	RUS
	China Radio Int.	MLI	100	NAf NAf
	Radio Free Asia (BBG)	XXX		CHN Tibetan
	Voice of America	ARM	100	CHN Tibetan
	Voice of America	STP	100	2-6 Af Shona/Ndebele 2-6 Af
	Voice of America	STP	100	2-6 Af
	Voice of America	STP	100	Af
	Xinjiang PBS	CHN		CHN Kyrgyz CHN Kyrgyz
11980	AWR Guam (KDSR)	GUM	100	J CHN J
	AWR Guam (KDSR)	GUM	100	J J
	BBC World Service	CYP	250	2-6 Eu Croatian
	BBC World Service	OMA	250	CHN
	China Radio Int.	CHN	150	SEA
	Dniprovska khvvylia	UKR	0.1	1,7 UKR Ukrainian/Russian
	HCJB	EQA	100	Am German (Low)
	HCJB	EQA	100	SAm
	R. Rossii	RUS	240	RUS
	Radio Liberty (BBG)	PHL	250	CAs Uzbek
	V.O. Russia	RUS	240	Eu Eu Polish
11985	All India Radio	IND	250	CLN Tamil
	All India Radio	IND	250	CLN Sinhala
	All India Radio	IND	500	WAs Kannada
	Central Broadc. System	TWN	100	2-6 CHN MNG Mongolian CHN
	Deutsche Welle	ATG	250	Oc
	Radio Farda (BBG)	D	100	ME Farsi
	Radio Sawa (BBG)	MOR	250	NAf
	Radio Taipei Int.	TWN	100	RUS CHN
	V.O. Russia	RUS	500	ME,Eu Turkish Eu Greek
	V.O. Russia	RUS	500	ME,Eu
	Voice of America	MOR	250	NAf ME Farsi
11990	Radio Budapest	HNG	250	SAm Hungarian
	Radio Budapest	HNG	250	2 SAm Hungarian
	Radio Kuwait	KWT	500	Eu,NAm Eu,NAm
	Radio Liberty (BBG)	GRC	250	RUS Tatar/Bashkir
	Radio Liberty (BBG)	MOR	250	RUS Tatar/Bashkir
	Radio Prague	CZE	100	Eu
	Radio Slovakia Int.	SVK	250	SAm Eu,As
	Voice of America	MRA	500	CHN
	Voice of America	RUS	200	CHN
	Voice of America	STP	100	Af Hausa
	YLE Radio Finland	FIN	500	NAm,SAm Finnish
11995	FEBA	UAE	250	CAs Pashto
	FEBA	UAE	250	CAs Dari

Legend: *Arabic* *Chinese* *English* *French* *German* *Japanese* *Mandarin*
Russian *Spanish* *Other*

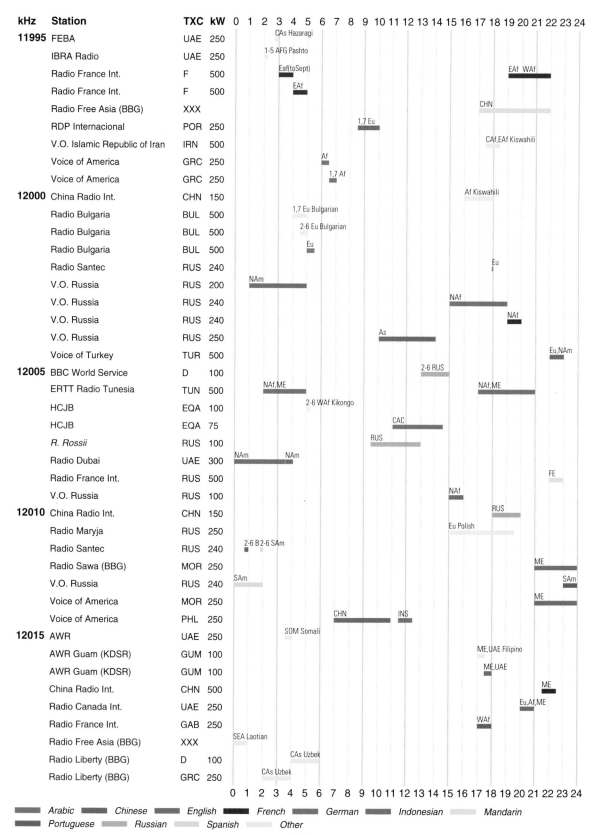

kHz	Station	TXC	kW
11995	FEBA	UAE	250
	IBRA Radio	UAE	250
	Radio France Int.	F	500
	Radio France Int.	F	500
	Radio Free Asia (BBG)	XXX	
	RDP Internacional	POR	250
	V.O. Islamic Republic of Iran	IRN	500
	Voice of America	GRC	250
	Voice of America	GRC	250
12000	China Radio Int.	CHN	150
	Radio Bulgaria	BUL	500
	Radio Bulgaria	BUL	500
	Radio Bulgaria	BUL	500
	Radio Santec	RUS	240
	V.O. Russia	RUS	200
	V.O. Russia	RUS	240
	V.O. Russia	RUS	240
	V.O. Russia	RUS	250
	Voice of Turkey	TUR	500
12005	BBC World Service	D	100
	ERTT Radio Tunesia	TUN	500
	HCJB	EQA	100
	HCJB	EQA	75
	R. Rossii	RUS	100
	Radio Dubai	UAE	300
	Radio France Int.	RUS	500
	V.O. Russia	RUS	100
12010	China Radio Int.	CHN	150
	Radio Maryja	RUS	250
	Radio Santec	RUS	240
	Radio Sawa (BBG)	MOR	250
	V.O. Russia	RUS	240
	Voice of America	MOR	250
	Voice of America	PHL	250
12015	AWR	UAE	250
	AWR Guam (KDSR)	GUM	100
	AWR Guam (KDSR)	GUM	100
	China Radio Int.	CHN	500
	Radio Canada Int.	UAE	250
	Radio France Int.	GAB	250
	Radio Free Asia (BBG)	XXX	
	Radio Liberty (BBG)	D	100
	Radio Liberty (BBG)	GRC	250

Legend: Arabic, Chinese, English, French, German, Indonesian, Mandarin, Portuguese, Russian, Spanish, Other

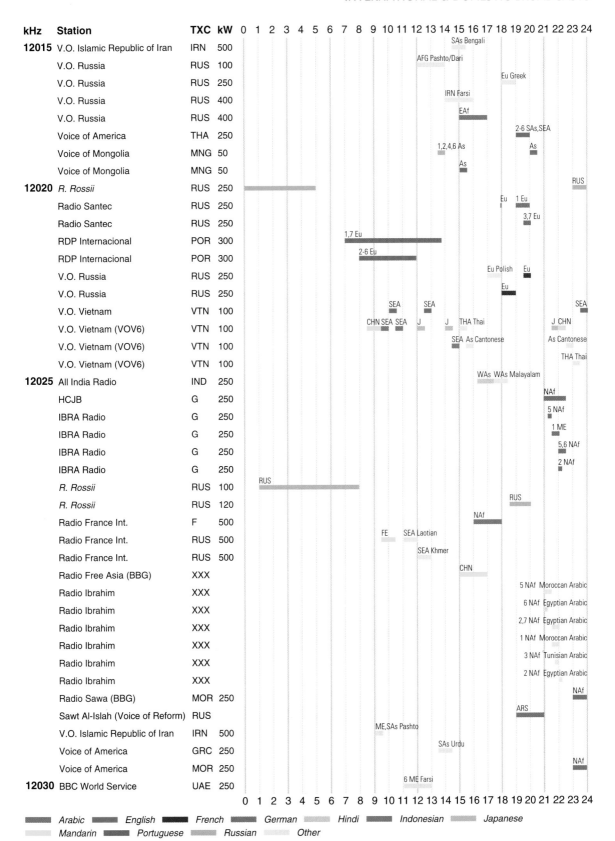

kHz	Station	TXC	kW
12015	V.O. Islamic Republic of Iran	IRN	500
	V.O. Russia	RUS	100
	V.O. Russia	RUS	250
	V.O. Russia	RUS	400
	V.O. Russia	RUS	400
	Voice of America	THA	250
	Voice of Mongolia	MNG	50
	Voice of Mongolia	MNG	50
12020	*R. Rossii*	RUS	250
	Radio Santec	RUS	250
	Radio Santec	RUS	250
	RDP Internacional	POR	300
	RDP Internacional	POR	300
	V.O. Russia	RUS	250
	V.O. Russia	RUS	250
	V.O. Vietnam	VTN	100
	V.O. Vietnam (VOV6)	VTN	100
	V.O. Vietnam (VOV6)	VTN	100
	V.O. Vietnam (VOV6)	VTN	100
12025	All India Radio	IND	250
	HCJB	G	250
	IBRA Radio	G	250
	IBRA Radio	G	250
	IBRA Radio	G	250
	IBRA Radio	G	250
	R. Rossii	RUS	100
	R. Rossii	RUS	120
	Radio France Int.	F	500
	Radio France Int.	RUS	500
	Radio France Int.	RUS	500
	Radio Free Asia (BBG)	XXX	
	Radio Ibrahim	XXX	
	Radio Ibrahim	XXX	
	Radio Ibrahim	XXX	
	Radio Ibrahim	XXX	
	Radio Ibrahim	XXX	
	Radio Ibrahim	XXX	
	Radio Sawa (BBG)	MOR	250
	Sawt Al-Islah (Voice of Reform)	RUS	
	V.O. Islamic Republic of Iran	IRN	500
	Voice of America	GRC	250
	Voice of America	MOR	250
12030	BBC World Service	UAE	250

Arabic English French German Hindi Indonesian Japanese

Mandarin Portuguese Russian Other

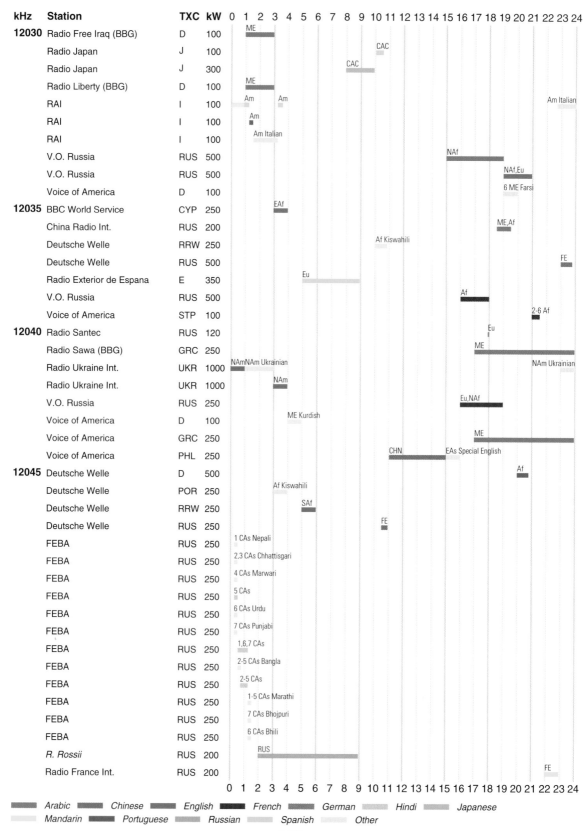

kHz	Station	TXC	kW
12030	Radio Free Iraq (BBG)	D	100
	Radio Japan	J	100
	Radio Japan	J	300
	Radio Liberty (BBG)	D	100
	RAI	I	100
	RAI	I	100
	RAI	I	100
	V.O. Russia	RUS	500
	V.O. Russia	RUS	500
	Voice of America	D	100
12035	BBC World Service	CYP	250
	China Radio Int.	RUS	200
	Deutsche Welle	RRW	250
	Deutsche Welle	RUS	500
	Radio Exterior de Espana	E	350
	V.O. Russia	RUS	500
	Voice of America	STP	100
12040	Radio Santec	RUS	120
	Radio Sawa (BBG)	GRC	250
	Radio Ukraine Int.	UKR	1000
	Radio Ukraine Int.	UKR	1000
	V.O. Russia	RUS	250
	Voice of America	D	100
	Voice of America	GRC	250
	Voice of America	PHL	250
12045	Deutsche Welle	D	500
	Deutsche Welle	POR	250
	Deutsche Welle	RRW	250
	Deutsche Welle	RUS	250
	FEBA	RUS	250
	FEBA	RUS	250
	FEBA	RUS	250
	FEBA	RUS	250
	FEBA	RUS	250
	FEBA	RUS	250
	FEBA	RUS	250
	FEBA	RUS	250
	FEBA	RUS	250
	FEBA	RUS	250
	FEBA	RUS	250
	FEBA	RUS	250
	R. Rossii	RUS	200
	Radio France Int.	RUS	200

Legend: Arabic, Chinese, English, French, German, Hindi, Japanese, Mandarin, Portuguese, Russian, Spanish, Other

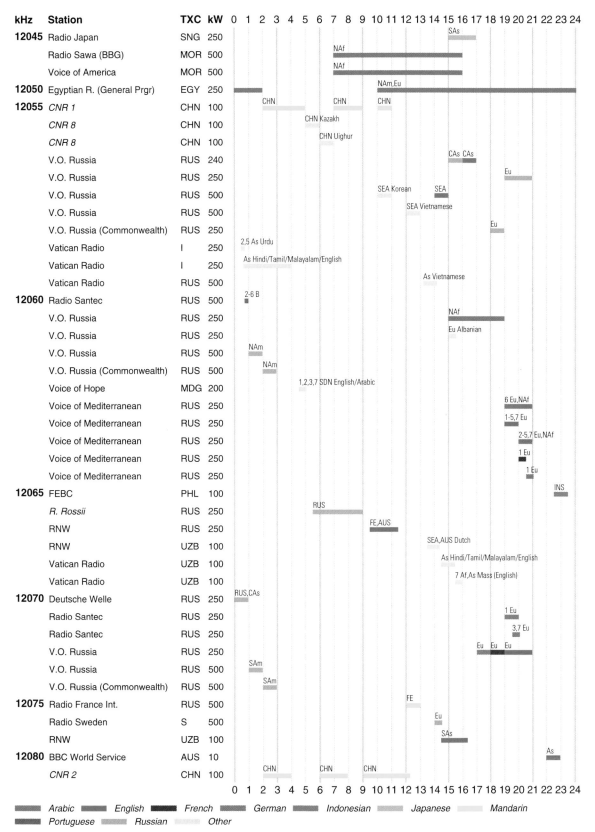

kHz	Station	TXC	kW
12045	Radio Japan	SNG	250
	Radio Sawa (BBG)	MOR	500
	Voice of America	MOR	500
12050	Egyptian R. (General Prgr)	EGY	250
12055	CNR 1	CHN	100
	CNR 8	CHN	100
	CNR 8	CHN	100
	V.O. Russia	RUS	240
	V.O. Russia	RUS	250
	V.O. Russia	RUS	500
	V.O. Russia	RUS	500
	V.O. Russia (Commonwealth)	RUS	250
	Vatican Radio	I	250
	Vatican Radio	I	250
	Vatican Radio	RUS	500
12060	Radio Santec	RUS	500
	V.O. Russia	RUS	250
	V.O. Russia	RUS	250
	V.O. Russia	RUS	500
	V.O. Russia (Commonwealth)	RUS	500
	Voice of Hope	MDG	200
	Voice of Mediterranean	RUS	250
	Voice of Mediterranean	RUS	250
	Voice of Mediterranean	RUS	250
	Voice of Mediterranean	RUS	250
	Voice of Mediterranean	RUS	250
12065	FEBC	PHL	100
	R. Rossii	RUS	250
	RNW	RUS	250
	RNW	UZB	100
	Vatican Radio	UZB	100
	Vatican Radio	UZB	100
12070	Deutsche Welle	RUS	250
	Radio Santec	RUS	250
	Radio Santec	RUS	250
	V.O. Russia	RUS	250
	V.O. Russia	RUS	500
	V.O. Russia (Commonwealth)	RUS	500
12075	Radio France Int.	RUS	500
	Radio Sweden	S	500
	RNW	UZB	100
12080	BBC World Service	AUS	10
	CNR 2	CHN	100

Legend: Arabic · English · French · German · Indonesian · Japanese · Mandarin · Portuguese · Russian · Other

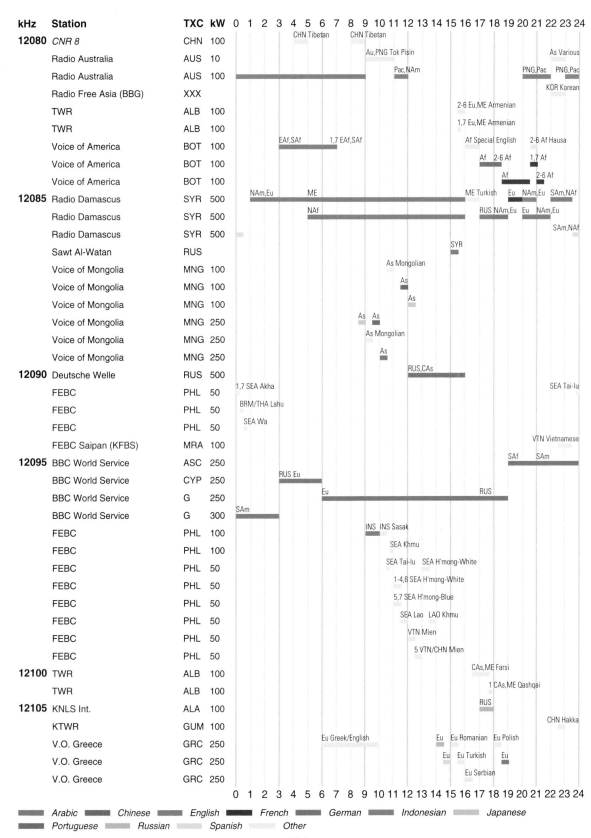

kHz	Station	TXC	kW																								
12080	*CNR 8*	CHN	100						CHN Tibetan			CHN Tibetan															
	Radio Australia	AUS	10									Au,PNG Tok Pisin												As Various			
	Radio Australia	AUS	100										Pac,NAm									PNG,Pac		PNG,Pac		
	Radio Free Asia (BBG)	XXX																						KOR Korean		
	TWR	ALB	100														2-6 Eu,ME Armenian									
	TWR	ALB	100														1,7 Eu,ME Armenian									
	Voice of America	BOT	100					EAf,SAf		1,7 EAf,SAf							Af Special English		2-6 Af Hausa							
	Voice of America	BOT	100													Af	2-6 Af		1,7 Af						
	Voice of America	BOT	100														Af		2-6 Af						
12085	Radio Damascus	SYR	500	NAm,Eu				ME						ME Turkish			Eu	NAm,Eu		SAm,NAf		
	Radio Damascus	SYR	500				NAf								RUS NAm,Eu		Eu	NAm,Eu			
	Radio Damascus	SYR	500																	SAm,NAf		
	Sawt Al-Watan	RUS												SYR								
	Voice of Mongolia	MNG	100								As Mongolian											
	Voice of Mongolia	MNG	100								As											
	Voice of Mongolia	MNG	100									As										
	Voice of Mongolia	MNG	250						As	As												
	Voice of Mongolia	MNG	250						As Mongolian													
	Voice of Mongolia	MNG	250							As												
12090	Deutsche Welle	RUS	500									RUS,CAs										
	FEBC	PHL	50	1,7 SEA Akha																SEA Tai-lu		
	FEBC	PHL	50	BRM/THA Lahu																		
	FEBC	PHL	50	SEA Wa																		
	FEBC Saipan (KFBS)	MRA	100																	VTN Vietnamese		
12095	BBC World Service	ASC	250															SAf	SAm			
	BBC World Service	CYP	250					RUS Eu														
	BBC World Service	G	250							Eu								RUS				
	BBC World Service	G	300	SAm																		
	FEBC	PHL	100								INS	INS Sasak										
	FEBC	PHL	100									SEA Khmu										
	FEBC	PHL	50								SEA Tai-lu	SEA H'mong-White									
	FEBC	PHL	50									1-4,6 SEA H'mong-White									
	FEBC	PHL	50									5,7 SEA H'mong-Blue									
	FEBC	PHL	50									SEA Lao	LAO Khmu								
	FEBC	PHL	50									VTN Mien									
	FEBC	PHL	50									5 VTN/CHN Mien									
12100	TWR	ALB	100													CAs,ME Farsi					
	TWR	ALB	100														1 CAs,ME Qashqai				
12105	KNLS Int.	ALA	100														RUS				
	KTWR	GUM	100																	CHN Hakka		
	V.O. Greece	GRC	250							Eu Greek/English					Eu	Eu Romanian	Eu Polish			
	V.O. Greece	GRC	250													Eu	Eu Turkish	Eu			
	V.O. Greece	GRC	250														Eu Serbian				

▬▬▬ Arabic	▬▬▬ Chinese	▬▬▬ English	▬▬▬ French	▬▬▬ German	▬▬▬ Indonesian	▬▬▬ Japanese
▬▬▬ Portuguese	▬▬▬ Russian	▬▬▬ Spanish	▬▬▬ Other			

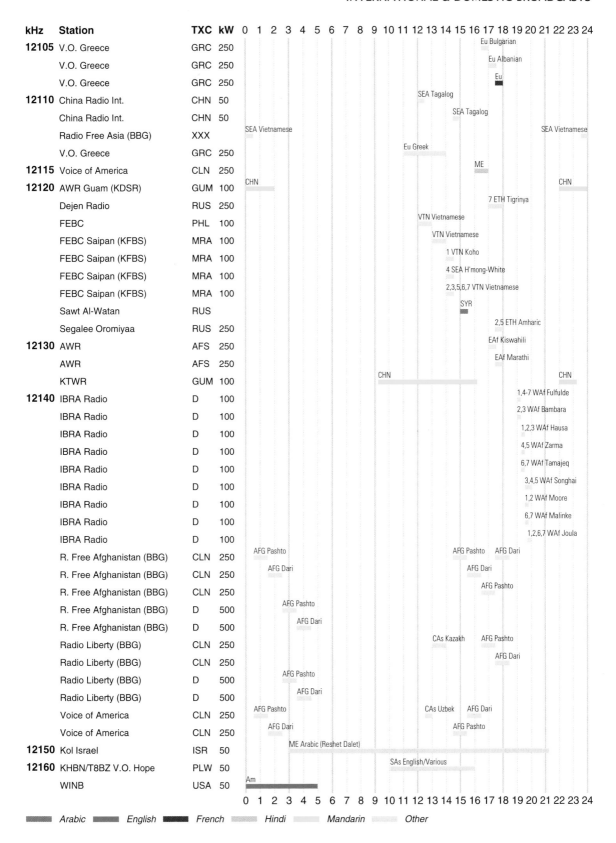

kHz	Station	TXC	kW	Broadcast (0–24 h)
12105	V.O. Greece	GRC	250	Eu Bulgarian
	V.O. Greece	GRC	250	Eu Albanian
	V.O. Greece	GRC	250	Eu
12110	China Radio Int.	CHN	50	SEA Tagalog
	China Radio Int.	CHN	50	SEA Tagalog
	Radio Free Asia (BBG)	XXX		SEA Vietnamese ... SEA Vietnamese
	V.O. Greece	GRC	250	Eu Greek
12115	Voice of America	CLN	250	ME
12120	AWR Guam (KDSR)	GUM	100	CHN ... CHN
	Dejen Radio	RUS	250	7 ETH Tigrinya
	FEBC	PHL	100	VTN Vietnamese
	FEBC Saipan (KFBS)	MRA	100	VTN Vietnamese
	FEBC Saipan (KFBS)	MRA	100	1 VTN Koho
	FEBC Saipan (KFBS)	MRA	100	4 SEA H'mong-White
	FEBC Saipan (KFBS)	MRA	100	2,3,5,6,7 VTN Vietnamese
	Sawt Al-Watan	RUS		SYR
	Segalee Oromiyaa	RUS	250	2,5 ETH Amharic
12130	AWR	AFS	250	EAf Kiswahili
	AWR	AFS	250	EAf Marathi
	KTWR	GUM	100	CHN ... CHN
12140	IBRA Radio	D	100	1,4-7 WAf Fulfulde
	IBRA Radio	D	100	2,3 WAf Bambara
	IBRA Radio	D	100	1,2,3 WAf Hausa
	IBRA Radio	D	100	4,5 WAf Zarma
	IBRA Radio	D	100	6,7 WAf Tamajeq
	IBRA Radio	D	100	3,4,5 WAf Songhai
	IBRA Radio	D	100	1,2 WAf Moore
	IBRA Radio	D	100	6,7 WAf Malinke
	IBRA Radio	D	100	1,2,6,7 WAf Joula
	R. Free Afghanistan (BBG)	CLN	250	AFG Pashto ... AFG Pashto ... AFG Dari
	R. Free Afghanistan (BBG)	CLN	250	AFG Dari ... AFG Dari
	R. Free Afghanistan (BBG)	CLN	250	AFG Pashto
	R. Free Afghanistan (BBG)	D	500	AFG Pashto
	R. Free Afghanistan (BBG)	D	500	AFG Dari
	Radio Liberty (BBG)	CLN	250	CAs Kazakh ... AFG Pashto
	Radio Liberty (BBG)	CLN	250	AFG Dari
	Radio Liberty (BBG)	D	500	AFG Pashto
	Radio Liberty (BBG)	D	500	AFG Dari
	Voice of America	CLN	250	AFG Pashto ... CAs Uzbek ... AFG Dari
	Voice of America	CLN	250	AFG Dari ... AFG Pashto
12150	Kol Israel	ISR	50	ME Arabic (Reshet Dalet)
12160	KHBN/T8BZ V.O. Hope	PLW	50	SAs English/Various
	WINB	USA	50	Am

Key: Arabic · English · French · Hindi · Mandarin · Other

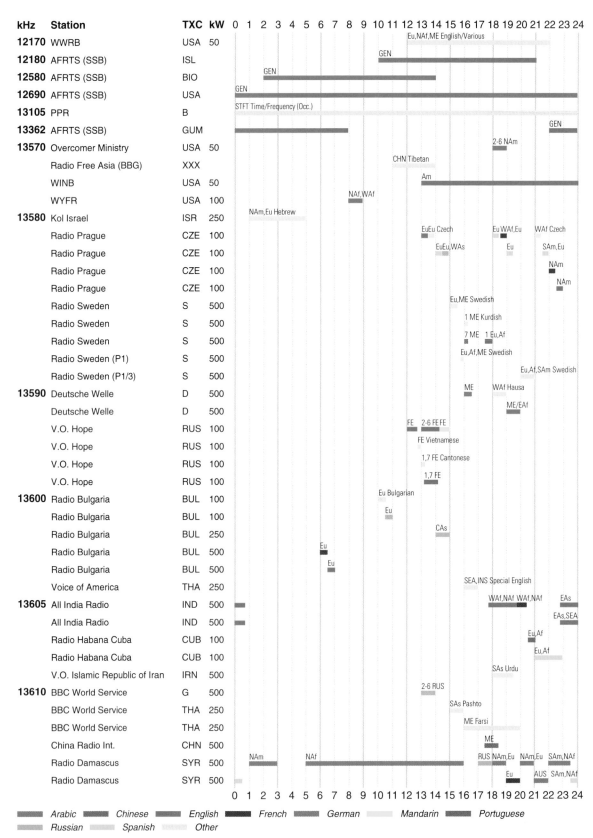

kHz	Station	TXC	kW
12170	WWRB	USA	50
12180	AFRTS (SSB)	ISL	
12580	AFRTS (SSB)	BIO	
12690	AFRTS (SSB)	USA	
13105	PPR	B	
13362	AFRTS (SSB)	GUM	
13570	Overcomer Ministry	USA	50
	Radio Free Asia (BBG)	XXX	
	WINB	USA	50
	WYFR	USA	100
13580	Kol Israel	ISR	250
	Radio Prague	CZE	100
	Radio Prague	CZE	100
	Radio Prague	CZE	100
	Radio Prague	CZE	100
	Radio Sweden	S	500
	Radio Sweden	S	500
	Radio Sweden	S	500
	Radio Sweden (P1)	S	500
	Radio Sweden (P1/3)	S	500
13590	Deutsche Welle	D	500
	Deutsche Welle	D	500
	V.O. Hope	RUS	100
	V.O. Hope	RUS	100
	V.O. Hope	RUS	100
	V.O. Hope	RUS	100
13600	Radio Bulgaria	BUL	100
	Radio Bulgaria	BUL	100
	Radio Bulgaria	BUL	250
	Radio Bulgaria	BUL	500
	Radio Bulgaria	BUL	500
	Voice of America	THA	250
13605	All India Radio	IND	500
	All India Radio	IND	500
	Radio Habana Cuba	CUB	100
	Radio Habana Cuba	CUB	100
	V.O. Islamic Republic of Iran	IRN	500
13610	BBC World Service	G	500
	BBC World Service	THA	250
	BBC World Service	THA	250
	China Radio Int.	CHN	500
	Radio Damascus	SYR	500
	Radio Damascus	SYR	500

Legend: Arabic, Chinese, English, French, German, Mandarin, Portuguese, Russian, Spanish, Other

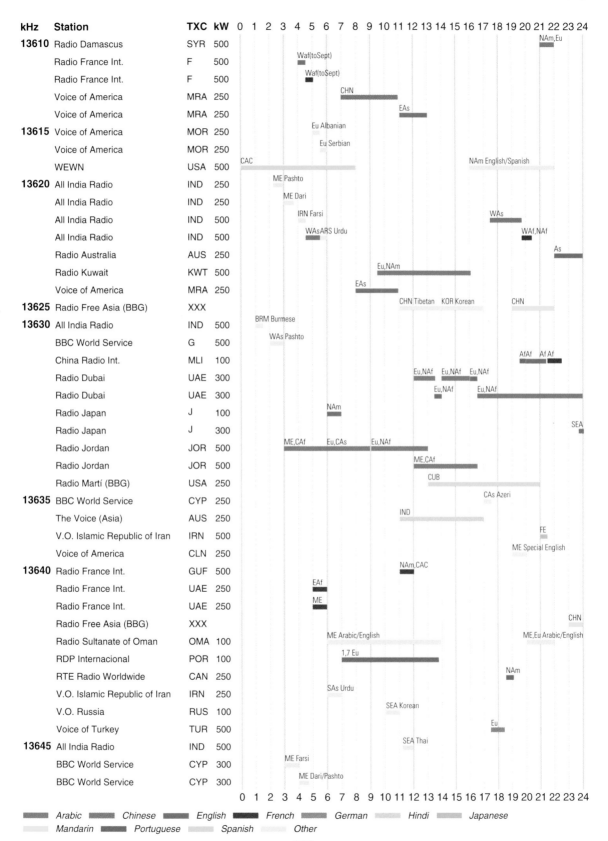

kHz	Station	TXC	kW
13610	Radio Damascus	SYR	500
	Radio France Int.	F	500
	Radio France Int.	F	500
	Voice of America	MRA	250
	Voice of America	MRA	250
13615	Voice of America	MOR	250
	Voice of America	MOR	250
	WEWN	USA	500
13620	All India Radio	IND	250
	All India Radio	IND	250
	All India Radio	IND	500
	All India Radio	IND	500
	Radio Australia	AUS	250
	Radio Kuwait	KWT	500
	Voice of America	MRA	250
13625	Radio Free Asia (BBG)	XXX	
13630	All India Radio	IND	500
	BBC World Service	G	500
	China Radio Int.	MLI	100
	Radio Dubai	UAE	300
	Radio Dubai	UAE	300
	Radio Japan	J	100
	Radio Japan	J	300
	Radio Jordan	JOR	500
	Radio Jordan	JOR	500
	Radio Martí (BBG)	USA	250
13635	BBC World Service	CYP	250
	The Voice (Asia)	AUS	250
	V.O. Islamic Republic of Iran	IRN	500
	Voice of America	CLN	250
13640	Radio France Int.	GUF	500
	Radio France Int.	UAE	250
	Radio France Int.	UAE	250
	Radio Free Asia (BBG)	XXX	
	Radio Sultanate of Oman	OMA	100
	RDP Internacional	POR	100
	RTE Radio Worldwide	CAN	250
	V.O. Islamic Republic of Iran	IRN	250
	V.O. Russia	RUS	100
	Voice of Turkey	TUR	500
13645	All India Radio	IND	500
	BBC World Service	CYP	300
	BBC World Service	CYP	300

Arabic · Chinese · English · French · German · Hindi · Japanese · Mandarin · Portuguese · Spanish · Other

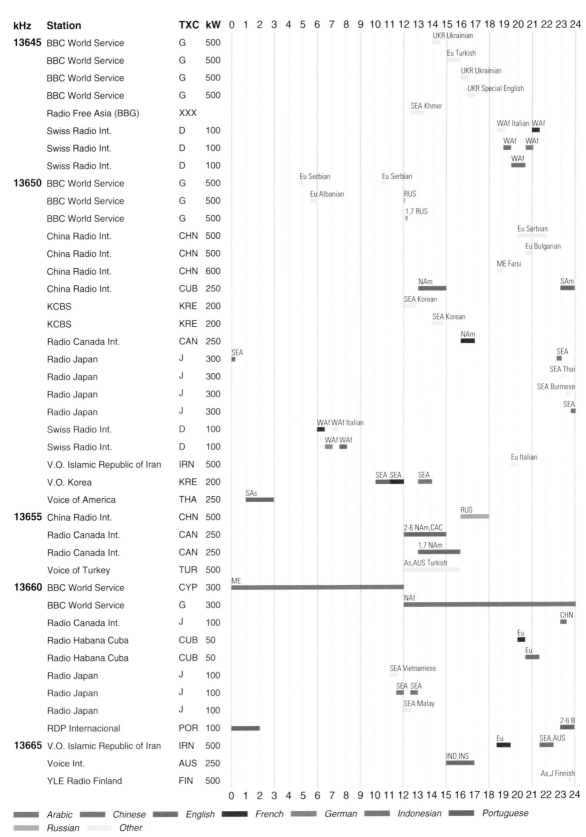

kHz	Station	TXC	kW	Timeline annotations
13645	BBC World Service	G	500	UKR Ukrainian
	BBC World Service	G	500	Eu Turkish
	BBC World Service	G	500	UKR Ukrainian
	BBC World Service	G	500	UKR Special English
	Radio Free Asia (BBG)	XXX		SEA Khmer
	Swiss Radio Int.	D	100	WAf Italian WAf
	Swiss Radio Int.	D	100	WAf WAf
	Swiss Radio Int.	D	100	WAf
13650	BBC World Service	G	500	Eu Serbian / Eu Serbian
	BBC World Service	G	500	Eu Albanian / RUS
	BBC World Service	G	500	1,7 RUS
	China Radio Int.	CHN	500	Eu Serbian
	China Radio Int.	CHN	500	Eu Bulgarian
	China Radio Int.	CHN	600	ME Farsi
	China Radio Int.	CUB	250	NAm / SAm
	KCBS	KRE	200	SEA Korean
	KCBS	KRE	200	SEA Korean
	Radio Canada Int.	CAN	250	NAm
	Radio Japan	J	300	SEA / SEA
	Radio Japan	J	300	SEA Thai
	Radio Japan	J	300	SEA Burmese
	Radio Japan	J	300	SEA
	Swiss Radio Int.	D	100	WAf WAf Italian
	Swiss Radio Int.	D	100	WAf WAf
	V.O. Islamic Republic of Iran	IRN	500	Eu Italian
	V.O. Korea	KRE	200	SEA SEA / SEA
	Voice of America	THA	250	SAs
13655	China Radio Int.	CHN	500	RUS
	Radio Canada Int.	CAN	250	2-6 NAm,CAC
	Radio Canada Int.	CAN	250	1,7 NAm
	Voice of Turkey	TUR	500	As,AUS Turkish
13660	BBC World Service	CYP	300	ME
	BBC World Service	G	300	NAf
	Radio Canada Int.	J	100	CHN
	Radio Habana Cuba	CUB	50	Eu
	Radio Habana Cuba	CUB	50	Eu
	Radio Japan	J	100	SEA Vietnamese
	Radio Japan	J	100	SEA SEA
	Radio Japan	J	100	SEA Malay
	RDP Internacional	POR	100	2-6 B
13665	V.O. Islamic Republic of Iran	IRN	500	Eu / SEA,AUS
	Voice Int.	AUS	250	IND,INS
	YLE Radio Finland	FIN	500	As,J Finnish

Legend: ▬ Arabic ▬ Chinese ▬ English ▬ French ▬ German ▬ Indonesian ▬ Portuguese ▬ Russian ▬ Other

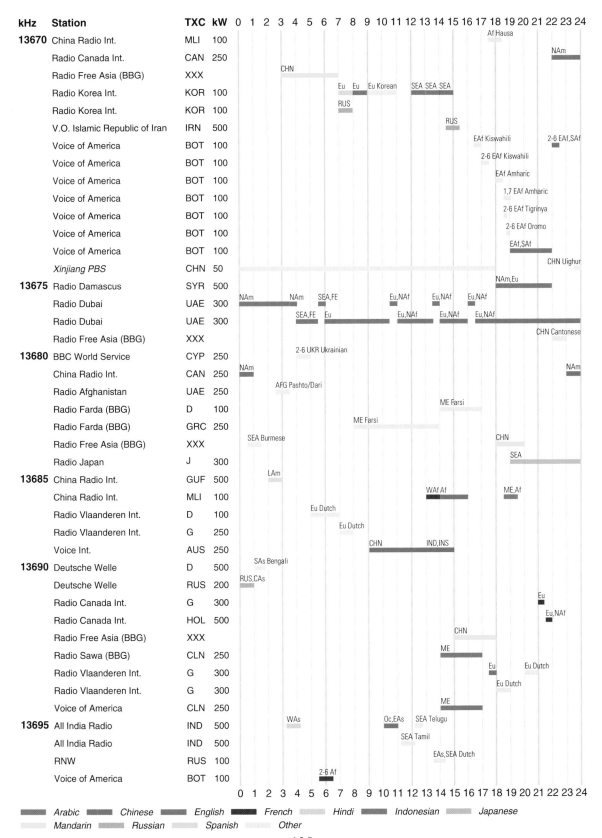

kHz	Station	TXC	kW
13670	China Radio Int.	MLI	100
	Radio Canada Int.	CAN	250
	Radio Free Asia (BBG)	XXX	
	Radio Korea Int.	KOR	100
	Radio Korea Int.	KOR	100
	V.O. Islamic Republic of Iran	IRN	500
	Voice of America	BOT	100
	Voice of America	BOT	100
	Voice of America	BOT	100
	Voice of America	BOT	100
	Voice of America	BOT	100
	Voice of America	BOT	100
	Voice of America	BOT	100
	Xinjiang PBS	CHN	50
13675	Radio Damascus	SYR	500
	Radio Dubai	UAE	300
	Radio Dubai	UAE	300
	Radio Free Asia (BBG)	XXX	
13680	BBC World Service	CYP	250
	China Radio Int.	CAN	250
	Radio Afghanistan	UAE	250
	Radio Farda (BBG)	D	100
	Radio Farda (BBG)	GRC	250
	Radio Free Asia (BBG)	XXX	
	Radio Japan	J	300
13685	China Radio Int.	GUF	500
	China Radio Int.	MLI	100
	Radio Vlaanderen Int.	D	100
	Radio Vlaanderen Int.	G	250
	Voice Int.	AUS	250
13690	Deutsche Welle	D	500
	Deutsche Welle	RUS	200
	Radio Canada Int.	G	300
	Radio Canada Int.	HOL	500
	Radio Free Asia (BBG)	XXX	
	Radio Sawa (BBG)	CLN	250
	Radio Vlaanderen Int.	G	300
	Radio Vlaanderen Int.	G	300
	Voice of America	CLN	250
13695	All India Radio	IND	500
	All India Radio	IND	500
	RNW	RUS	100
	Voice of America	BOT	100

Legend: Arabic, Chinese, English, French, Hindi, Indonesian, Japanese, Mandarin, Russian, Spanish, Other

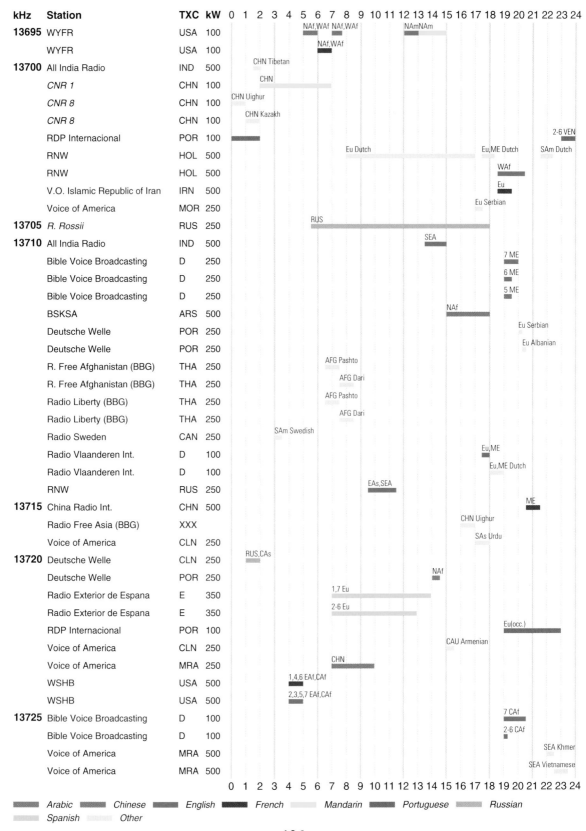

kHz	Station	TXC	kW
13695	WYFR	USA	100
	WYFR	USA	100
13700	All India Radio	IND	500
	CNR 1	CHN	100
	CNR 8	CHN	100
	CNR 8	CHN	100
	RDP Internacional	POR	100
	RNW	HOL	500
	RNW	HOL	500
	V.O. Islamic Republic of Iran	IRN	500
	Voice of America	MOR	250
13705	*R. Rossii*	RUS	250
13710	All India Radio	IND	500
	Bible Voice Broadcasting	D	250
	Bible Voice Broadcasting	D	250
	Bible Voice Broadcasting	D	250
	BSKSA	ARS	500
	Deutsche Welle	POR	250
	Deutsche Welle	POR	250
	R. Free Afghanistan (BBG)	THA	250
	R. Free Afghanistan (BBG)	THA	250
	Radio Liberty (BBG)	THA	250
	Radio Liberty (BBG)	THA	250
	Radio Sweden	CAN	250
	Radio Vlaanderen Int.	D	100
	Radio Vlaanderen Int.	D	100
	RNW	RUS	250
13715	China Radio Int.	CHN	500
	Radio Free Asia (BBG)	XXX	
	Voice of America	CLN	250
13720	Deutsche Welle	CLN	250
	Deutsche Welle	POR	250
	Radio Exterior de Espana	E	350
	Radio Exterior de Espana	E	350
	RDP Internacional	POR	100
	Voice of America	CLN	250
	Voice of America	MRA	250
	WSHB	USA	500
	WSHB	USA	500
13725	Bible Voice Broadcasting	D	100
	Bible Voice Broadcasting	D	100
	Voice of America	MRA	500
	Voice of America	MRA	500

Legend: Arabic · Chinese · English · French · Mandarin · Portuguese · Russian · Spanish · Other

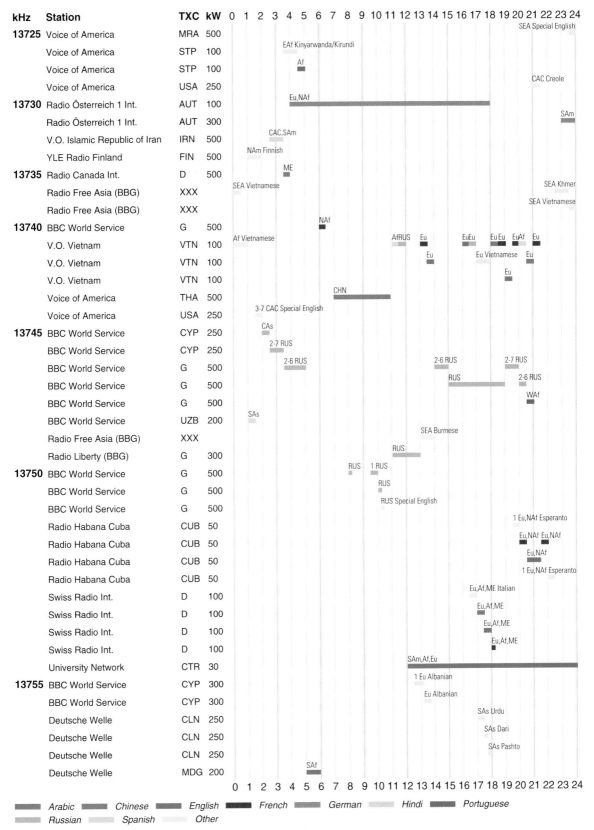

kHz	Station	TXC	kW	Schedule (0-24)
13725	Voice of America	MRA	500	SEA Special English
	Voice of America	STP	100	EAf Kinyarwanda/Kirundi
	Voice of America	STP	100	Af
	Voice of America	USA	250	CAC Creole
13730	Radio Österreich 1 Int.	AUT	100	Eu,NAf
	Radio Österreich 1 Int.	AUT	300	SAm
	V.O. Islamic Republic of Iran	IRN	500	CAC,SAm
	YLE Radio Finland	FIN	500	NAm Finnish
13735	Radio Canada Int.	D	500	ME
	Radio Free Asia (BBG)	XXX		SEA Vietnamese / SEA Khmer
	Radio Free Asia (BBG)	XXX		SEA Vietnamese
13740	BBC World Service	G	500	NAf
	V.O. Vietnam	VTN	100	Af Vietnamese / AfRUS Eu EuEu EuEu EuAf Eu
	V.O. Vietnam	VTN	100	Eu / Eu Vietnamese Eu
	V.O. Vietnam	VTN	100	Eu
	Voice of America	THA	500	CHN
	Voice of America	USA	250	3-7 CAC Special English
13745	BBC World Service	CYP	250	CAs
	BBC World Service	CYP	250	2-7 RUS
	BBC World Service	G	500	2-6 RUS / 2-6 RUS 2-7 RUS
	BBC World Service	G	500	RUS 2-6 RUS
	BBC World Service	G	500	WAf
	BBC World Service	UZB	200	SAs
	Radio Free Asia (BBG)	XXX		SEA Burmese
	Radio Liberty (BBG)	G	300	RUS
13750	BBC World Service	G	500	RUS 1 RUS
	BBC World Service	G	500	RUS
	BBC World Service	G	500	RUS Special English
	Radio Habana Cuba	CUB	50	1 Eu,NAf Esperanto
	Radio Habana Cuba	CUB	50	Eu,NAf Eu,NAf
	Radio Habana Cuba	CUB	50	Eu,NAf
	Radio Habana Cuba	CUB	50	1 Eu,NAf Esperanto
	Swiss Radio Int.	D	100	Eu,Af,ME Italian
	Swiss Radio Int.	D	100	Eu,Af,ME
	Swiss Radio Int.	D	100	Eu,Af,ME
	Swiss Radio Int.	D	100	Eu,Af,ME
	University Network	CTR	30	SAm,Af,Eu
13755	BBC World Service	CYP	300	1 Eu Albanian
	BBC World Service	CYP	300	Eu Albanian
	Deutsche Welle	CLN	250	SAs Urdu
	Deutsche Welle	CLN	250	SAs Dari
	Deutsche Welle	CLN	250	SAs Pashto
	Deutsche Welle	MDG	200	SAf

Legend: *Arabic* — *Chinese* — *English* — *French* — *German* — *Hindi* — *Portuguese* — *Russian* — *Spanish* — *Other*

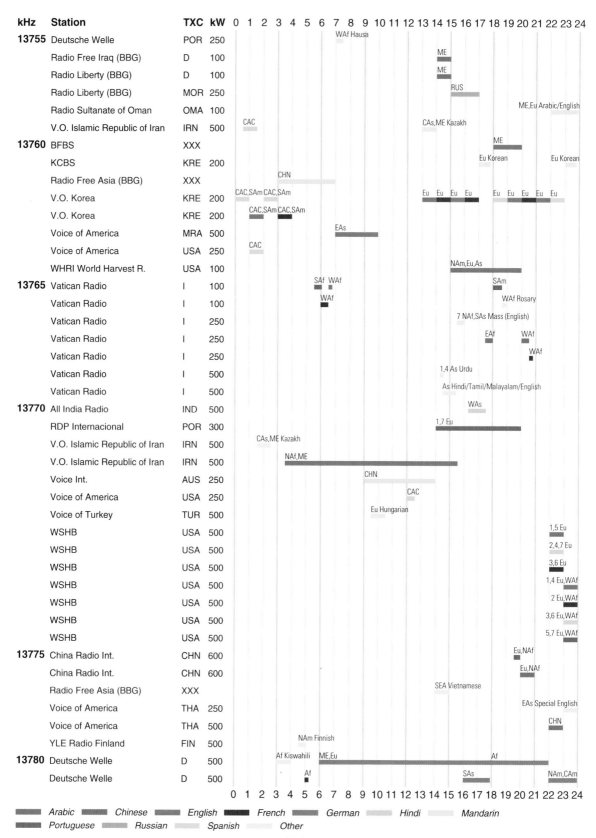

kHz	Station	TXC	kW	Notes
13755	Deutsche Welle	POR	250	WAf Hausa
	Radio Free Iraq (BBG)	D	100	ME
	Radio Liberty (BBG)	D	100	ME
	Radio Liberty (BBG)	MOR	250	RUS
	Radio Sultanate of Oman	OMA	100	ME,Eu Arabic/English
	V.O. Islamic Republic of Iran	IRN	500	CAC CAs,ME Kazakh
13760	BFBS	XXX		ME
	KCBS	KRE	200	Eu Korean Eu Korean
	Radio Free Asia (BBG)	XXX		CHN
	V.O. Korea	KRE	200	CAC,SAm CAC,SAm Eu Eu Eu Eu Eu Eu Eu Eu Eu
	V.O. Korea	KRE	200	CAC,SAm CAC,SAm
	Voice of America	MRA	500	EAs
	Voice of America	USA	250	CAC
	WHRI World Harvest R.	USA	100	NAm,Eu,As
13765	Vatican Radio	I	100	SAf WAf SAm
	Vatican Radio	I	100	WAf WAf Rosary
	Vatican Radio	I	250	7 NAf,SAs Mass (English)
	Vatican Radio	I	250	EAf WAf
	Vatican Radio	I	250	WAf
	Vatican Radio	I	500	1,4 As Urdu
	Vatican Radio	I	500	As Hindi/Tamil/Malayalam/English
13770	All India Radio	IND	500	WAs
	RDP Internacional	POR	300	1,7 Eu
	V.O. Islamic Republic of Iran	IRN	500	CAs,ME Kazakh
	V.O. Islamic Republic of Iran	IRN	500	NAf,ME
	Voice Int.	AUS	250	CHN
	Voice of America	USA	250	CAC
	Voice of Turkey	TUR	500	Eu Hungarian
	WSHB	USA	500	1,5 Eu
	WSHB	USA	500	2,4,7 Eu
	WSHB	USA	500	3,6 Eu
	WSHB	USA	500	1,4 Eu,WAf
	WSHB	USA	500	2 Eu,WAf
	WSHB	USA	500	3,6 Eu,WAf
	WSHB	USA	500	5,7 Eu,WAf
13775	China Radio Int.	CHN	600	Eu,NAf
	China Radio Int.	CHN	600	Eu,NAf
	Radio Free Asia (BBG)	XXX		SEA Vietnamese
	Voice of America	THA	250	EAs Special English
	Voice of America	THA	500	CHN
	YLE Radio Finland	FIN	500	NAm Finnish
13780	Deutsche Welle	D	500	Af Kiswahili ME,Eu Af
	Deutsche Welle	D	500	Af SAs NAm,CAm

Arabic   Chinese   English   French   German   Hindi   Mandarin
Portuguese   Russian   Spanish   Other

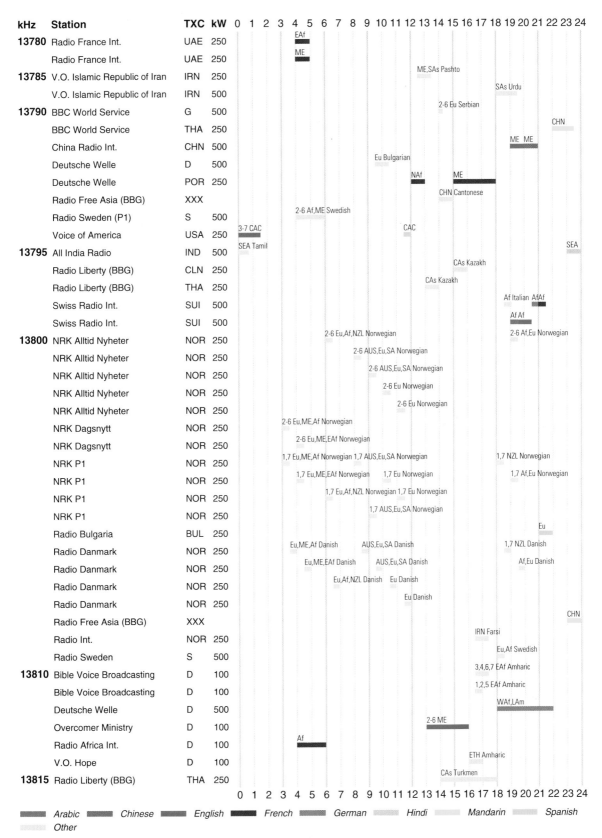

kHz	Station	TXC	kW
13780	Radio France Int.	UAE	250
	Radio France Int.	UAE	250
13785	V.O. Islamic Republic of Iran	IRN	250
	V.O. Islamic Republic of Iran	IRN	500
13790	BBC World Service	G	500
	BBC World Service	THA	250
	China Radio Int.	CHN	500
	Deutsche Welle	D	500
	Deutsche Welle	POR	250
	Radio Free Asia (BBG)	XXX	
	Radio Sweden (P1)	S	500
	Voice of America	USA	250
13795	All India Radio	IND	500
	Radio Liberty (BBG)	CLN	250
	Radio Liberty (BBG)	THA	250
	Swiss Radio Int.	SUI	500
	Swiss Radio Int.	SUI	500
13800	NRK Alltid Nyheter	NOR	250
	NRK Alltid Nyheter	NOR	250
	NRK Alltid Nyheter	NOR	250
	NRK Alltid Nyheter	NOR	250
	NRK Alltid Nyheter	NOR	250
	NRK Dagsnytt	NOR	250
	NRK Dagsnytt	NOR	250
	NRK P1	NOR	250
	NRK P1	NOR	250
	NRK P1	NOR	250
	NRK P1	NOR	250
	Radio Bulgaria	BUL	250
	Radio Danmark	NOR	250
	Radio Danmark	NOR	250
	Radio Danmark	NOR	250
	Radio Danmark	NOR	250
	Radio Free Asia (BBG)	XXX	
	Radio Int.	NOR	250
	Radio Sweden	S	500
13810	Bible Voice Broadcasting	D	100
	Bible Voice Broadcasting	D	100
	Deutsche Welle	D	500
	Overcomer Ministry	D	100
	Radio Africa Int.	D	100
	V.O. Hope	D	100
13815	Radio Liberty (BBG)	THA	250

Arabic   Chinese   English   French   German   Hindi   Mandarin   Spanish
Other

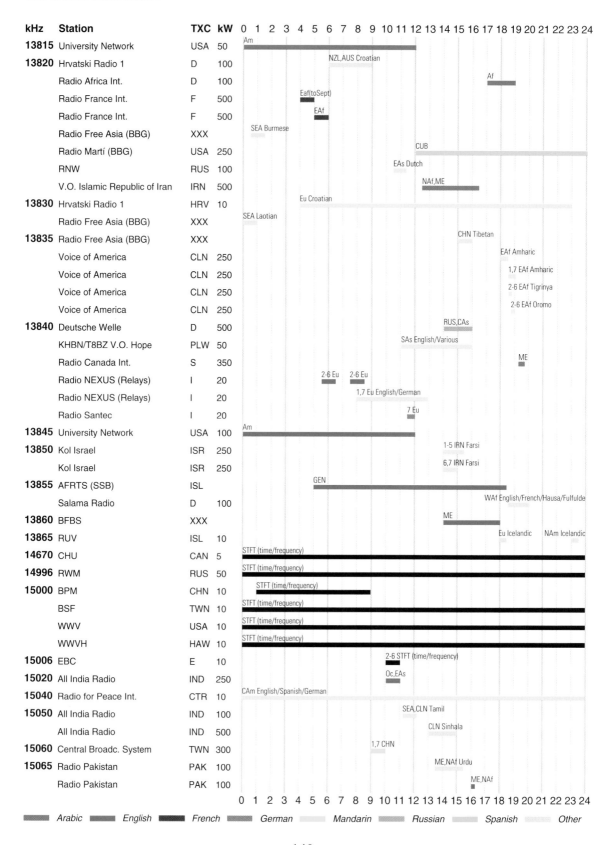

kHz	Station	TXC	kW
13815	University Network	USA	50
13820	Hrvatski Radio 1	D	100
	Radio Africa Int.	D	100
	Radio France Int.	F	500
	Radio France Int.	F	500
	Radio Free Asia (BBG)	XXX	
	Radio Martí (BBG)	USA	250
	RNW	RUS	100
	V.O. Islamic Republic of Iran	IRN	500
13830	Hrvatski Radio 1	HRV	10
	Radio Free Asia (BBG)	XXX	
13835	Radio Free Asia (BBG)	XXX	
	Voice of America	CLN	250
	Voice of America	CLN	250
	Voice of America	CLN	250
	Voice of America	CLN	250
13840	Deutsche Welle	D	500
	KHBN/T8BZ V.O. Hope	PLW	50
	Radio Canada Int.	S	350
	Radio NEXUS (Relays)	I	20
	Radio NEXUS (Relays)	I	20
	Radio Santec	I	20
13845	University Network	USA	100
13850	Kol Israel	ISR	250
	Kol Israel	ISR	250
13855	AFRTS (SSB)	ISL	
	Salama Radio	D	100
13860	BFBS	XXX	
13865	RUV	ISL	10
14670	CHU	CAN	5
14996	RWM	RUS	50
15000	BPM	CHN	10
	BSF	TWN	10
	WWV	USA	10
	WWVH	HAW	10
15006	EBC	E	10
15020	All India Radio	IND	250
15040	Radio for Peace Int.	CTR	10
15050	All India Radio	IND	100
	All India Radio	IND	500
15060	Central Broadc. System	TWN	300
15065	Radio Pakistan	PAK	100
	Radio Pakistan	PAK	100

Arabic   English   French   German   Mandarin   Russian   Spanish   Other

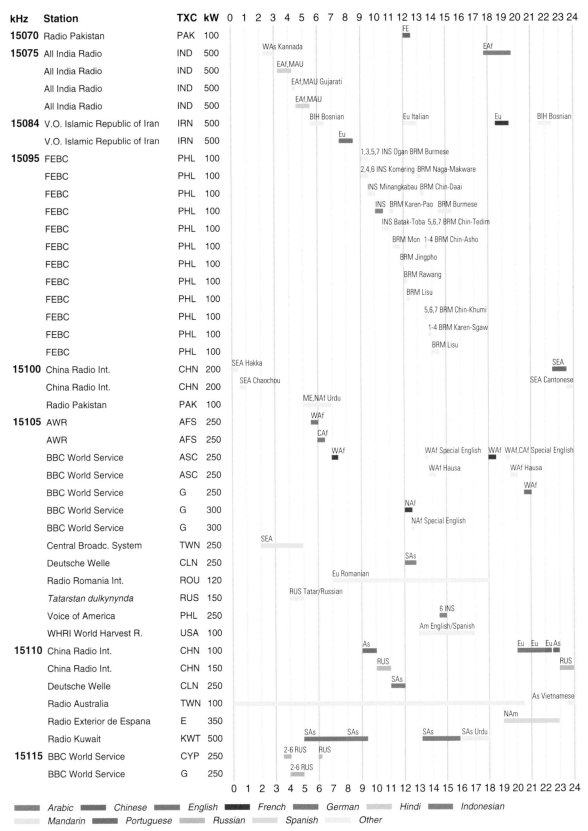

kHz	Station	TXC	kW
15070	Radio Pakistan	PAK	100
15075	All India Radio	IND	500
	All India Radio	IND	500
	All India Radio	IND	500
	All India Radio	IND	500
15084	V.O. Islamic Republic of Iran	IRN	500
	V.O. Islamic Republic of Iran	IRN	500
15095	FEBC	PHL	100
	FEBC	PHL	100
	FEBC	PHL	100
	FEBC	PHL	100
	FEBC	PHL	100
	FEBC	PHL	100
	FEBC	PHL	100
	FEBC	PHL	100
	FEBC	PHL	100
	FEBC	PHL	100
	FEBC	PHL	100
	FEBC	PHL	100
15100	China Radio Int.	CHN	200
	China Radio Int.	CHN	200
	Radio Pakistan	PAK	100
15105	AWR	AFS	250
	AWR	AFS	250
	BBC World Service	ASC	250
	BBC World Service	ASC	250
	BBC World Service	G	250
	BBC World Service	G	300
	BBC World Service	G	300
	Central Broadc. System	TWN	250
	Deutsche Welle	CLN	250
	Radio Romania Int.	ROU	120
	Tatarstan dulkynynda	RUS	150
	Voice of America	PHL	250
	WHRI World Harvest R.	USA	100
15110	China Radio Int.	CHN	100
	China Radio Int.	CHN	150
	Deutsche Welle	CLN	250
	Radio Australia	TWN	100
	Radio Exterior de Espana	E	350
	Radio Kuwait	KWT	500
15115	BBC World Service	CYP	250
	BBC World Service	G	250

Legend (languages):
Arabic · Chinese · English · French · German · Hindi · Indonesian
Mandarin · Portuguese · Russian · Spanish · Other

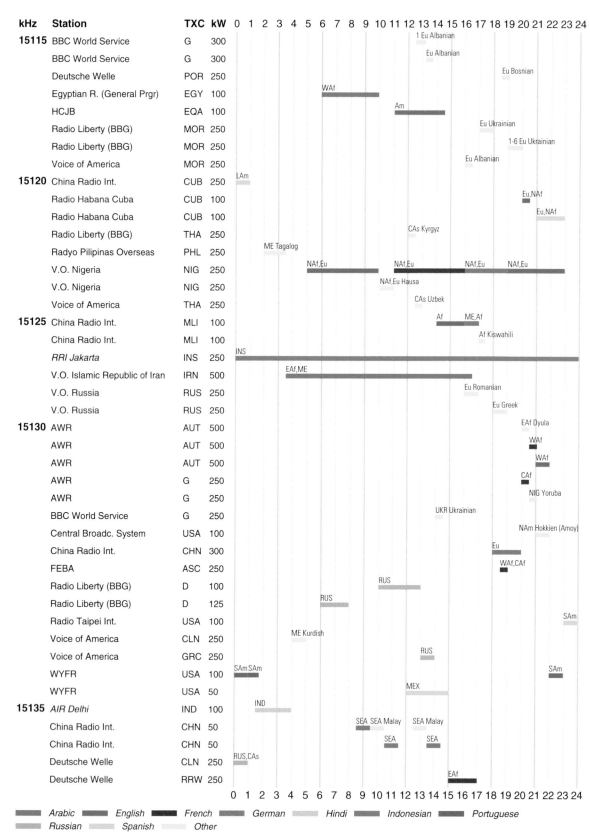

kHz	Station	TXC	kW	0 1 2 3 4 5 6 7 8 9 10 11 12 13 14 15 16 17 18 19 20 21 22 23 24
15115	BBC World Service	G	300	1 Eu Albanian
	BBC World Service	G	300	Eu Albanian
	Deutsche Welle	POR	250	Eu Bosnian
	Egyptian R. (General Prgr)	EGY	100	WAf
	HCJB	EQA	100	Am
	Radio Liberty (BBG)	MOR	250	Eu Ukrainian
	Radio Liberty (BBG)	MOR	250	1-6 Eu Ukrainian
	Voice of America	MOR	250	Eu Albanian
15120	China Radio Int.	CUB	250	LAm
	Radio Habana Cuba	CUB	100	Eu,NAf
	Radio Habana Cuba	CUB	100	Eu,NAf
	Radio Liberty (BBG)	THA	250	CAs Kyrgyz
	Radyo Pilipinas Overseas	PHL	250	ME Tagalog
	V.O. Nigeria	NIG	250	NAf,Eu NAf,Eu NAf,Eu NAf,Eu
	V.O. Nigeria	NIG	250	NAf,Eu Hausa
	Voice of America	THA	250	CAs Uzbek
15125	China Radio Int.	MLI	100	Af ME,Af
	China Radio Int.	MLI	100	Af Kiswahili
	RRI Jakarta	INS	250	INS
	V.O. Islamic Republic of Iran	IRN	500	EAf,ME
	V.O. Russia	RUS	250	Eu Romanian
	V.O. Russia	RUS	250	Eu Greek
15130	AWR	AUT	500	EAf Dyula
	AWR	AUT	500	WAf
	AWR	AUT	500	WAf
	AWR	G	250	CAf
	AWR	G	250	NIG Yoruba
	BBC World Service	G	250	UKR Ukrainian
	Central Broadc. System	USA	100	NAm Hokkien (Amoy)
	China Radio Int.	CHN	300	Eu
	FEBA	ASC	250	WAf,CAf
	Radio Liberty (BBG)	D	100	RUS
	Radio Liberty (BBG)	D	125	RUS
	Radio Taipei Int.	USA	100	SAm
	Voice of America	CLN	250	ME Kurdish
	Voice of America	GRC	250	RUS
	WYFR	USA	100	SAm SAm SAm
	WYFR	USA	50	MEX
15135	*AIR Delhi*	IND	100	IND
	China Radio Int.	CHN	50	SEA SEA Malay SEA Malay
	China Radio Int.	CHN	50	SEA SEA
	Deutsche Welle	CLN	250	RUS,CAs
	Deutsche Welle	RRW	250	EAf

Legend: Arabic · English · French · German · Hindi · Indonesian · Portuguese · Russian · Spanish · Other

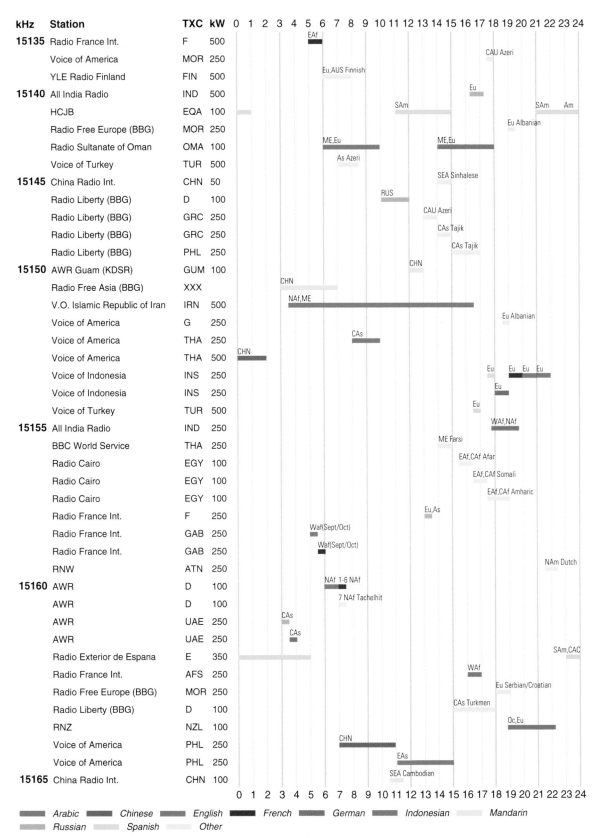

kHz	Station	TXC	kW
15135	Radio France Int.	F	500
	Voice of America	MOR	250
	YLE Radio Finland	FIN	500
15140	All India Radio	IND	500
	HCJB	EQA	100
	Radio Free Europe (BBG)	MOR	250
	Radio Sultanate of Oman	OMA	100
	Voice of Turkey	TUR	500
15145	China Radio Int.	CHN	50
	Radio Liberty (BBG)	D	100
	Radio Liberty (BBG)	GRC	250
	Radio Liberty (BBG)	GRC	250
	Radio Liberty (BBG)	PHL	250
15150	AWR Guam (KDSR)	GUM	100
	Radio Free Asia (BBG)	XXX	
	V.O. Islamic Republic of Iran	IRN	500
	Voice of America	G	250
	Voice of America	THA	250
	Voice of America	THA	500
	Voice of Indonesia	INS	250
	Voice of Indonesia	INS	250
	Voice of Turkey	TUR	500
15155	All India Radio	IND	250
	BBC World Service	THA	250
	Radio Cairo	EGY	100
	Radio Cairo	EGY	100
	Radio Cairo	EGY	100
	Radio France Int.	F	250
	Radio France Int.	GAB	250
	Radio France Int.	GAB	250
	RNW	ATN	250
15160	AWR	D	100
	AWR	D	100
	AWR	UAE	250
	AWR	UAE	250
	Radio Exterior de Espana	E	350
	Radio France Int.	AFS	250
	Radio Free Europe (BBG)	MOR	250
	Radio Liberty (BBG)	D	100
	RNZ	NZL	100
	Voice of America	PHL	250
	Voice of America	PHL	250
15165	China Radio Int.	CHN	100

Legend: Arabic Chinese English French German Indonesian Mandarin Russian Spanish Other

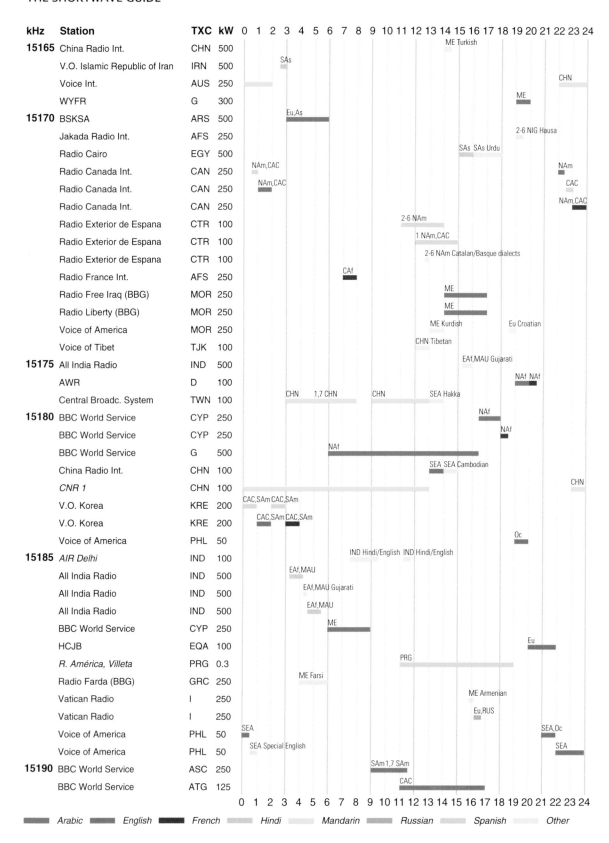

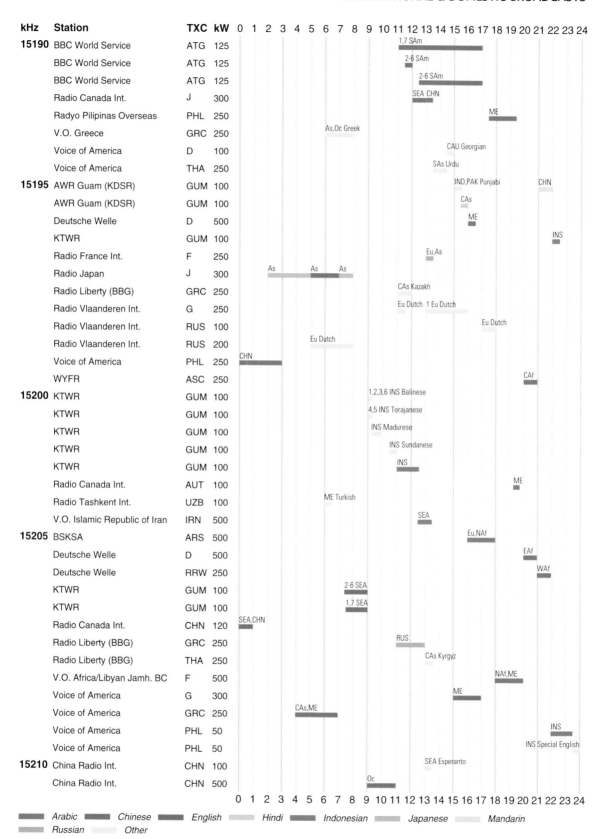

kHz	Station	TXC	kW
15190	BBC World Service	ATG	125
	BBC World Service	ATG	125
	BBC World Service	ATG	125
	Radio Canada Int.	J	300
	Radyo Pilipinas Overseas	PHL	250
	V.O. Greece	GRC	250
	Voice of America	D	100
	Voice of America	THA	250
15195	AWR Guam (KDSR)	GUM	100
	AWR Guam (KDSR)	GUM	100
	Deutsche Welle	D	500
	KTWR	GUM	100
	Radio France Int.	F	250
	Radio Japan	J	300
	Radio Liberty (BBG)	GRC	250
	Radio Vlaanderen Int.	G	250
	Radio Vlaanderen Int.	RUS	100
	Radio Vlaanderen Int.	RUS	200
	Voice of America	PHL	250
	WYFR	ASC	250
15200	KTWR	GUM	100
	KTWR	GUM	100
	KTWR	GUM	100
	KTWR	GUM	100
	KTWR	GUM	100
	Radio Canada Int.	AUT	100
	Radio Tashkent Int.	UZB	100
	V.O. Islamic Republic of Iran	IRN	500
15205	BSKSA	ARS	500
	Deutsche Welle	D	500
	Deutsche Welle	RRW	250
	KTWR	GUM	100
	KTWR	GUM	100
	Radio Canada Int.	CHN	120
	Radio Liberty (BBG)	GRC	250
	Radio Liberty (BBG)	THA	250
	V.O. Africa/Libyan Jamh. BC	F	500
	Voice of America	G	300
	Voice of America	GRC	250
	Voice of America	PHL	50
	Voice of America	PHL	50
15210	China Radio Int.	CHN	100
	China Radio Int.	CHN	500

Legend: *Arabic* *Chinese* *English* *Hindi* *Indonesian* *Japanese* *Mandarin* *Russian* *Other*

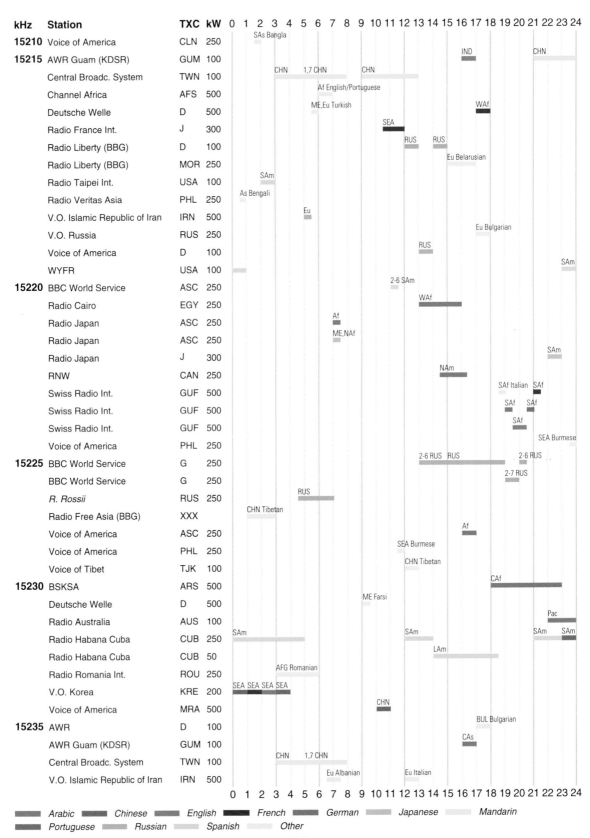

kHz	Station	TXC	kW
15210	Voice of America	CLN	250
15215	AWR Guam (KDSR)	GUM	100
	Central Broadc. System	TWN	100
	Channel Africa	AFS	500
	Deutsche Welle	D	500
	Radio France Int.	J	300
	Radio Liberty (BBG)	D	100
	Radio Liberty (BBG)	MOR	250
	Radio Taipei Int.	USA	100
	Radio Veritas Asia	PHL	250
	V.O. Islamic Republic of Iran	IRN	500
	V.O. Russia	RUS	250
	Voice of America	D	100
	WYFR	USA	100
15220	BBC World Service	ASC	250
	Radio Cairo	EGY	250
	Radio Japan	ASC	250
	Radio Japan	ASC	250
	Radio Japan	J	300
	RNW	CAN	250
	Swiss Radio Int.	GUF	500
	Swiss Radio Int.	GUF	500
	Swiss Radio Int.	GUF	500
	Voice of America	PHL	250
15225	BBC World Service	G	250
	BBC World Service	G	250
	R. Rossii	RUS	250
	Radio Free Asia (BBG)	XXX	
	Voice of America	ASC	250
	Voice of America	PHL	250
	Voice of Tibet	TJK	100
15230	BSKSA	ARS	500
	Deutsche Welle	D	500
	Radio Australia	AUS	100
	Radio Habana Cuba	CUB	250
	Radio Habana Cuba	CUB	50
	Radio Romania Int.	ROU	250
	V.O. Korea	KRE	200
	Voice of America	MRA	500
15235	AWR	D	100
	AWR Guam (KDSR)	GUM	100
	Central Broadc. System	TWN	100
	V.O. Islamic Republic of Iran	IRN	500

Legend: Arabic, Chinese, English, French, German, Japanese, Mandarin, Portuguese, Russian, Spanish, Other

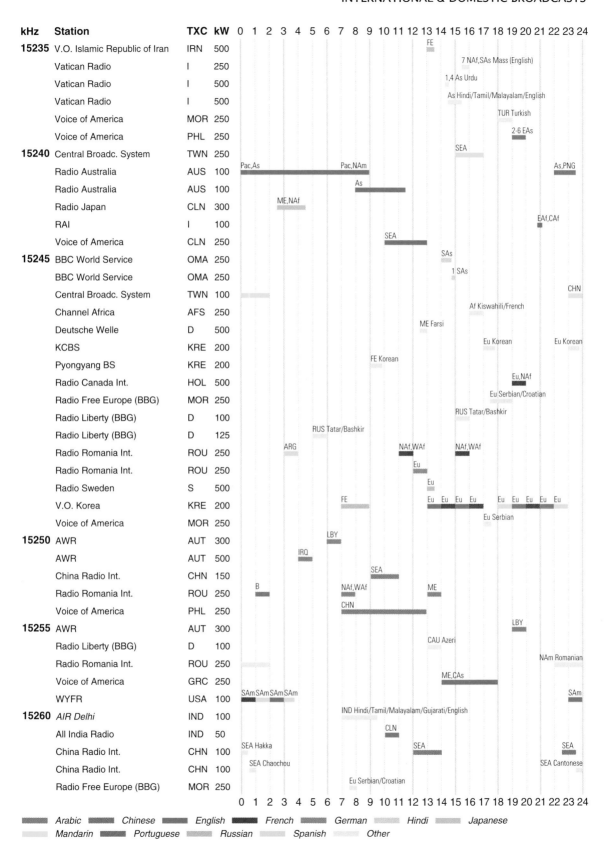

kHz	Station	TXC	kW	Notes (timeline 0–24)
15235	V.O. Islamic Republic of Iran	IRN	500	FE
	Vatican Radio	I	250	7 NAf,SAs Mass (English)
	Vatican Radio	I	500	1,4 As Urdu
	Vatican Radio	I	500	As Hindi/Tamil/Malayalam/English
	Voice of America	MOR	250	TUR Turkish
	Voice of America	PHL	250	2-6 EAs
15240	Central Broadc. System	TWN	250	SEA
	Radio Australia	AUS	100	Pac,As · Pac,NAm · As,PNG
	Radio Australia	AUS	100	As
	Radio Japan	CLN	300	ME,NAf
	RAI	I	100	EAf,CAf
	Voice of America	CLN	250	SEA
15245	BBC World Service	OMA	250	SAs
	BBC World Service	OMA	250	1 SAs
	Central Broadc. System	TWN	100	CHN
	Channel Africa	AFS	250	Af Kiswahili/French
	Deutsche Welle	D	500	ME Farsi
	KCBS	KRE	200	Eu Korean · Eu Korean
	Pyongyang BS	KRE	200	FE Korean
	Radio Canada Int.	HOL	500	Eu,NAf
	Radio Free Europe (BBG)	MOR	250	Eu Serbian/Croatian
	Radio Liberty (BBG)	D	100	RUS Tatar/Bashkir
	Radio Liberty (BBG)	D	125	RUS Tatar/Bashkir
	Radio Romania Int.	ROU	250	ARG · NAf,WAf · NAf,WAf
	Radio Romania Int.	ROU	250	Eu
	Radio Sweden	S	500	Eu
	V.O. Korea	KRE	200	FE · Eu Eu Eu Eu · Eu Eu Eu Eu Eu
	Voice of America	MOR	250	Eu Serbian
15250	AWR	AUT	300	LBY
	AWR	AUT	500	IRQ
	China Radio Int.	CHN	150	SEA
	Radio Romania Int.	ROU	250	B · NAf,WAf · ME
	Voice of America	PHL	250	CHN
15255	AWR	AUT	300	LBY
	Radio Liberty (BBG)	D	100	CAU Azeri
	Radio Romania Int.	ROU	250	NAm Romanian
	Voice of America	GRC	250	ME,CAs
	WYFR	USA	100	SAm SAm SAm SAm · SAm
15260	AIR Delhi	IND	100	IND Hindi/Tamil/Malayalam/Gujarati/English
	All India Radio	IND	50	CLN
	China Radio Int.	CHN	100	SEA Hakka · SEA · SEA
	China Radio Int.	CHN	100	SEA Chaochou · SEA Cantonese
	Radio Free Europe (BBG)	MOR	250	Eu Serbian/Croatian

Legend: Arabic · Chinese · English · French · German · Hindi · Japanese · Mandarin · Portuguese · Russian · Spanish · Other

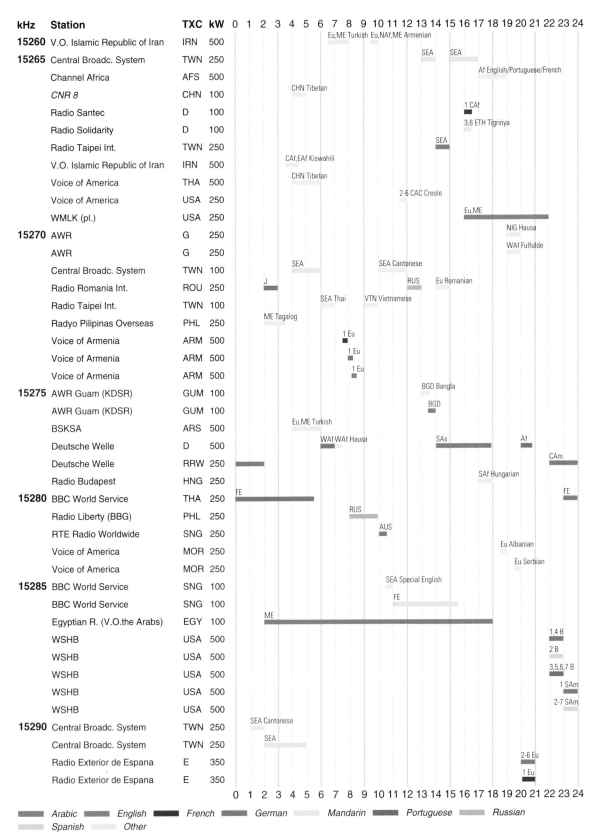

kHz	Station	TXC	kW	Schedule
15260	V.O. Islamic Republic of Iran	IRN	500	Eu,ME Turkish — Eu,NAf,ME Armenian
15265	Central Broadc. System	TWN	250	SEA — SEA
	Channel Africa	AFS	500	Af English/Portuguese/French
	CNR 8	CHN	100	CHN Tibetan
	Radio Santec	D	100	1 CAf
	Radio Solidarity	D	100	3,6 ETH Tigrinya
	Radio Taipei Int.	TWN	250	SEA
	V.O. Islamic Republic of Iran	IRN	500	CAf,EAf Kiswahili
	Voice of America	THA	500	CHN Tibetan
	Voice of America	USA	250	2-6 CAC Creole
	WMLK (pl.)	USA	250	Eu,ME
15270	AWR	G	250	NIG Hausa
	AWR	G	250	WAf Fulfulde
	Central Broadc. System	TWN	100	SEA — SEA Cantonese
	Radio Romania Int.	ROU	250	J — RUS — Eu Romanian
	Radio Taipei Int.	TWN	100	SEA Thai — VTN Vietnamese
	Radyo Pilipinas Overseas	PHL	250	ME Tagalog
	Voice of Armenia	ARM	500	1 Eu
	Voice of Armenia	ARM	500	1 Eu
	Voice of Armenia	ARM	500	1 Eu
15275	AWR Guam (KDSR)	GUM	100	BGD Bangla
	AWR Guam (KDSR)	GUM	100	BGD
	BSKSA	ARS	500	Eu,ME Turkish
	Deutsche Welle	D	500	WAf WAf Hausa — SAs — Af
	Deutsche Welle	RRW	250	CAm
	Radio Budapest	HNG	250	SAf Hungarian
15280	BBC World Service	THA	250	FE — FE
	Radio Liberty (BBG)	PHL	250	RUS
	RTE Radio Worldwide	SNG	250	AUS
	Voice of America	MOR	250	Eu Albanian
	Voice of America	MOR	250	Eu Serbian
15285	BBC World Service	SNG	100	SEA Special English
	BBC World Service	SNG	100	FE
	Egyptian R. (V.O.the Arabs)	EGY	100	ME
	WSHB	USA	500	1,4 B
	WSHB	USA	500	2 B
	WSHB	USA	500	3,5,6,7 B
	WSHB	USA	500	1 SAm
	WSHB	USA	500	2-7 SAm
15290	Central Broadc. System	TWN	250	SEA Cantonese
	Central Broadc. System	TWN	250	SEA
	Radio Exterior de Espana	E	350	2-6 Eu
	Radio Exterior de Espana	E	350	1 Eu

Arabic   English   French   German   Mandarin   Portuguese   Russian
Spanish   Other

148

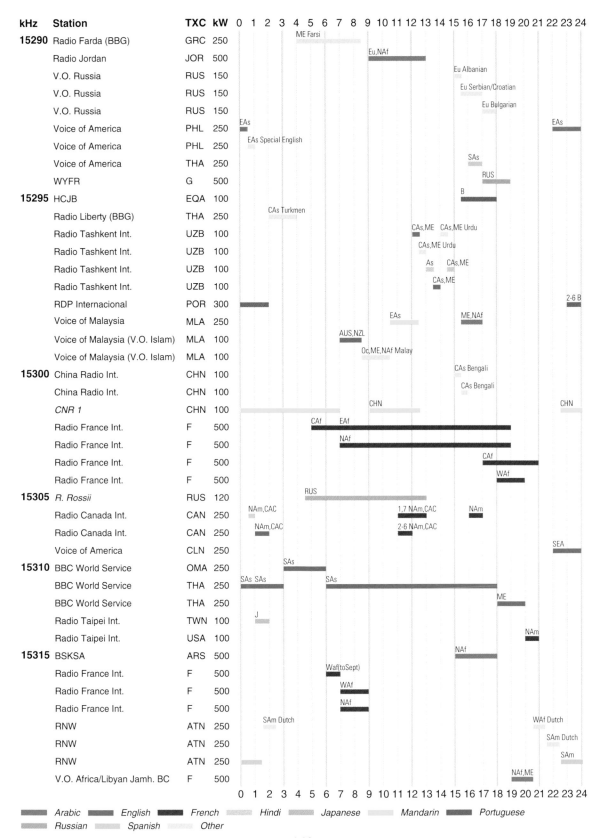

kHz	Station	TXC	kW
15290	Radio Farda (BBG)	GRC	250
	Radio Jordan	JOR	500
	V.O. Russia	RUS	150
	V.O. Russia	RUS	150
	V.O. Russia	RUS	150
	Voice of America	PHL	250
	Voice of America	PHL	250
	Voice of America	THA	250
	WYFR	G	500
15295	HCJB	EQA	100
	Radio Liberty (BBG)	THA	250
	Radio Tashkent Int.	UZB	100
	Radio Tashkent Int.	UZB	100
	Radio Tashkent Int.	UZB	100
	Radio Tashkent Int.	UZB	100
	RDP Internacional	POR	300
	Voice of Malaysia	MLA	250
	Voice of Malaysia (V.O. Islam)	MLA	100
	Voice of Malaysia (V.O. Islam)	MLA	100
15300	China Radio Int.	CHN	100
	China Radio Int.	CHN	100
	CNR 1	CHN	100
	Radio France Int.	F	500
	Radio France Int.	F	500
	Radio France Int.	F	500
	Radio France Int.	F	500
15305	*R. Rossii*	RUS	120
	Radio Canada Int.	CAN	250
	Radio Canada Int.	CAN	250
	Voice of America	CLN	250
15310	BBC World Service	OMA	250
	BBC World Service	THA	250
	BBC World Service	THA	250
	Radio Taipei Int.	TWN	100
	Radio Taipei Int.	USA	100
15315	BSKSA	ARS	500
	Radio France Int.	F	500
	Radio France Int.	F	500
	Radio France Int.	F	500
	RNW	ATN	250
	RNW	ATN	250
	RNW	ATN	250
	V.O. Africa/Libyan Jamh. BC	F	500

Legend:
Arabic, English, French, Hindi, Japanese, Mandarin, Portuguese, Russian, Spanish, Other

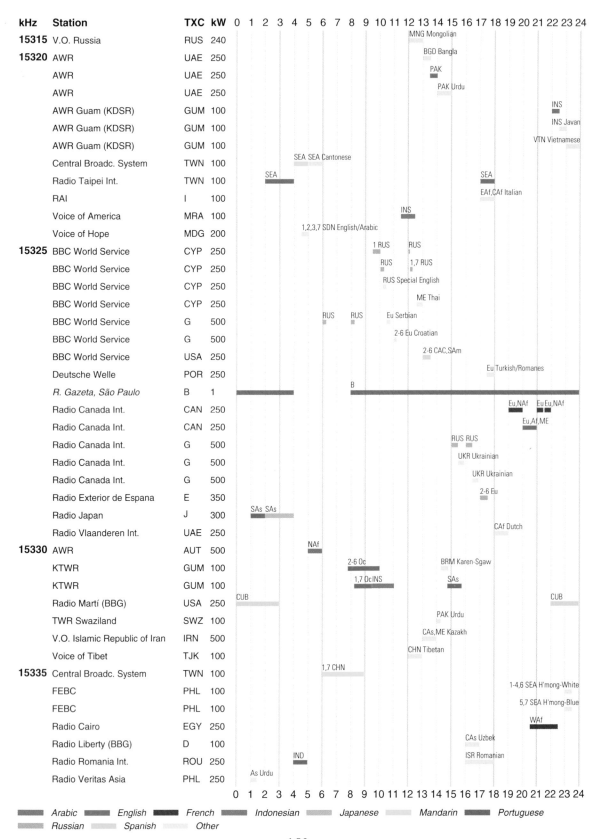

kHz	Station	TXC	kW	Notes
15315	V.O. Russia	RUS	240	MNG Mongolian
15320	AWR	UAE	250	BGD Bangla
	AWR	UAE	250	PAK
	AWR	UAE	250	PAK Urdu
	AWR Guam (KDSR)	GUM	100	INS
	AWR Guam (KDSR)	GUM	100	INS Javan
	AWR Guam (KDSR)	GUM	100	VTN Vietnamese
	Central Broadc. System	TWN	100	SEA SEA Cantonese
	Radio Taipei Int.	TWN	100	SEA SEA
	RAI	I	100	EAf,CAf Italian
	Voice of America	MRA	100	INS
	Voice of Hope	MDG	200	1,2,3,7 SDN English/Arabic
15325	BBC World Service	CYP	250	1 RUS RUS
	BBC World Service	CYP	250	RUS 1,7 RUS
	BBC World Service	CYP	250	RUS Special English
	BBC World Service	CYP	250	ME Thai
	BBC World Service	G	500	RUS RUS Eu Serbian
	BBC World Service	G	500	2-6 Eu Croatian
	BBC World Service	USA	250	2-6 CAC,SAm
	Deutsche Welle	POR	250	Eu Turkish/Romanes
	R. Gazeta, São Paulo	B	1	B
	Radio Canada Int.	CAN	250	Eu,NAf Eu Eu,NAf
	Radio Canada Int.	CAN	250	Eu,Af,ME
	Radio Canada Int.	G	500	RUS RUS
	Radio Canada Int.	G	500	UKR Ukrainian
	Radio Canada Int.	G	500	UKR Ukrainian
	Radio Exterior de Espana	E	350	2-6 Eu
	Radio Japan	J	300	SAs SAs
	Radio Vlaanderen Int.	UAE	250	CAf Dutch
15330	AWR	AUT	500	NAf
	KTWR	GUM	100	2-6 Oc BRM Karen-Sgaw
	KTWR	GUM	100	1,7 OcINS SAs
	Radio Martí (BBG)	USA	250	CUB CUB
	TWR Swaziland	SWZ	100	PAK Urdu
	V.O. Islamic Republic of Iran	IRN	500	CAs,ME Kazakh
	Voice of Tibet	TJK	100	CHN Tibetan
15335	Central Broadc. System	TWN	100	1,7 CHN
	FEBC	PHL	100	1-4,6 SEA H'mong-White
	FEBC	PHL	100	5,7 SEA H'mong-Blue
	Radio Cairo	EGY	250	WAf
	Radio Liberty (BBG)	D	100	CAs Uzbek
	Radio Romania Int.	ROU	250	IND ISR Romanian
	Radio Veritas Asia	PHL	250	As Urdu

Legend: ▬ *Arabic* ▬ *English* ▬ *French* ▬ *Indonesian* ▬ *Japanese* ▬ *Mandarin* ▬ *Portuguese* ▬ *Russian* ▬ *Spanish* ▬ *Other*

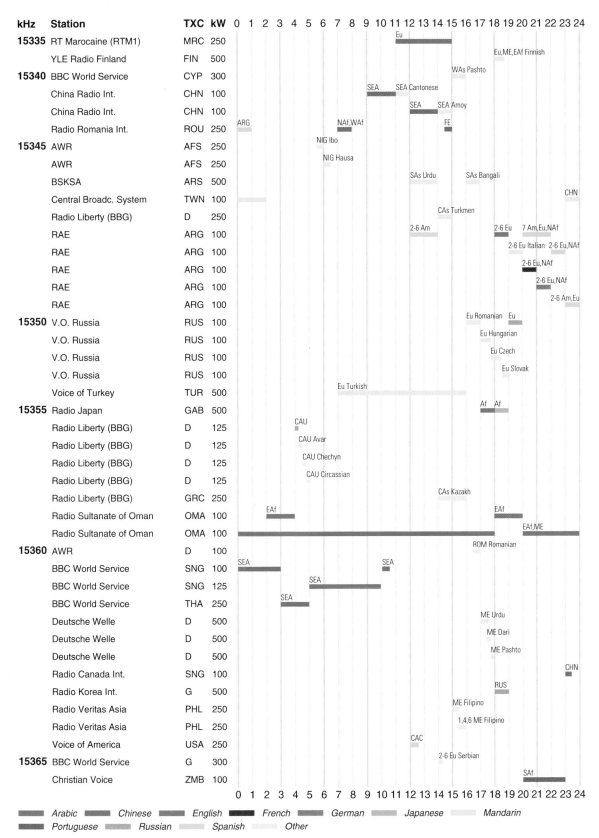

kHz	Station	TXC	kW
15335	RT Marocaine (RTM1)	MRC	250
	YLE Radio Finland	FIN	500
15340	BBC World Service	CYP	300
	China Radio Int.	CHN	100
	China Radio Int.	CHN	100
	Radio Romania Int.	ROU	250
15345	AWR	AFS	250
	AWR	AFS	250
	BSKSA	ARS	500
	Central Broadc. System	TWN	100
	Radio Liberty (BBG)	D	250
	RAE	ARG	100
	RAE	ARG	100
	RAE	ARG	100
	RAE	ARG	100
	RAE	ARG	100
15350	V.O. Russia	RUS	100
	V.O. Russia	RUS	100
	V.O. Russia	RUS	100
	V.O. Russia	RUS	100
	Voice of Turkey	TUR	500
15355	Radio Japan	GAB	500
	Radio Liberty (BBG)	D	125
	Radio Liberty (BBG)	D	125
	Radio Liberty (BBG)	D	125
	Radio Liberty (BBG)	D	125
	Radio Liberty (BBG)	GRC	250
	Radio Sultanate of Oman	OMA	100
	Radio Sultanate of Oman	OMA	100
15360	AWR	D	100
	BBC World Service	SNG	100
	BBC World Service	SNG	125
	BBC World Service	THA	250
	Deutsche Welle	D	500
	Deutsche Welle	D	500
	Deutsche Welle	D	500
	Radio Canada Int.	SNG	100
	Radio Korea Int.	G	500
	Radio Veritas Asia	PHL	250
	Radio Veritas Asia	PHL	250
	Voice of America	USA	250
15365	BBC World Service	G	300
	Christian Voice	ZMB	100

Legend: *Arabic* · *Chinese* · *English* · **French** · *German* · *Japanese* · *Mandarin* · **Portuguese** · *Russian* · *Spanish* · *Other*

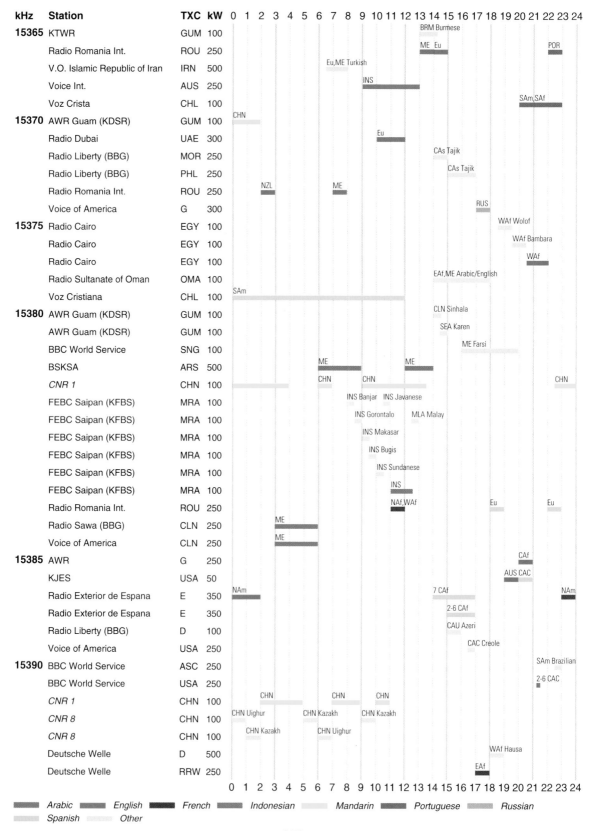

kHz	Station	TXC	kW
15365	KTWR	GUM	100
	Radio Romania Int.	ROU	250
	V.O. Islamic Republic of Iran	IRN	500
	Voice Int.	AUS	250
	Voz Crista	CHL	100
15370	AWR Guam (KDSR)	GUM	100
	Radio Dubai	UAE	300
	Radio Liberty (BBG)	MOR	250
	Radio Liberty (BBG)	PHL	250
	Radio Romania Int.	ROU	250
	Voice of America	G	300
15375	Radio Cairo	EGY	100
	Radio Cairo	EGY	100
	Radio Cairo	EGY	100
	Radio Sultanate of Oman	OMA	100
	Voz Cristiana	CHL	100
15380	AWR Guam (KDSR)	GUM	100
	AWR Guam (KDSR)	GUM	100
	BBC World Service	SNG	100
	BSKSA	ARS	500
	CNR 1	CHN	100
	FEBC Saipan (KFBS)	MRA	100
	FEBC Saipan (KFBS)	MRA	100
	FEBC Saipan (KFBS)	MRA	100
	FEBC Saipan (KFBS)	MRA	100
	FEBC Saipan (KFBS)	MRA	100
	FEBC Saipan (KFBS)	MRA	100
	Radio Romania Int.	ROU	250
	Radio Sawa (BBG)	CLN	250
	Voice of America	CLN	250
15385	AWR	G	250
	KJES	USA	50
	Radio Exterior de Espana	E	350
	Radio Exterior de Espana	E	350
	Radio Liberty (BBG)	D	100
	Voice of America	USA	250
15390	BBC World Service	ASC	250
	BBC World Service	USA	250
	CNR 1	CHN	100
	CNR 8	CHN	100
	CNR 8	CHN	100
	Deutsche Welle	D	500
	Deutsche Welle	RRW	250

Legend: Arabic, English, French, Indonesian, Mandarin, Portuguese, Russian, Spanish, Other

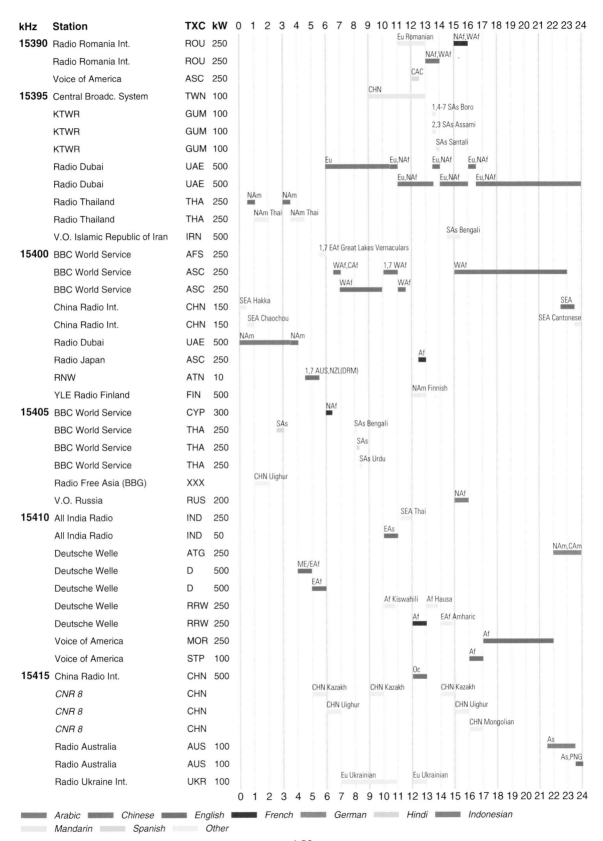

kHz	Station	TXC	kW
15390	Radio Romania Int.	ROU	250
	Radio Romania Int.	ROU	250
	Voice of America	ASC	250
15395	Central Broadc. System	TWN	100
	KTWR	GUM	100
	KTWR	GUM	100
	KTWR	GUM	100
	Radio Dubai	UAE	500
	Radio Dubai	UAE	500
	Radio Thailand	THA	250
	Radio Thailand	THA	250
	V.O. Islamic Republic of Iran	IRN	500
15400	BBC World Service	AFS	250
	BBC World Service	ASC	250
	BBC World Service	ASC	250
	China Radio Int.	CHN	150
	China Radio Int.	CHN	150
	Radio Dubai	UAE	500
	Radio Japan	ASC	250
	RNW	ATN	10
	YLE Radio Finland	FIN	500
15405	BBC World Service	CYP	300
	BBC World Service	THA	250
	BBC World Service	THA	250
	BBC World Service	THA	250
	Radio Free Asia (BBG)	XXX	
	V.O. Russia	RUS	200
15410	All India Radio	IND	250
	All India Radio	IND	50
	Deutsche Welle	ATG	250
	Deutsche Welle	D	500
	Deutsche Welle	D	500
	Deutsche Welle	RRW	250
	Deutsche Welle	RRW	250
	Voice of America	MOR	250
	Voice of America	STP	100
15415	China Radio Int.	CHN	500
	CNR 8	CHN	
	CNR 8	CHN	
	CNR 8	CHN	
	Radio Australia	AUS	100
	Radio Australia	AUS	100
	Radio Ukraine Int.	UKR	100

Chart annotations (by row):

- Radio Romania Int. (ROU 250): Eu Romanian; NAf,WAf
- Radio Romania Int. (ROU 250): NAf,WAf
- Voice of America (ASC 250): CAC
- Central Broadc. System (TWN 100): CHN
- KTWR (GUM 100): 1,4-7 SAs Boro
- KTWR (GUM 100): 2,3 SAs Assami
- KTWR (GUM 100): SAs Santali
- Radio Dubai (UAE 500): Eu; Eu,NAf; Eu,NAf; Eu,NAf
- Radio Dubai (UAE 500): Eu,NAf; Eu,NAf; Eu,NAf
- Radio Thailand (THA 250): NAm; NAm
- Radio Thailand (THA 250): NAm Thai; NAm Thai
- V.O. Islamic Republic of Iran (IRN 500): SAs Bengali
- BBC World Service (AFS 250): 1,7 EAf Great Lakes Vernaculars
- BBC World Service (ASC 250): WAf,CAf; 1,7 WAf; WAf
- BBC World Service (ASC 250): WAf; WAf
- China Radio Int. (CHN 150): SEA Hakka; SEA
- China Radio Int. (CHN 150): SEA Chaochou; SEA Cantonese
- Radio Dubai (UAE 500): NAm; NAm
- Radio Japan (ASC 250): Af
- RNW (ATN 10): 1,7 AUS,NZL(DRM)
- YLE Radio Finland (FIN 500): NAm Finnish
- BBC World Service (CYP 300): NAf
- BBC World Service (THA 250): SAs; SAs Bengali
- BBC World Service (THA 250): SAs
- BBC World Service (THA 250): SAs Urdu
- Radio Free Asia (BBG) (XXX): CHN Uighur
- V.O. Russia (RUS 200): NAf
- All India Radio (IND 250): SEA Thai
- All India Radio (IND 50): EAs
- Deutsche Welle (ATG 250): NAm,CAm
- Deutsche Welle (D 500): ME/EAf
- Deutsche Welle (D 500): EAf
- Deutsche Welle (RRW 250): Af Kiswahili; Af Hausa
- Deutsche Welle (RRW 250): Af; EAf Amharic
- Voice of America (MOR 250): Af
- Voice of America (STP 100): Af
- China Radio Int. (CHN 500): Oc
- CNR 8 (CHN): CHN Kazakh; CHN Kazakh; CHN Kazakh
- CNR 8 (CHN): CHN Uighur; CHN Uighur
- CNR 8 (CHN): CHN Mongolian
- Radio Australia (AUS 100): As
- Radio Australia (AUS 100): As,PNG
- Radio Ukraine Int. (UKR 100): Eu Ukrainian; Eu Ukrainian

Legend: Arabic · Chinese · English · French · German · Hindi · Indonesian · Mandarin · Spanish · Other

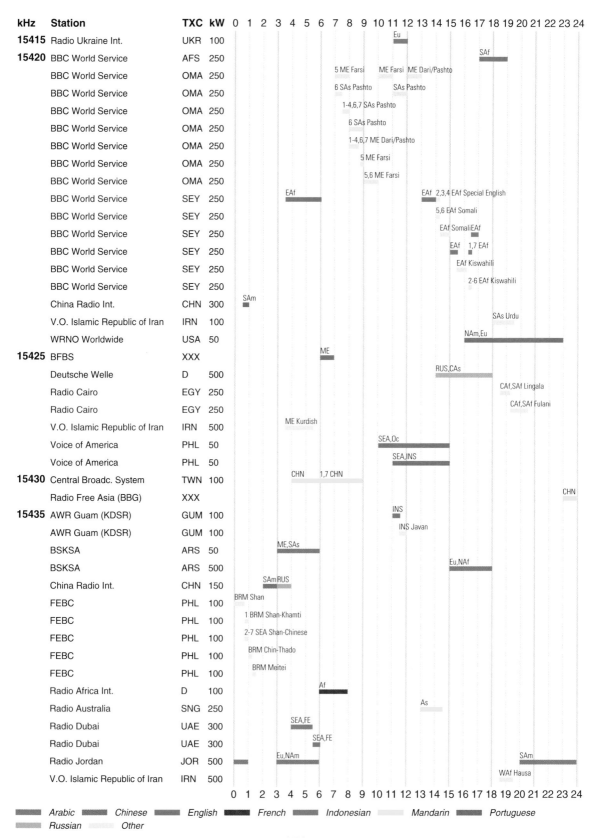

kHz	Station	TXC	kW
15415	Radio Ukraine Int.	UKR	100
15420	BBC World Service	AFS	250
	BBC World Service	OMA	250
	BBC World Service	OMA	250
	BBC World Service	OMA	250
	BBC World Service	OMA	250
	BBC World Service	OMA	250
	BBC World Service	OMA	250
	BBC World Service	OMA	250
	BBC World Service	SEY	250
	BBC World Service	SEY	250
	BBC World Service	SEY	250
	BBC World Service	SEY	250
	BBC World Service	SEY	250
	BBC World Service	SEY	250
	China Radio Int.	CHN	300
	V.O. Islamic Republic of Iran	IRN	100
	WRNO Worldwide	USA	50
15425	BFBS	XXX	
	Deutsche Welle	D	500
	Radio Cairo	EGY	250
	Radio Cairo	EGY	250
	V.O. Islamic Republic of Iran	IRN	500
	Voice of America	PHL	50
	Voice of America	PHL	50
15430	Central Broadc. System	TWN	100
	Radio Free Asia (BBG)	XXX	
15435	AWR Guam (KDSR)	GUM	100
	AWR Guam (KDSR)	GUM	100
	BSKSA	ARS	50
	BSKSA	ARS	500
	China Radio Int.	CHN	150
	FEBC	PHL	100
	FEBC	PHL	100
	FEBC	PHL	100
	FEBC	PHL	100
	FEBC	PHL	100
	Radio Africa Int.	D	100
	Radio Australia	SNG	250
	Radio Dubai	UAE	300
	Radio Dubai	UAE	300
	Radio Jordan	JOR	500
	V.O. Islamic Republic of Iran	IRN	500

Legend: Arabic, Chinese, English, French, Indonesian, Mandarin, Portuguese, Russian, Other

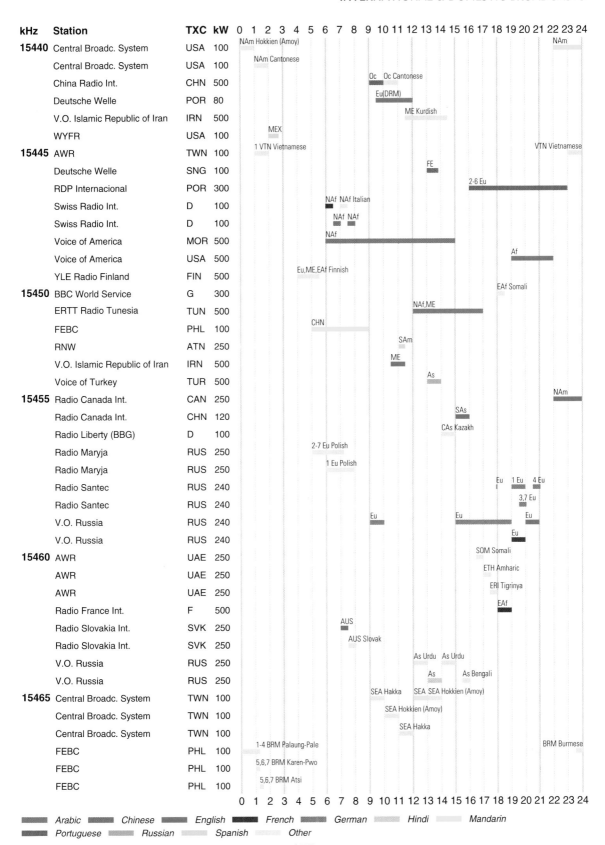

kHz	Station	TXC	kW
15440	Central Broadc. System	USA	100
	Central Broadc. System	USA	100
	China Radio Int.	CHN	500
	Deutsche Welle	POR	80
	V.O. Islamic Republic of Iran	IRN	500
	WYFR	USA	100
15445	AWR	TWN	100
	Deutsche Welle	SNG	100
	RDP Internacional	POR	300
	Swiss Radio Int.	D	100
	Swiss Radio Int.	D	100
	Voice of America	MOR	500
	Voice of America	USA	500
	YLE Radio Finland	FIN	500
15450	BBC World Service	G	300
	ERTT Radio Tunesia	TUN	500
	FEBC	PHL	100
	RNW	ATN	250
	V.O. Islamic Republic of Iran	IRN	500
	Voice of Turkey	TUR	500
15455	Radio Canada Int.	CAN	250
	Radio Canada Int.	CHN	120
	Radio Liberty (BBG)	D	100
	Radio Maryja	RUS	250
	Radio Maryja	RUS	250
	Radio Santec	RUS	240
	Radio Santec	RUS	240
	V.O. Russia	RUS	240
	V.O. Russia	RUS	240
15460	AWR	UAE	250
	AWR	UAE	250
	AWR	UAE	250
	Radio France Int.	F	500
	Radio Slovakia Int.	SVK	250
	Radio Slovakia Int.	SVK	250
	V.O. Russia	RUS	250
	V.O. Russia	RUS	250
15465	Central Broadc. System	TWN	100
	Central Broadc. System	TWN	100
	Central Broadc. System	TWN	100
	FEBC	PHL	100
	FEBC	PHL	100
	FEBC	PHL	100

Legend: Arabic, Chinese, English, French, German, Hindi, Mandarin, Portuguese, Russian, Spanish, Other

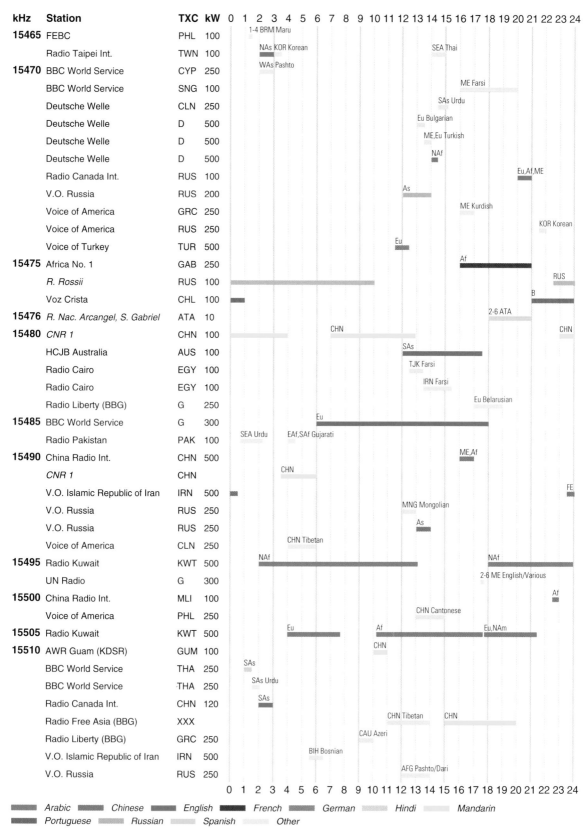

kHz	Station	TXC	kW																								
15465	FEBC	PHL	100	1-4 BRM Maru																					
	Radio Taipei Int.	TWN	100	NAs KOR Korean										SEA Thai							
15470	BBC World Service	CYP	250	WAs Pashto																	
	BBC World Service	SNG	100														ME Farsi				
	Deutsche Welle	CLN	250											SAs Urdu							
	Deutsche Welle	D	500									Eu Bulgarian								
	Deutsche Welle	D	500									ME,Eu Turkish							
	Deutsche Welle	D	500									NAf								
	Radio Canada Int.	RUS	100															Eu,Af,ME			
	V.O. Russia	RUS	200								As									
	Voice of America	GRC	250											ME Kurdish					
	Voice of America	RUS	250															KOR Korean
	Voice of Turkey	TUR	500								Eu								
15475	Africa No. 1	GAB	250												Af				
	R. Rossii	RUS	100																RUS
	Voz Crista	CHL	100															B	
15476	R. Nac. Arcangel, S. Gabriel	ATA	10												2-6 ATA			
15480	CNR 1	CHN	100						CHN									CHN
	HCJB Australia	AUS	100							SAs								
	Radio Cairo	EGY	100							TJK Farsi							
	Radio Cairo	EGY	100								IRN Farsi						
	Radio Liberty (BBG)	G	250										Eu Belarusian			
15485	BBC World Service	G	300					Eu									
	Radio Pakistan	PAK	100	SEA Urdu		EAf,SAf Gujarati									
15490	China Radio Int.	CHN	500										ME,Af			
	CNR 1	CHN						CHN								
	V.O. Islamic Republic of Iran	IRN	500													FE
	V.O. Russia	RUS	250								MNG Mongolian				
	V.O. Russia	RUS	250								As				
	Voice of America	CLN	250					CHN Tibetan						
15495	Radio Kuwait	KWT	500	NAf											NAf	
	UN Radio	G	300										2-6 ME English/Various	
15500	China Radio Int.	MLI	100													Af
	Voice of America	PHL	250									CHN Cantonese		
15505	Radio Kuwait	KWT	500				Eu				Af			Eu,NAm	
15510	AWR Guam (KDSR)	GUM	100							CHN				
	BBC World Service	THA	250	SAs										
	BBC World Service	THA	250	SAs Urdu									
	Radio Canada Int.	CHN	120	SAs									
	Radio Free Asia (BBG)	XXX							CHN Tibetan	CHN		
	Radio Liberty (BBG)	GRC	250					CAU Azeri			
	V.O. Islamic Republic of Iran	IRN	500				BIH Bosnian			
	V.O. Russia	RUS	250						AFG Pashto/Dari		

Arabic Chinese English French German Hindi Mandarin
Portuguese Russian Spanish Other

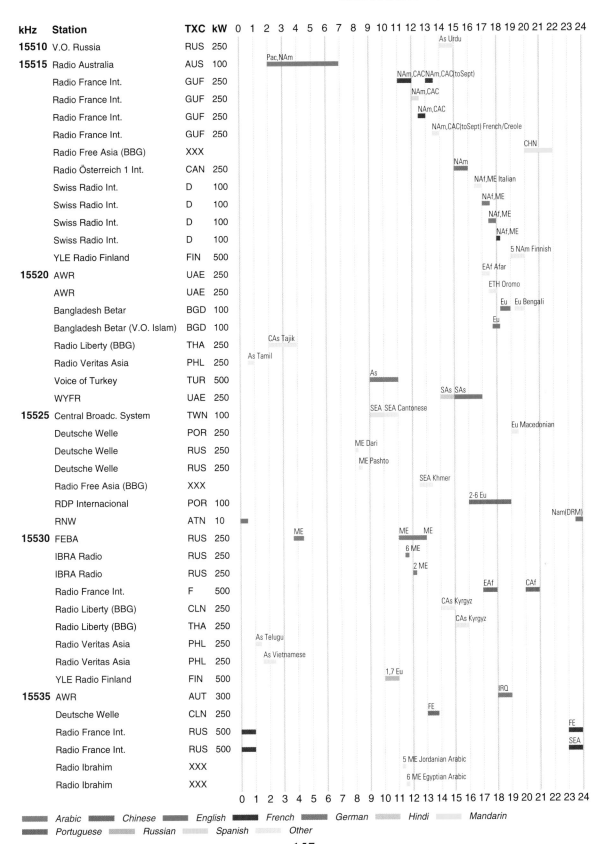

kHz	Station	TXC	kW
15510	V.O. Russia	RUS	250
15515	Radio Australia	AUS	100
	Radio France Int.	GUF	250
	Radio France Int.	GUF	250
	Radio France Int.	GUF	250
	Radio France Int.	GUF	250
	Radio Free Asia (BBG)	XXX	
	Radio Österreich 1 Int.	CAN	250
	Swiss Radio Int.	D	100
	Swiss Radio Int.	D	100
	Swiss Radio Int.	D	100
	Swiss Radio Int.	D	100
	YLE Radio Finland	FIN	500
15520	AWR	UAE	250
	AWR	UAE	250
	Bangladesh Betar	BGD	100
	Bangladesh Betar (V.O. Islam)	BGD	100
	Radio Liberty (BBG)	THA	250
	Radio Veritas Asia	PHL	250
	Voice of Turkey	TUR	500
	WYFR	UAE	250
15525	Central Broadc. System	TWN	100
	Deutsche Welle	POR	250
	Deutsche Welle	RUS	250
	Deutsche Welle	RUS	250
	Radio Free Asia (BBG)	XXX	
	RDP Internacional	POR	100
	RNW	ATN	10
15530	FEBA	RUS	250
	IBRA Radio	RUS	250
	IBRA Radio	RUS	250
	Radio France Int.	F	500
	Radio Liberty (BBG)	CLN	250
	Radio Liberty (BBG)	THA	250
	Radio Veritas Asia	PHL	250
	Radio Veritas Asia	PHL	250
	YLE Radio Finland	FIN	500
15535	AWR	AUT	300
	Deutsche Welle	CLN	250
	Radio France Int.	RUS	500
	Radio France Int.	RUS	500
	Radio Ibrahim	XXX	
	Radio Ibrahim	XXX	

Legend:
Arabic, Chinese, English, French, German, Hindi, Mandarin, Portuguese, Russian, Spanish, Other

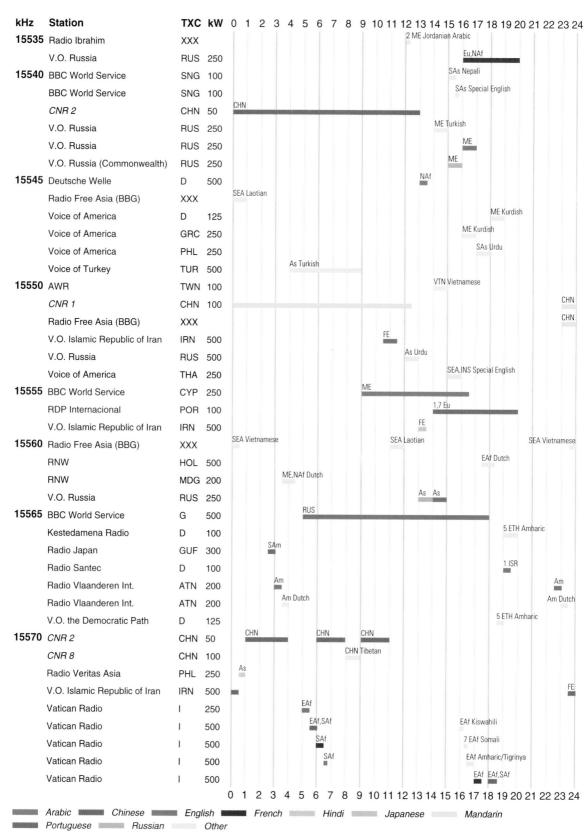

kHz	Station	TXC	kW	Timeline notes
15535	Radio Ibrahim	XXX		2 ME Jordanian Arabic
	V.O. Russia	RUS	250	Eu,NAf
15540	BBC World Service	SNG	100	SAs Nepali
	BBC World Service	SNG	100	SAs Special English
	CNR 2	CHN	50	CHN
	V.O. Russia	RUS	250	ME Turkish
	V.O. Russia	RUS	250	ME
	V.O. Russia (Commonwealth)	RUS	250	ME
15545	Deutsche Welle	D	500	NAf
	Radio Free Asia (BBG)	XXX		SEA Laotian
	Voice of America	D	125	ME Kurdish
	Voice of America	GRC	250	ME Kurdish
	Voice of America	PHL	250	SAs Urdu
	Voice of Turkey	TUR	500	As Turkish
15550	AWR	TWN	100	VTN Vietnamese
	CNR 1	CHN	100	CHN
	Radio Free Asia (BBG)	XXX		CHN
	V.O. Islamic Republic of Iran	IRN	500	FE
	V.O. Russia	RUS	500	As Urdu
	Voice of America	THA	250	SEA,INS Special English
15555	BBC World Service	CYP	250	ME
	RDP Internacional	POR	100	1,7 Eu
	V.O. Islamic Republic of Iran	IRN	500	FE
15560	Radio Free Asia (BBG)	XXX		SEA Vietnamese / SEA Laotian / SEA Vietnamese
	RNW	HOL	500	EAf Dutch
	RNW	MDG	200	ME,NAf Dutch
	V.O. Russia	RUS	250	As As
15565	BBC World Service	G	500	RUS
	Kestedamena Radio	D	100	5 ETH Amharic
	Radio Japan	GUF	300	SAm
	Radio Santec	D	100	1 ISR
	Radio Vlaanderen Int.	ATN	200	Am / Am
	Radio Vlaanderen Int.	ATN	200	Am Dutch / Am Dutch
	V.O. the Democratic Path	D	125	5 ETH Amharic
15570	*CNR 2*	CHN	50	CHN / CHN / CHN
	CNR 8	CHN	100	CHN Tibetan
	Radio Veritas Asia	PHL	250	As
	V.O. Islamic Republic of Iran	IRN	500	FE
	Vatican Radio	I	250	EAf
	Vatican Radio	I	500	EAf,SAf / EAf Kiswahili
	Vatican Radio	I	500	SAf / 7 EAf Somali
	Vatican Radio	I	500	SAf / EAf Amharic/Tigrinya
	Vatican Radio	I	500	EAf EAf,SAf

Legend: *Arabic*, *Chinese*, *English*, *French*, *Hindi*, *Japanese*, *Mandarin*, *Portuguese*, *Russian*, *Other*

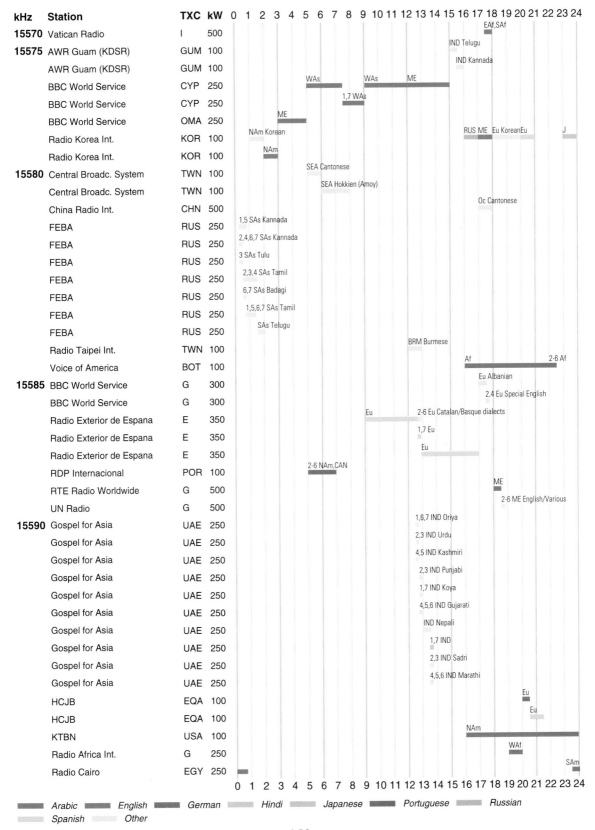

kHz	Station	TXC	kW
15570	Vatican Radio	I	500
15575	AWR Guam (KDSR)	GUM	100
	AWR Guam (KDSR)	GUM	100
	BBC World Service	CYP	250
	BBC World Service	CYP	250
	BBC World Service	OMA	250
	Radio Korea Int.	KOR	100
	Radio Korea Int.	KOR	100
15580	Central Broadc. System	TWN	100
	Central Broadc. System	TWN	100
	China Radio Int.	CHN	500
	FEBA	RUS	250
	FEBA	RUS	250
	FEBA	RUS	250
	FEBA	RUS	250
	FEBA	RUS	250
	FEBA	RUS	250
	FEBA	RUS	250
	Radio Taipei Int.	TWN	100
	Voice of America	BOT	100
15585	BBC World Service	G	300
	BBC World Service	G	300
	Radio Exterior de Espana	E	350
	Radio Exterior de Espana	E	350
	Radio Exterior de Espana	E	350
	RDP Internacional	POR	100
	RTE Radio Worldwide	G	500
	UN Radio	G	500
15590	Gospel for Asia	UAE	250
	Gospel for Asia	UAE	250
	Gospel for Asia	UAE	250
	Gospel for Asia	UAE	250
	Gospel for Asia	UAE	250
	Gospel for Asia	UAE	250
	Gospel for Asia	UAE	250
	Gospel for Asia	UAE	250
	Gospel for Asia	UAE	250
	Gospel for Asia	UAE	250
	HCJB	EQA	100
	HCJB	EQA	100
	KTBN	USA	100
	Radio Africa Int.	G	250
	Radio Cairo	EGY	250

Program annotations (by station, reading across the time scale 0–24):

- Vatican Radio: EAf,SAf
- AWR Guam (KDSR): IND Telugu
- AWR Guam (KDSR): IND Kannada
- BBC World Service (CYP): WAs — WAs — ME
- BBC World Service (CYP): 1,7 WAs
- BBC World Service (OMA): ME
- Radio Korea Int. (KOR): NAm Korean / RUS ME Eu KoreanEu / J
- Radio Korea Int. (KOR): NAm
- Central Broadc. System (TWN): SEA Cantonese
- Central Broadc. System (TWN): SEA Hokkien (Amoy)
- China Radio Int. (CHN): Oc Cantonese
- FEBA: 1,5 SAs Kannada
- FEBA: 2,4,6,7 SAs Kannada
- FEBA: 3 SAs Tulu
- FEBA: 2,3,4 SAs Tamil
- FEBA: 6,7 SAs Badagi
- FEBA: 1,5,6,7 SAs Tamil
- FEBA: SAs Telugu
- Radio Taipei Int.: BRM Burmese
- Voice of America (BOT): Af — 2-6 Af
- BBC World Service (G): Eu Albanian
- BBC World Service (G): 2,4 Eu Special English
- Radio Exterior de Espana: Eu — 2-6 Eu Catalan/Basque dialects
- Radio Exterior de Espana: 1,7 Eu
- Radio Exterior de Espana: Eu
- RDP Internacional: 2-6 NAm,CAN
- RTE Radio Worldwide: ME
- UN Radio: 2-6 ME English/Various
- Gospel for Asia: 1,6,7 IND Oriya
- Gospel for Asia: 2,3 IND Urdu
- Gospel for Asia: 4,5 IND Kashmiri
- Gospel for Asia: 2,3 IND Punjabi
- Gospel for Asia: 1,7 IND Koya
- Gospel for Asia: 4,5,6 IND Gujarati
- Gospel for Asia: IND Nepali
- Gospel for Asia: 1,7 IND
- Gospel for Asia: 2,3 IND Sadri
- Gospel for Asia: 4,5,6 IND Marathi
- HCJB: Eu
- HCJB: Eu
- KTBN: NAm
- Radio Africa Int.: WAf
- Radio Cairo: SAm

Legend: Arabic · English · German · Hindi · Japanese · Portuguese · Russian · Spanish · Other

159

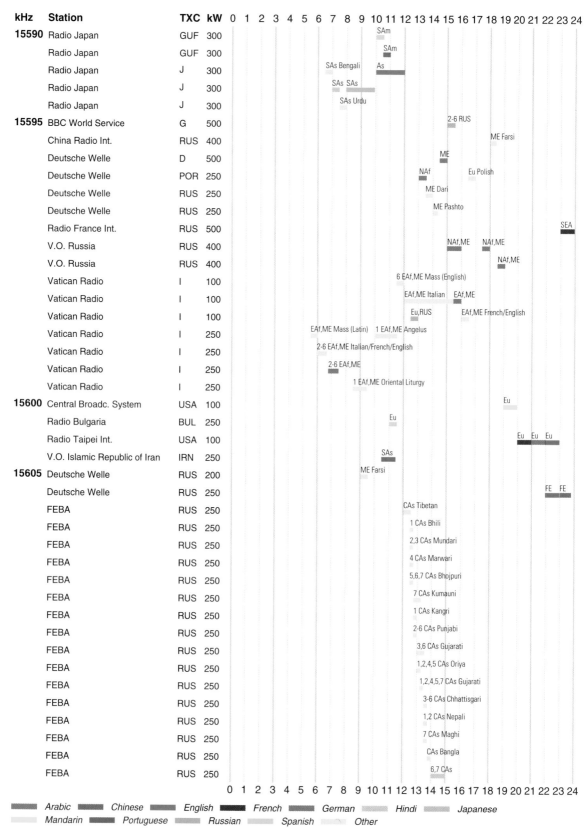

kHz	Station	TXC	kW																								
15590	Radio Japan	GUF	300										SAm													
	Radio Japan	GUF	300										SAm													
	Radio Japan	J	300							SAs Bengali		As														
	Radio Japan	J	300							SAs SAs															
	Radio Japan	J	300							SAs Urdu															
15595	BBC World Service	G	500												2-6 RUS											
	China Radio Int.	RUS	400															ME Farsi							
	Deutsche Welle	D	500												ME											
	Deutsche Welle	POR	250											NAf			Eu Polish							
	Deutsche Welle	RUS	250											ME Dari										
	Deutsche Welle	RUS	250											ME Pashto										
	Radio France Int.	RUS	500																					SEA	
	V.O. Russia	RUS	400												NAf,ME		NAf,ME							
	V.O. Russia	RUS	400															NAf,ME						
	Vatican Radio	I	100									6 EAf,ME Mass (English)												
	Vatican Radio	I	100									EAf,ME Italian	EAf,ME										
	Vatican Radio	I	100									Eu,RUS		EAf,ME French/English							
	Vatican Radio	I	250					EAf,ME Mass (Latin)	1 EAf,ME Angelus													
	Vatican Radio	I	250					2-6 EAf,ME Italian/French/English													
	Vatican Radio	I	250					2-6 EAf,ME														
	Vatican Radio	I	250						1 EAf,ME Oriental Liturgy											
15600	Central Broadc. System	USA	100																	Eu				
	Radio Bulgaria	BUL	250								Eu												
	Radio Taipei Int.	USA	100																		Eu Eu Eu		
	V.O. Islamic Republic of Iran	IRN	250							SAs												
15605	Deutsche Welle	RUS	200							ME Farsi												
	Deutsche Welle	RUS	250																				FE FE	
	FEBA	RUS	250									CAs Tibetan										
	FEBA	RUS	250									1 CAs Bhili										
	FEBA	RUS	250									2,3 CAs Mundari										
	FEBA	RUS	250									4 CAs Marwari										
	FEBA	RUS	250									5,6,7 CAs Bhojpuri									
	FEBA	RUS	250									7 CAs Kumauni									
	FEBA	RUS	250									1 CAs Kangri									
	FEBA	RUS	250									2-6 CAs Punjabi								
	FEBA	RUS	250									3,6 CAs Gujarati								
	FEBA	RUS	250									1,2,4,5 CAs Oriya								
	FEBA	RUS	250									1,2,4,5,7 CAs Gujarati							
	FEBA	RUS	250									3-6 CAs Chhattisgari							
	FEBA	RUS	250									1,2 CAs Nepali								
	FEBA	RUS	250									7 CAs Maghi								
	FEBA	RUS	250									CAs Bangla								
	FEBA	RUS	250									6,7 CAs								

Arabic Chinese English French German Hindi Japanese
Mandarin Portuguese Russian Spanish Other

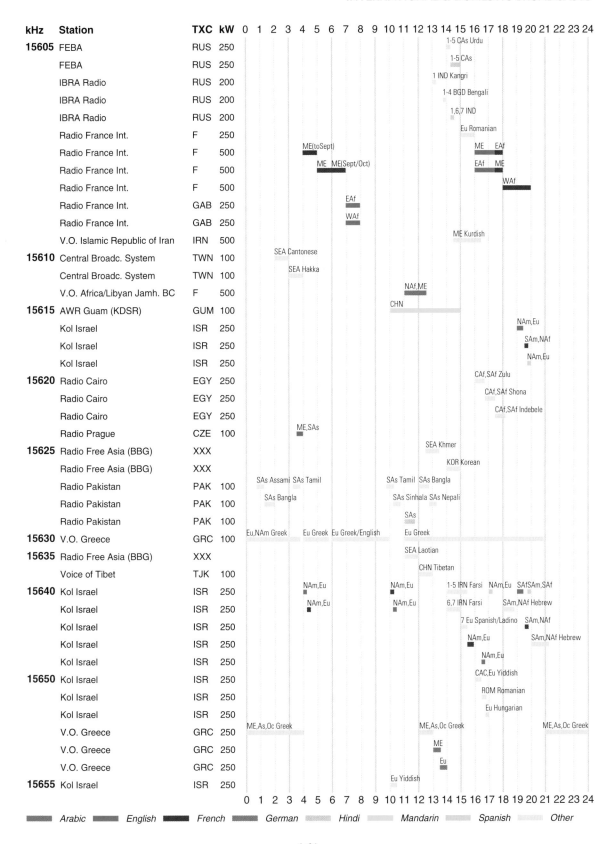

kHz	Station	TXC	kW
15605	FEBA	RUS	250
	FEBA	RUS	250
	IBRA Radio	RUS	200
	IBRA Radio	RUS	200
	IBRA Radio	RUS	200
	Radio France Int.	F	250
	Radio France Int.	F	500
	Radio France Int.	F	500
	Radio France Int.	F	500
	Radio France Int.	GAB	250
	Radio France Int.	GAB	250
	V.O. Islamic Republic of Iran	IRN	500
15610	Central Broadc. System	TWN	100
	Central Broadc. System	TWN	100
	V.O. Africa/Libyan Jamh. BC	F	500
15615	AWR Guam (KDSR)	GUM	100
	Kol Israel	ISR	250
	Kol Israel	ISR	250
	Kol Israel	ISR	250
15620	Radio Cairo	EGY	250
	Radio Cairo	EGY	250
	Radio Cairo	EGY	250
	Radio Prague	CZE	100
15625	Radio Free Asia (BBG)	XXX	
	Radio Free Asia (BBG)	XXX	
	Radio Pakistan	PAK	100
	Radio Pakistan	PAK	100
	Radio Pakistan	PAK	100
15630	V.O. Greece	GRC	100
15635	Radio Free Asia (BBG)	XXX	
	Voice of Tibet	TJK	100
15640	Kol Israel	ISR	250
	Kol Israel	ISR	250
	Kol Israel	ISR	250
	Kol Israel	ISR	250
	Kol Israel	ISR	250
15650	Kol Israel	ISR	250
	Kol Israel	ISR	250
	Kol Israel	ISR	250
	V.O. Greece	GRC	250
	V.O. Greece	GRC	250
	V.O. Greece	GRC	250
15655	Kol Israel	ISR	250

Schedule annotations (by row):

- 15605 FEBA RUS 250: 1-5 CAs Urdu
- FEBA RUS 250: 1-5 CAs
- IBRA Radio RUS 200: 1 IND Kangri
- IBRA Radio RUS 200: 1-4 BGD Bengali
- IBRA Radio RUS 200: 1,6,7 IND
- Radio France Int. F 250: Eu Romanian
- Radio France Int. F 500: ME(toSept); ME; EAf
- Radio France Int. F 500: ME ME(Sept/Oct); EAf; ME
- Radio France Int. F 500: WAf
- Radio France Int. GAB 250: EAf
- Radio France Int. GAB 250: WAf
- V.O. Islamic Republic of Iran IRN 500: ME Kurdish
- 15610 Central Broadc. System TWN 100: SEA Cantonese
- Central Broadc. System TWN 100: SEA Hakka
- V.O. Africa/Libyan Jamh. BC F 500: NAf,ME
- 15615 AWR Guam (KDSR) GUM 100: CHN
- Kol Israel ISR 250: NAm,Eu
- Kol Israel ISR 250: SAm,NAf
- Kol Israel ISR 250: NAm,Eu
- 15620 Radio Cairo EGY 250: CAf,SAf Zulu
- Radio Cairo EGY 250: CAf,SAf Shona
- Radio Cairo EGY 250: CAf,SAf Indebele
- Radio Prague CZE 100: ME,SAs
- 15625 Radio Free Asia (BBG) XXX: SEA Khmer
- Radio Free Asia (BBG) XXX: KOR Korean
- Radio Pakistan PAK 100: SAs Assami SAs Tamil; SAs Tamil SAs Bangla
- Radio Pakistan PAK 100: SAs Bangla; SAs Sinhala SAs Nepali
- Radio Pakistan PAK 100: SAs
- 15630 V.O. Greece GRC 100: Eu,NAm Greek; Eu Greek; Eu Greek/English; Eu Greek
- 15635 Radio Free Asia (BBG) XXX: SEA Laotian
- Voice of Tibet TJK 100: CHN Tibetan
- 15640 Kol Israel ISR 250: NAm,Eu; NAm,Eu; 1-5 IRN Farsi; NAm,Eu; SAfSAm,SAf
- Kol Israel ISR 250: NAm,Eu; NAm,Eu; 6,7 IRN Farsi; SAm,NAf Hebrew
- Kol Israel ISR 250: 7 Eu Spanish/Ladino; SAm,NAf
- Kol Israel ISR 250: NAm,Eu; SAm,NAf Hebrew
- Kol Israel ISR 250: NAm,Eu
- 15650 Kol Israel ISR 250: CAC,Eu Yiddish
- Kol Israel ISR 250: ROM Romanian
- Kol Israel ISR 250: Eu Hungarian
- V.O. Greece GRC 250: ME,As,Oc Greek; ME,As,Oc Greek; ME,As,Oc Greek
- V.O. Greece GRC 250: ME
- V.O. Greece GRC 250: Eu
- 15655 Kol Israel ISR 250: Eu Yiddish

Legend: Arabic — English — French — German — Hindi — Mandarin — Spanish — Other

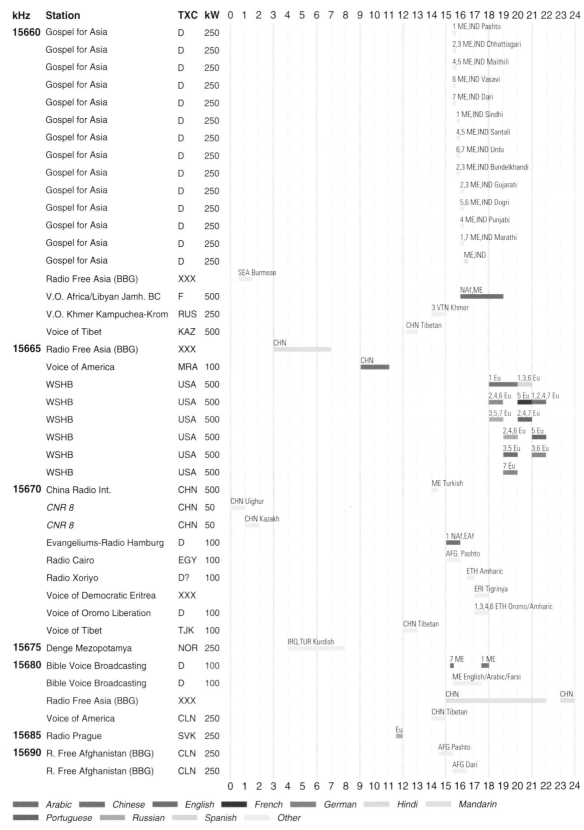

kHz	Station	TXC	kW																							
				0	1	2	3	4	5	6	7	8	9	10	11	12	13	14	15	16	17	18	19	20	21	22	23	24
15660	Gospel for Asia	D	250																	1 ME,IND Pashto
	Gospel for Asia	D	250																	2,3 ME,IND Chhattisgari
	Gospel for Asia	D	250																	4,5 ME,IND Maithili
	Gospel for Asia	D	250																	6 ME,IND Vasavi
	Gospel for Asia	D	250																	7 ME,IND Dari
	Gospel for Asia	D	250																	1 ME,IND Sindhi
	Gospel for Asia	D	250																	4,5 ME,IND Santali
	Gospel for Asia	D	250																	6,7 ME,IND Urdu
	Gospel for Asia	D	250																	2,3 ME,IND Bundelkhandi
	Gospel for Asia	D	250																	2,3 ME,IND Gujarati
	Gospel for Asia	D	250																	5,6 ME,IND Dogri
	Gospel for Asia	D	250																	4 ME,IND Punjabi
	Gospel for Asia	D	250																	1,7 ME,IND Marathi
	Gospel for Asia	D	250																	ME,IND
	Radio Free Asia (BBG)	XXX			SEA Burmese
	V.O. Africa/Libyan Jamh. BC	F	500																	NAf,ME
	V.O. Khmer Kampuchea-Krom	RUS	250														3 VTN Khmer
	Voice of Tibet	KAZ	500													CHN Tibetan
15665	Radio Free Asia (BBG)	XXX						CHN
	Voice of America	MRA	100											CHN
	WSHB	USA	500																	1 Eu			1,3,6 Eu
	WSHB	USA	500																	2,4,6 Eu		5 Eu	1,2,4,7 Eu
	WSHB	USA	500																	3,5,7 Eu	2,4,7 Eu
	WSHB	USA	500																		2,4,6 Eu	5 Eu
	WSHB	USA	500																		3,5 Eu	3,6 Eu
	WSHB	USA	500																		7 Eu
15670	China Radio Int.	CHN	500															ME Turkish
	CNR 8	CHN	50		CHN Uighur
	CNR 8	CHN	50			CHN Kazakh
	Evangeliums-Radio Hamburg	D	100															1 NAf,EAf
	Radio Cairo	EGY	100															AFG. Pashto
	Radio Xoriyo	D?	100																ETH Amharic
	Voice of Democratic Eritrea	XXX																	ERI Tigrinya
	Voice of Oromo Liberation	D	100																	1,3,4,6 ETH Oromo/Amharic
	Voice of Tibet	TJK	100													CHN Tibetan
15675	Denge Mezopotamya	NOR	250						IRQ,TUR Kurdish
15680	Bible Voice Broadcasting	D	100																7 ME		1 ME
	Bible Voice Broadcasting	D	100																ME English/Arabic/Farsi
	Radio Free Asia (BBG)	XXX																	CHN							CHN
	Voice of America	CLN	250														CHN Tibetan
15685	Radio Prague	SVK	250												Eu
15690	R. Free Afghanistan (BBG)	CLN	250															AFG Pashto
	R. Free Afghanistan (BBG)	CLN	250															AFG Dari

Legend: Arabic · Chinese · English · French · German · Hindi · Mandarin · Portuguese · Russian · Spanish · Other

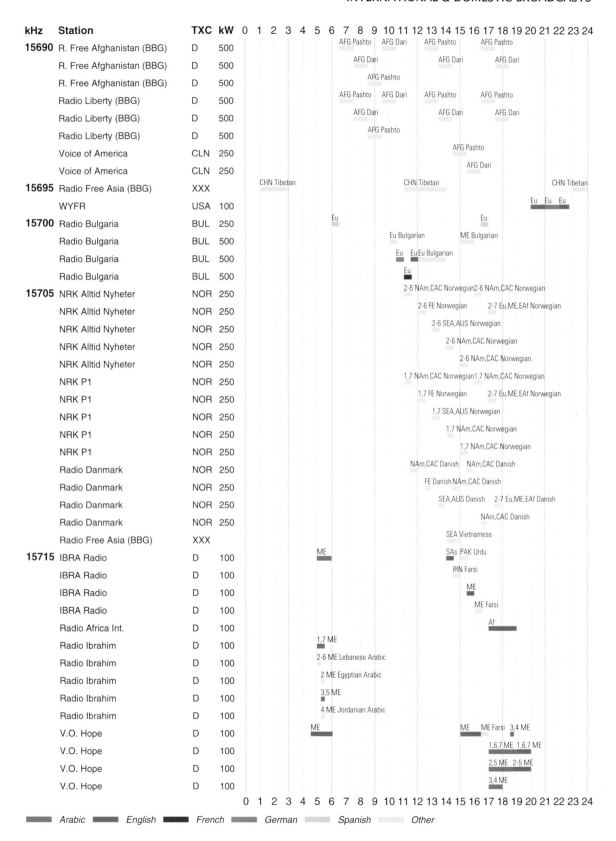

kHz	Station	TXC	kW	0 1 2 3 4 5 6 7 8 9 10 11 12 13 14 15 16 17 18 19 20 21 22 23 24
15690	R. Free Afghanistan (BBG)	D	500	AFG Pashto · AFG Dari · AFG Pashto · AFG Pashto
	R. Free Afghanistan (BBG)	D	500	AFG Dari · AFG Dari · AFG Dari
	R. Free Afghanistan (BBG)	D	500	AFG Pashto
	Radio Liberty (BBG)	D	500	AFG Pashto · AFG Dari · AFG Pashto · AFG Pashto
	Radio Liberty (BBG)	D	500	AFG Dari · AFG Dari · AFG Dari
	Radio Liberty (BBG)	D	500	AFG Pashto
	Voice of America	CLN	250	AFG Pashto
	Voice of America	CLN	250	AFG Dari
15695	Radio Free Asia (BBG)	XXX		CHN Tibetan · CHN Tibetan · CHN Tibetan
	WYFR	USA	100	Eu Eu Eu
15700	Radio Bulgaria	BUL	250	Eu · Eu
	Radio Bulgaria	BUL	500	Eu Bulgarian · ME Bulgarian
	Radio Bulgaria	BUL	500	Eu Eu Eu Bulgarian
	Radio Bulgaria	BUL	500	Eu
15705	NRK Alltid Nyheter	NOR	250	2-6 NAm,CAC Norwegian 2-6 NAm,CAC Norwegian
	NRK Alltid Nyheter	NOR	250	2-6 FE Norwegian · 2-7 Eu,ME,EAf Norwegian
	NRK Alltid Nyheter	NOR	250	2-6 SEA,AUS Norwegian
	NRK Alltid Nyheter	NOR	250	2-6 NAm,CAC Norwegian
	NRK Alltid Nyheter	NOR	250	2-6 NAm,CAC Norwegian
	NRK P1	NOR	250	1,7 NAm,CAC Norwegian 1,7 NAm,CAC Norwegian
	NRK P1	NOR	250	1,7 FE Norwegian · 2-7 Eu,ME,EAf Norwegian
	NRK P1	NOR	250	1,7 SEA,AUS Norwegian
	NRK P1	NOR	250	1,7 NAm,CAC Norwegian
	NRK P1	NOR	250	1,7 NAm,CAC Norwegian
	Radio Danmark	NOR	250	NAm,CAC Danish · NAm,CAC Danish
	Radio Danmark	NOR	250	FE Danish NAm,CAC Danish
	Radio Danmark	NOR	250	SEA,AUS Danish · 2-7 Eu,ME,EAf Danish
	Radio Danmark	NOR	250	NAm,CAC Danish
	Radio Free Asia (BBG)	XXX		SEA Vietnamese
15715	IBRA Radio	D	100	ME · SAs PAK Urdu
	IBRA Radio	D	100	IRN Farsi
	IBRA Radio	D	100	ME
	IBRA Radio	D	100	ME Farsi
	Radio Africa Int.	D	100	Af
	Radio Ibrahim	D	100	1,7 ME
	Radio Ibrahim	D	100	2-6 ME Lebanese Arabic
	Radio Ibrahim	D	100	2 ME Egyptian Arabic
	Radio Ibrahim	D	100	3,5 ME
	Radio Ibrahim	D	100	4 ME Jordanian Arabic
	V.O. Hope	D	100	ME · ME ME Farsi 3,4 ME
	V.O. Hope	D	100	1,6,7 ME 1,6,7 ME
	V.O. Hope	D	100	2,5 ME 2-5 ME
	V.O. Hope	D	100	3,4 ME

0 1 2 3 4 5 6 7 8 9 10 11 12 13 14 15 16 17 18 19 20 21 22 23 24

Arabic   English   French   German   Spanish   Other

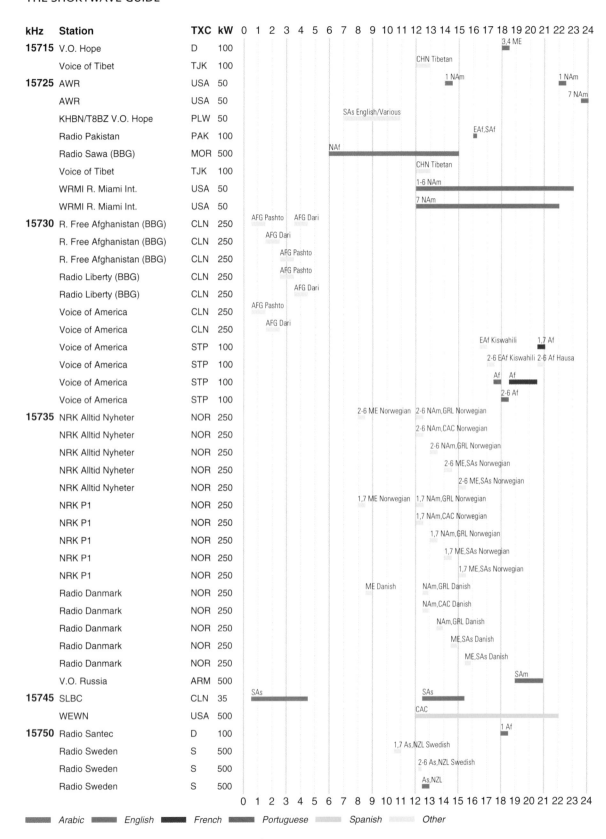

kHz	Station	TXC	kW
15715	V.O. Hope	D	100
	Voice of Tibet	TJK	100
15725	AWR	USA	50
	AWR	USA	50
	KHBN/T8BZ V.O. Hope	PLW	50
	Radio Pakistan	PAK	100
	Radio Sawa (BBG)	MOR	500
	Voice of Tibet	TJK	100
	WRMI R. Miami Int.	USA	50
	WRMI R. Miami Int.	USA	50
15730	R. Free Afghanistan (BBG)	CLN	250
	R. Free Afghanistan (BBG)	CLN	250
	R. Free Afghanistan (BBG)	CLN	250
	Radio Liberty (BBG)	CLN	250
	Radio Liberty (BBG)	CLN	250
	Voice of America	CLN	250
	Voice of America	CLN	250
	Voice of America	STP	100
	Voice of America	STP	100
	Voice of America	STP	100
	Voice of America	STP	100
15735	NRK Alltid Nyheter	NOR	250
	NRK Alltid Nyheter	NOR	250
	NRK Alltid Nyheter	NOR	250
	NRK Alltid Nyheter	NOR	250
	NRK Alltid Nyheter	NOR	250
	NRK P1	NOR	250
	NRK P1	NOR	250
	NRK P1	NOR	250
	NRK P1	NOR	250
	NRK P1	NOR	250
	Radio Danmark	NOR	250
	Radio Danmark	NOR	250
	Radio Danmark	NOR	250
	Radio Danmark	NOR	250
	Radio Danmark	NOR	250
	V.O. Russia	ARM	500
15745	SLBC	CLN	35
	WEWN	USA	500
15750	Radio Santec	D	100
	Radio Sweden	S	500
	Radio Sweden	S	500
	Radio Sweden	S	500

Legend: Arabic · English · French · Portuguese · Spanish · Other

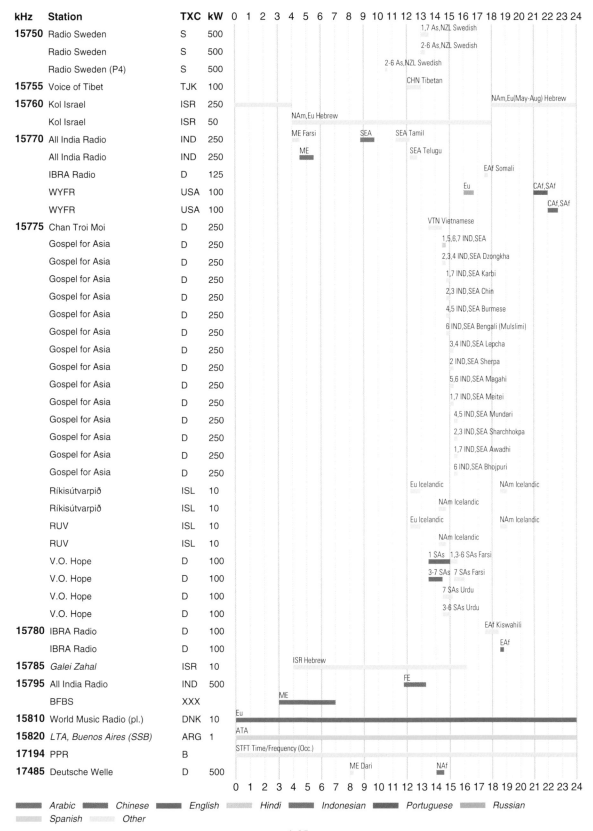

kHz	Station	TXC	kW
15750	Radio Sweden	S	500
	Radio Sweden	S	500
	Radio Sweden (P4)	S	500
15755	Voice of Tibet	TJK	100
15760	Kol Israel	ISR	250
	Kol Israel	ISR	50
15770	All India Radio	IND	250
	All India Radio	IND	250
	IBRA Radio	D	125
	WYFR	USA	100
	WYFR	USA	100
15775	Chan Troi Moi	D	250
	Gospel for Asia	D	250
	Gospel for Asia	D	250
	Gospel for Asia	D	250
	Gospel for Asia	D	250
	Gospel for Asia	D	250
	Gospel for Asia	D	250
	Gospel for Asia	D	250
	Gospel for Asia	D	250
	Gospel for Asia	D	250
	Gospel for Asia	D	250
	Gospel for Asia	D	250
	Gospel for Asia	D	250
	Gospel for Asia	D	250
	Gospel for Asia	D	250
	Ríkisútvarpið	ISL	10
	Ríkisútvarpið	ISL	10
	RUV	ISL	10
	RUV	ISL	10
	V.O. Hope	D	100
	V.O. Hope	D	100
	V.O. Hope	D	100
	V.O. Hope	D	100
15780	IBRA Radio	D	100
	IBRA Radio	D	100
15785	*Galei Zahal*	ISR	10
15795	All India Radio	IND	500
	BFBS	XXX	
15810	World Music Radio (pl.)	DNK	10
15820	*LTA, Buenos Aires (SSB)*	ARG	1
17194	PPR	B	
17485	Deutsche Welle	D	500

Chart annotations (left to right by time):
- 1,7 As,NZL Swedish
- 2-6 As,NZL Swedish
- 2-6 As,NZL Swedish
- CHN Tibetan
- NAm,Eu(May-Aug) Hebrew
- NAm,Eu Hebrew
- ME Farsi / SEA / SEA Tamil
- ME / SEA Telugu
- EAf Somali
- Eu / CAf,SAf
- CAf,SAf
- VTN Vietnamese
- 1,5,6,7 IND,SEA
- 2,3,4 IND,SEA Dzongkha
- 1,7 IND,SEA Karbi
- 2,3 IND,SEA Chin
- 4,5 IND,SEA Burmese
- 6 IND,SEA Bengali (Mulslimi)
- 3,4 IND,SEA Lepcha
- 2 IND,SEA Sherpa
- 5,6 IND,SEA Magahi
- 1,7 IND,SEA Meitei
- 4,5 IND,SEA Mundari
- 2,3 IND,SEA Sharchhokpa
- 1,7 IND,SEA Awadhi
- 6 IND,SEA Bhojpuri
- Eu Icelandic / NAm Icelandic
- NAm Icelandic
- Eu Icelandic / NAm Icelandic
- NAm Icelandic
- 1 SAs / 1,3-6 SAs Farsi
- 3-7 SAs / 7 SAs Farsi
- 7 SAs Urdu
- 3-6 SAs Urdu
- EAf Kiswahili
- EAf
- ISR Hebrew
- FE
- ME
- Eu
- ATA
- STFT Time/Frequency (Occ.)
- ME Dari / NAf

Legend:
- Arabic
- Chinese
- English
- Hindi
- Indonesian
- Portuguese
- Russian
- Spanish
- Other

THE SHORTWAVE GUIDE

kHz	Station	TXC	kW	0 1 2 3 4 5 6 7 8 9 10 11 12 13 14 15 16 17 18 19 20 21 22 23 24
17485	Deutsche Welle	D	500	ME Pashto · · ME
	Deutsche Welle	KAZ	500	SEA,Oc
	Radio Free Asia (BBG)	XXX		CHN Tibetan
	Radio Prague	CZE	100	CAf CAf Czech
	Radio Prague	CZE	100	CAf
	Radio Sweden (P1)	S	500	Eu,Af,SAm Swedish
17495	Democratic Voice of Burma	MDG	50	BRM Burmese
	IBC Tamil Radio	MDG	50	SEA Tamil
	Radio Free Asia (BBG)	XXX		CHN
	Radio Pakistan	PAK	100	SAs Tamil
	Radio Pakistan	PAK	100	SAs Sinhala
	V.O. Russia	TJK	500	SEA
	WBCQ	USA	50	NAm
17500	NRK Alltid Nyheter	NOR	250	2-6 FE,NZL Norwegian
	NRK P1	NOR	250	1,7 FE,NZL Norwegian
	Radio Bulgaria	BUL	250	Eu
	Radio Bulgaria	BUL	500	SAf Bulgarian
	Radio Danmark	NOR	250	FE,NZL Danish
17505	NRK Alltid Nyheter	NOR	250	2-6 NAm,CAC Norwegian
	NRK P1	NOR	250	1,7 NAm,CAC Norwegian
	Radio Danmark	NOR	250	NAm,CAC Danish
	Radio Sweden	S	500	1,7 As,AUS Swedish · · ME Swedish
	Radio Sweden	S	500	AUS SAs,AUS
	Radio Sweden	S	500	1,7 As,AUS Swedish
	Radio Sweden	S	500	2-6 As,AUS Swedish
	Radio Sweden	S	500	2-6 As,AUS Swedish
	Radio Sweden	S	500	ME,As
	Radio Sweden	S	500	ME,As Swedish
	Radio Sweden (P1)	S	500	1 Af,ME Swedish
	Radio Sweden (P4)	S	500	7 Af,ME Swedish · · 2-6 As,AUS Swedish
17510	All India Radio	IND	250	SEA Oc
	KWHR World Harvest R.	HAW	100	1-5,7 FE,SEA · · FE,SEA
	KWHR World Harvest R.	HAW	100	6 FE,SEA · · 1-6 FE,SEA
	KWHR World Harvest R.	HAW	100	FE,SEA · · 7 FE,SEA
	Radio France Int.	F	500	NAm
	Radio Free Asia (BBG)	XXX		CHN Tibetan
	Radio Sedaye Iran	F	250	IRN Farsi
17515	Radio France Int.	F	500	IND
	Radio France Int.	F	500	ME
	Vatican Radio	I	100	1 Eu,RUS Oriental Liturgy
	Vatican Radio	I	250	6 EAf,ME Mass (English)
	Vatican Radio	I	500	1-6 As · · EAf Kiswahili
	Vatican Radio	I	500	7 As Mass (Chinese) · · EAf,SAf

0 1 2 3 4 5 6 7 8 9 10 11 12 13 14 15 16 17 18 19 20 21 22 23 24

Arabic *Chinese* *English* *French* *German* *Indonesian* *Japanese*
Mandarin *Spanish* *Other*

166

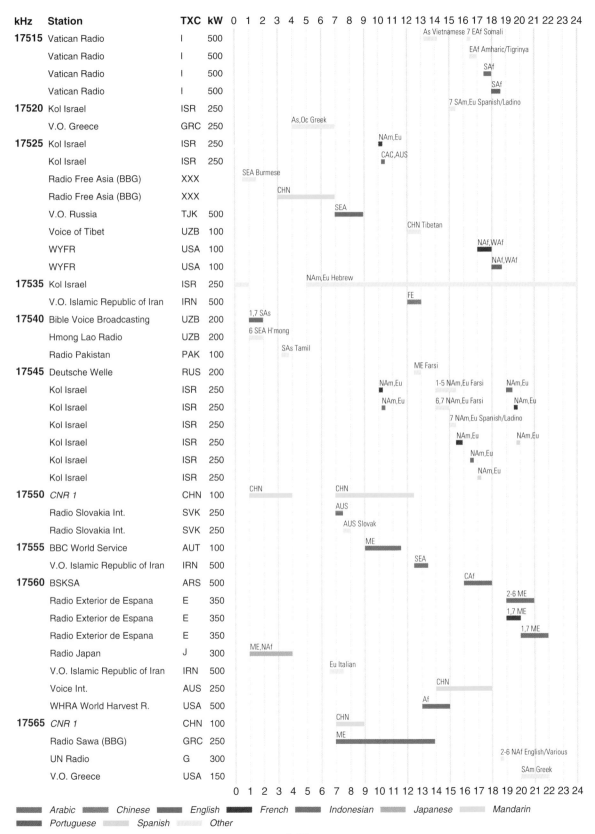

kHz	Station	TXC	kW
17515	Vatican Radio	I	500
	Vatican Radio	I	500
	Vatican Radio	I	500
	Vatican Radio	I	500
17520	Kol Israel	ISR	250
	V.O. Greece	GRC	250
17525	Kol Israel	ISR	250
	Kol Israel	ISR	250
	Radio Free Asia (BBG)	XXX	
	Radio Free Asia (BBG)	XXX	
	V.O. Russia	TJK	500
	Voice of Tibet	UZB	100
	WYFR	USA	100
	WYFR	USA	100
17535	Kol Israel	ISR	250
	V.O. Islamic Republic of Iran	IRN	500
17540	Bible Voice Broadcasting	UZB	200
	Hmong Lao Radio	UZB	200
	Radio Pakistan	PAK	100
17545	Deutsche Welle	RUS	200
	Kol Israel	ISR	250
	Kol Israel	ISR	250
	Kol Israel	ISR	250
	Kol Israel	ISR	250
	Kol Israel	ISR	250
	Kol Israel	ISR	250
17550	*CNR 1*	CHN	100
	Radio Slovakia Int.	SVK	250
	Radio Slovakia Int.	SVK	250
17555	BBC World Service	AUT	100
	V.O. Islamic Republic of Iran	IRN	500
17560	BSKSA	ARS	500
	Radio Exterior de Espana	E	350
	Radio Exterior de Espana	E	350
	Radio Exterior de Espana	E	350
	Radio Japan	J	300
	V.O. Islamic Republic of Iran	IRN	500
	Voice Int.	AUS	250
	WHRA World Harvest R.	USA	500
17565	*CNR 1*	CHN	100
	Radio Sawa (BBG)	GRC	250
	UN Radio	G	300
	V.O. Greece	USA	150

Schedule labels (by row):
- Vatican Radio: As Vietnamese 7 EAf Somali
- Vatican Radio: EAf Amharic/Tigrinya
- Vatican Radio: SAf
- Vatican Radio: SAf
- Kol Israel: 7 SAm,Eu Spanish/Ladino
- V.O. Greece: As,Oc Greek
- Kol Israel: NAm,Eu
- Kol Israel: CAC,AUS
- Radio Free Asia: SEA Burmese
- Radio Free Asia: CHN
- V.O. Russia: SEA
- Voice of Tibet: CHN Tibetan
- WYFR: NAf,WAf
- WYFR: NAf,WAf
- Kol Israel: NAm,Eu Hebrew
- V.O. Islamic Republic of Iran: FE
- Bible Voice Broadcasting: 1,7 SAs
- Hmong Lao Radio: 6 SEA H'mong
- Radio Pakistan: SAs Tamil
- Deutsche Welle: ME Farsi
- Kol Israel: NAm,Eu / 1-5 NAm,Eu Farsi / NAm,Eu
- Kol Israel: NAm,Eu / 6,7 NAm,Eu Farsi / NAm,Eu
- Kol Israel: 7 NAm,Eu Spanish/Ladino
- Kol Israel: NAm,Eu / NAm,Eu
- Kol Israel: NAm,Eu
- Kol Israel: NAm,Eu
- CNR 1: CHN / CHN
- Radio Slovakia Int.: AUS
- Radio Slovakia Int.: AUS Slovak
- BBC World Service: ME
- V.O. Islamic Republic of Iran: SEA
- BSKSA: CAf
- Radio Exterior de Espana: 2-6 ME
- Radio Exterior de Espana: 1,7 ME
- Radio Exterior de Espana: 1,7 ME
- Radio Japan: ME,NAf
- V.O. Islamic Republic of Iran: Eu Italian
- Voice Int.: CHN
- WHRA World Harvest R.: Af
- CNR 1: CHN
- Radio Sawa (BBG): ME
- UN Radio: 2-6 NAf English/Various
- V.O. Greece: SAm Greek

Legend: Arabic · Chinese · English · French · Indonesian · Japanese · Mandarin · Portuguese · Spanish · Other

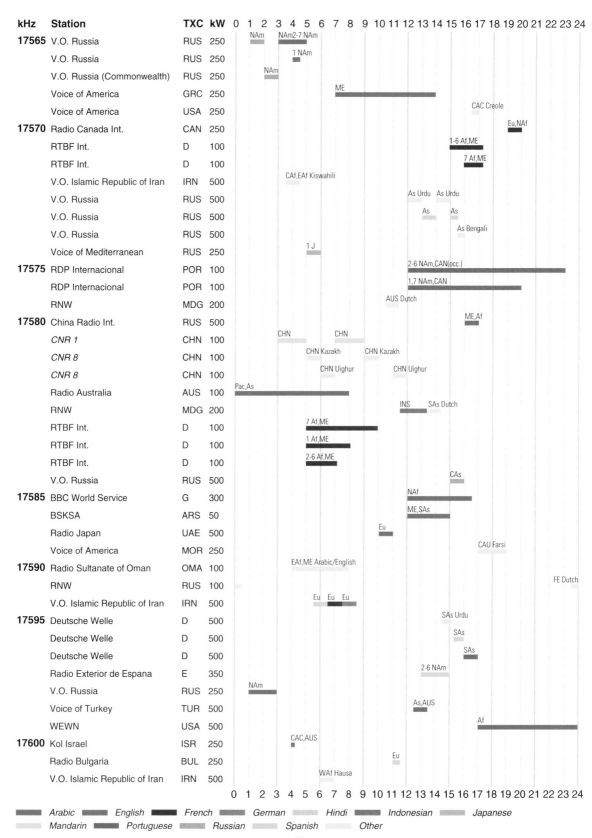

kHz	Station	TXC	kW	Schedule (0–24 h)
17565	V.O. Russia	RUS	250	NAm; NAm2-7 NAm
	V.O. Russia	RUS	250	1 NAm
	V.O. Russia (Commonwealth)	RUS	250	NAm
	Voice of America	GRC	250	ME
	Voice of America	USA	250	CAC Creole
17570	Radio Canada Int.	CAN	250	Eu,NAf
	RTBF Int.	D	100	1-6 Af,ME
	RTBF Int.	D	100	7 Af,ME
	V.O. Islamic Republic of Iran	IRN	500	CAf,EAf Kiswahili
	V.O. Russia	RUS	500	As Urdu As Urdu
	V.O. Russia	RUS	500	As As
	V.O. Russia	RUS	500	As Bengali
	Voice of Mediterranean	RUS	250	1 J
17575	RDP Internacional	POR	100	2-6 NAm,CAN(occ.)
	RDP Internacional	POR	100	1,7 NAm,CAN
	RNW	MDG	200	AUS Dutch
17580	China Radio Int.	RUS	500	ME,Af
	CNR 1	CHN	100	CHN CHN
	CNR 8	CHN	100	CHN Kazakh CHN Kazakh
	CNR 8	CHN	100	CHN Uighur CHN Uighur
	Radio Australia	AUS	100	Pac,As
	RNW	MDG	200	INS SAs Dutch
	RTBF Int.	D	100	7 Af,ME
	RTBF Int.	D	100	1 Af,ME
	RTBF Int.	D	100	2-6 Af,ME
	V.O. Russia	RUS	500	CAs
17585	BBC World Service	G	300	NAf
	BSKSA	ARS	50	ME,SAs
	Radio Japan	UAE	500	Eu
	Voice of America	MOR	250	CAU Farsi
17590	Radio Sultanate of Oman	OMA	100	EAf,ME Arabic/English
	RNW	RUS	100	FE Dutch
	V.O. Islamic Republic of Iran	IRN	500	Eu Eu Eu
17595	Deutsche Welle	D	500	SAs Urdu
	Deutsche Welle	D	500	SAs
	Deutsche Welle	D	500	SAs
	Radio Exterior de Espana	E	350	2-6 NAm
	V.O. Russia	RUS	250	NAm
	Voice of Turkey	TUR	500	As,AUS
	WEWN	USA	500	Af
17600	Kol Israel	ISR	250	CAC,AUS
	Radio Bulgaria	BUL	250	Eu
	V.O. Islamic Republic of Iran	IRN	500	WAf Hausa

Legend: Arabic · English · French · German · Hindi · Indonesian · Japanese · Mandarin · Portuguese · Russian · Spanish · Other

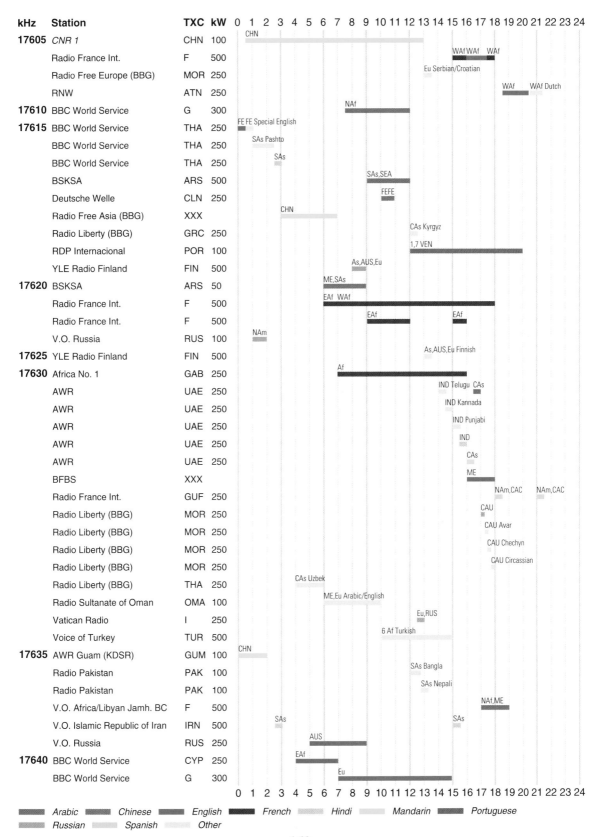

kHz	Station	TXC	kW
17605	CNR 1	CHN	100
	Radio France Int.	F	500
	Radio Free Europe (BBG)	MOR	250
	RNW	ATN	250
17610	BBC World Service	G	300
17615	BBC World Service	THA	250
	BBC World Service	THA	250
	BBC World Service	THA	250
	BSKSA	ARS	500
	Deutsche Welle	CLN	250
	Radio Free Asia (BBG)	XXX	
	Radio Liberty (BBG)	GRC	250
	RDP Internacional	POR	100
	YLE Radio Finland	FIN	500
17620	BSKSA	ARS	50
	Radio France Int.	F	500
	Radio France Int.	F	500
	V.O. Russia	RUS	100
17625	YLE Radio Finland	FIN	500
17630	Africa No. 1	GAB	250
	AWR	UAE	250
	AWR	UAE	250
	AWR	UAE	250
	AWR	UAE	250
	AWR	UAE	250
	BFBS	XXX	
	Radio France Int.	GUF	250
	Radio Liberty (BBG)	MOR	250
	Radio Liberty (BBG)	MOR	250
	Radio Liberty (BBG)	MOR	250
	Radio Liberty (BBG)	MOR	250
	Radio Liberty (BBG)	THA	250
	Radio Sultanate of Oman	OMA	100
	Vatican Radio	I	250
	Voice of Turkey	TUR	500
17635	AWR Guam (KDSR)	GUM	100
	Radio Pakistan	PAK	100
	Radio Pakistan	PAK	100
	V.O. Africa/Libyan Jamh. BC	F	500
	V.O. Islamic Republic of Iran	IRN	500
	V.O. Russia	RUS	250
17640	BBC World Service	CYP	250
	BBC World Service	G	300

Station labels and bars:
- CNR 1: CHN
- Radio France Int.: WAf WAf WAf
- Radio Free Europe (BBG): Eu Serbian/Croatian
- RNW: WAf WAf Dutch
- BBC World Service (17610): NAf
- BBC World Service (THA): FE FE Special English
- BBC World Service (THA): SAs Pashto
- BBC World Service (THA): SAs
- BSKSA: SAs,SEA
- Deutsche Welle: FE FE
- Radio Free Asia (BBG): CHN
- Radio Liberty (BBG): CAs Kyrgyz
- RDP Internacional: 1,7 VEN
- YLE Radio Finland: As,AUS,Eu
- BSKSA: ME,SAs
- Radio France Int.: EAf WAf
- Radio France Int.: EAf EAf
- V.O. Russia: NAm
- YLE Radio Finland: As,AUS,Eu Finnish
- Africa No. 1: Af
- AWR: IND Telugu CAs
- AWR: IND Kannada
- AWR: IND Punjabi
- AWR: IND
- AWR: CAs
- BFBS: ME
- Radio France Int.: NAm,CAC NAm,CAC
- Radio Liberty (BBG): CAU
- Radio Liberty (BBG): CAU Avar
- Radio Liberty (BBG): CAU Chechyn
- Radio Liberty (BBG): CAU Circassian
- Radio Liberty (BBG): CAs Uzbek
- Radio Sultanate of Oman: ME,Eu Arabic/English
- Vatican Radio: Eu,RUS
- Voice of Turkey: 6 Af Turkish
- AWR Guam (KDSR): CHN
- Radio Pakistan: SAs Bangla
- Radio Pakistan: SAs Nepali
- V.O. Africa/Libyan Jamh. BC: NAf,ME
- V.O. Islamic Republic of Iran: SAs SAs
- V.O. Russia: AUS
- BBC World Service (CYP): EAf
- BBC World Service (G): Eu

Legend:
Arabic | Chinese | English | French | Hindi | Mandarin | Portuguese
Russian | Spanish | Other

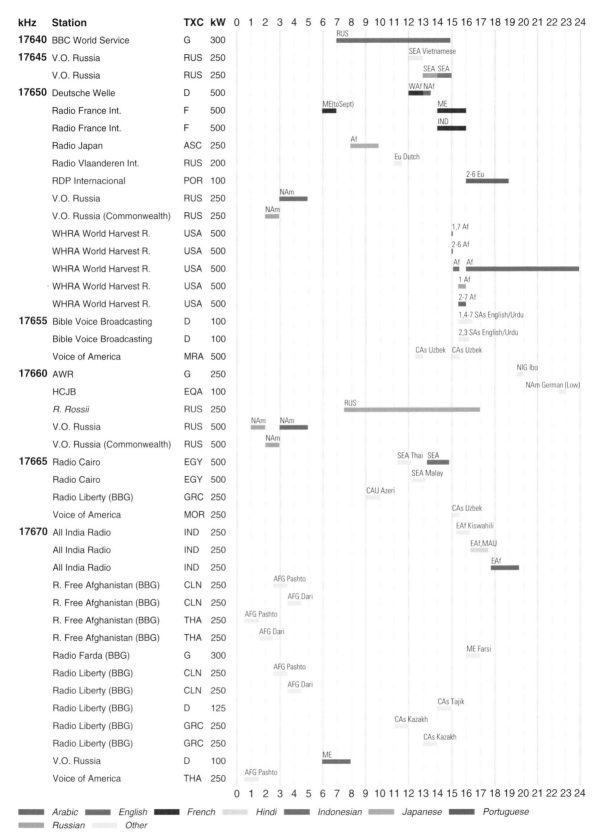

kHz	Station	TXC	kW
17640	BBC World Service	G	300
17645	V.O. Russia	RUS	250
	V.O. Russia	RUS	250
17650	Deutsche Welle	D	500
	Radio France Int.	F	500
	Radio France Int.	F	500
	Radio Japan	ASC	250
	Radio Vlaanderen Int.	RUS	200
	RDP Internacional	POR	100
	V.O. Russia	RUS	250
	V.O. Russia (Commonwealth)	RUS	250
	WHRA World Harvest R.	USA	500
	WHRA World Harvest R.	USA	500
	WHRA World Harvest R.	USA	500
	WHRA World Harvest R.	USA	500
	WHRA World Harvest R.	USA	500
17655	Bible Voice Broadcasting	D	100
	Bible Voice Broadcasting	D	100
	Voice of America	MRA	500
17660	AWR	G	250
	HCJB	EQA	100
	R. Rossii	RUS	250
	V.O. Russia	RUS	500
	V.O. Russia (Commonwealth)	RUS	500
17665	Radio Cairo	EGY	500
	Radio Cairo	EGY	500
	Radio Liberty (BBG)	GRC	250
	Voice of America	MOR	250
17670	All India Radio	IND	250
	All India Radio	IND	250
	All India Radio	IND	250
	R. Free Afghanistan (BBG)	CLN	250
	R. Free Afghanistan (BBG)	CLN	250
	R. Free Afghanistan (BBG)	THA	250
	R. Free Afghanistan (BBG)	THA	250
	Radio Farda (BBG)	G	300
	Radio Liberty (BBG)	CLN	250
	Radio Liberty (BBG)	CLN	250
	Radio Liberty (BBG)	D	125
	Radio Liberty (BBG)	GRC	250
	Radio Liberty (BBG)	GRC	250
	V.O. Russia	D	100
	Voice of America	THA	250

Legend: *Arabic* · *English* · *French* · *Hindi* · *Indonesian* · *Japanese* · *Portuguese* · *Russian* · *Other*

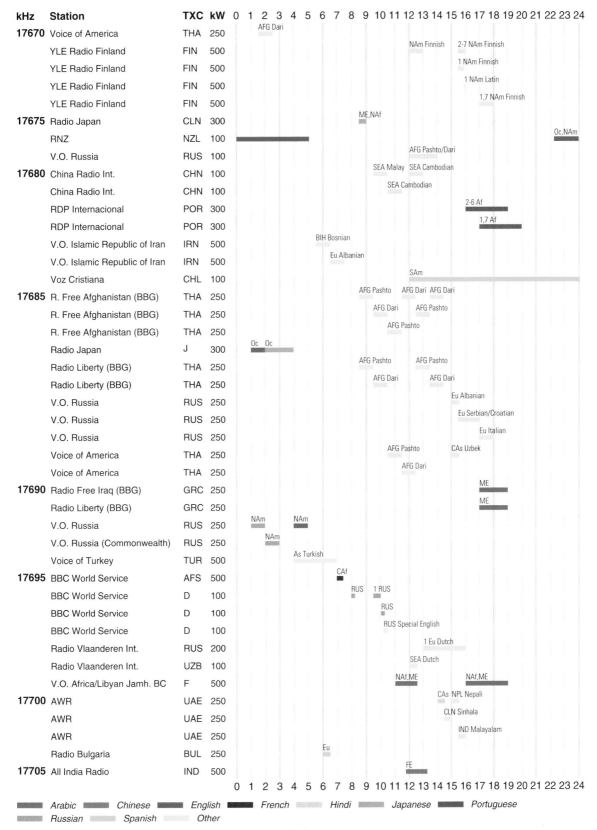

kHz	Station	TXC	kW																								
17670	Voice of America	THA	250	AFG Dari
	YLE Radio Finland	FIN	500	NAm Finnish / 2-7 NAm Finnish
	YLE Radio Finland	FIN	500	1 NAm Finnish
	YLE Radio Finland	FIN	500	1 NAm Latin
	YLE Radio Finland	FIN	500	1,7 NAm Finnish
17675	Radio Japan	CLN	300	ME,NAf
	RNZ	NZL	100	Oc,NAm
	V.O. Russia	RUS	100	AFG Pashto/Dari
17680	China Radio Int.	CHN	100	SEA Malay / SEA Cambodian
	China Radio Int.	CHN	100	SEA Cambodian
	RDP Internacional	POR	300	2-6 Af
	RDP Internacional	POR	300	1,7 Af
	V.O. Islamic Republic of Iran	IRN	500	BIH Bosnian
	V.O. Islamic Republic of Iran	IRN	500	Eu Albanian
	Voz Cristiana	CHL	100	SAm
17685	R. Free Afghanistan (BBG)	THA	250	AFG Pashto / AFG Dari / AFG Dari
	R. Free Afghanistan (BBG)	THA	250	AFG Dari / AFG Pashto
	R. Free Afghanistan (BBG)	THA	250	AFG Pashto
	Radio Japan	J	300	Oc Oc
	Radio Liberty (BBG)	THA	250	AFG Pashto / AFG Pashto
	Radio Liberty (BBG)	THA	250	AFG Dari / AFG Dari
	V.O. Russia	RUS	250	Eu Albanian
	V.O. Russia	RUS	250	Eu Serbian/Croatian
	V.O. Russia	RUS	250	Eu Italian
	Voice of America	THA	250	AFG Pashto / CAs Uzbek
	Voice of America	THA	250	AFG Dari
17690	Radio Free Iraq (BBG)	GRC	250	ME
	Radio Liberty (BBG)	GRC	250	ME
	V.O. Russia	RUS	250	NAm / NAm
	V.O. Russia (Commonwealth)	RUS	250	NAm
	Voice of Turkey	TUR	500	As Turkish
17695	BBC World Service	AFS	500	CAf
	BBC World Service	D	100	RUS / 1 RUS
	BBC World Service	D	100	RUS
	BBC World Service	D	100	RUS Special English
	Radio Vlaanderen Int.	RUS	200	1 Eu Dutch
	Radio Vlaanderen Int.	UZB	100	SEA Dutch
	V.O. Africa/Libyan Jamh. BC	F	500	NAf,ME / NAf,ME
17700	AWR	UAE	250	CAs NPL Nepali
	AWR	UAE	250	CLN Sinhala
	AWR	UAE	250	IND Malayalam
	Radio Bulgaria	BUL	250	Eu
17705	All India Radio	IND	500	FE

Legend: Arabic, Chinese, English, French, Hindi, Japanese, Portuguese, Russian, Spanish, Other

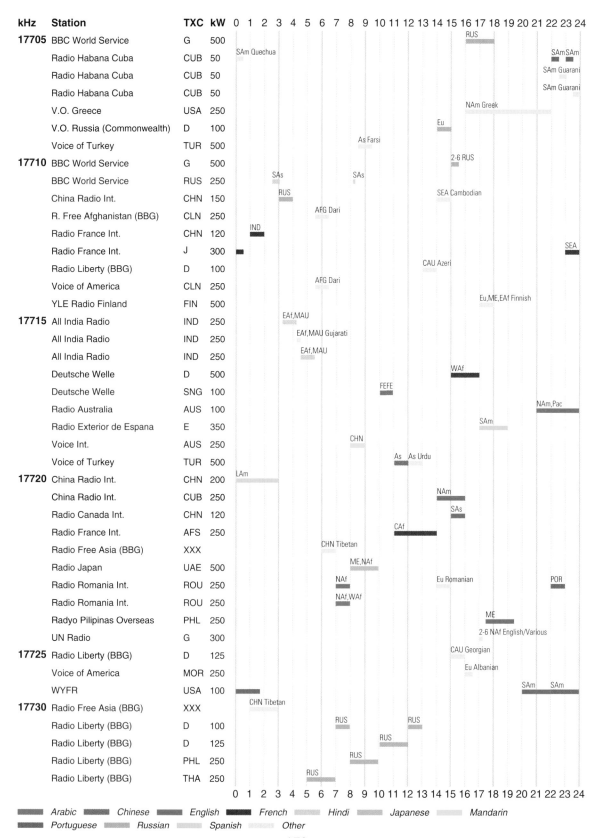

kHz	Station	TXC	kW	0 1 2 3 4 5 6 7 8 9 10 11 12 13 14 15 16 17 18 19 20 21 22 23 24
17705	BBC World Service	G	500	RUS
	Radio Habana Cuba	CUB	50	SAm Quechua · SAm SAm
	Radio Habana Cuba	CUB	50	SAm Guarani
	Radio Habana Cuba	CUB	50	SAm Guarani
	V.O. Greece	USA	250	NAm Greek
	V.O. Russia (Commonwealth)	D	100	Eu
	Voice of Turkey	TUR	500	As Farsi
17710	BBC World Service	G	500	2-6 RUS
	BBC World Service	RUS	250	SAs · SAs
	China Radio Int.	CHN	150	RUS · SEA Cambodian
	R. Free Afghanistan (BBG)	CLN	250	AFG Dari
	Radio France Int.	CHN	120	IND
	Radio France Int.	J	300	· SEA
	Radio Liberty (BBG)	D	100	CAU Azeri
	Voice of America	CLN	250	AFG Dari
	YLE Radio Finland	FIN	500	Eu,ME,EAf Finnish
17715	All India Radio	IND	250	EAf,MAU
	All India Radio	IND	250	EAf,MAU Gujarati
	All India Radio	IND	250	EAf,MAU
	Deutsche Welle	D	500	WAf
	Deutsche Welle	SNG	100	FEFE
	Radio Australia	AUS	100	NAm,Pac
	Radio Exterior de Espana	E	350	SAm
	Voice Int.	AUS	250	CHN
	Voice of Turkey	TUR	500	As As Urdu
17720	China Radio Int.	CHN	200	LAm
	China Radio Int.	CUB	250	NAm
	Radio Canada Int.	CHN	120	SAs
	Radio France Int.	AFS	250	CAf
	Radio Free Asia (BBG)	XXX		CHN Tibetan
	Radio Japan	UAE	500	ME,NAf
	Radio Romania Int.	ROU	250	NAf · Eu Romanian · POR
	Radio Romania Int.	ROU	250	NAf,WAf
	Radyo Pilipinas Overseas	PHL	250	ME
	UN Radio	G	300	2-6 NAf English/Various
17725	Radio Liberty (BBG)	D	125	CAU Georgian
	Voice of America	MOR	250	Eu Albanian
	WYFR	USA	100	· SAm SAm
17730	Radio Free Asia (BBG)	XXX		CHN Tibetan
	Radio Liberty (BBG)	D	100	RUS RUS
	Radio Liberty (BBG)	D	125	RUS
	Radio Liberty (BBG)	PHL	250	RUS
	Radio Liberty (BBG)	THA	250	RUS

Arabic	Chinese	English	French	Hindi	Japanese	Mandarin
Portuguese	Russian	Spanish	Other			

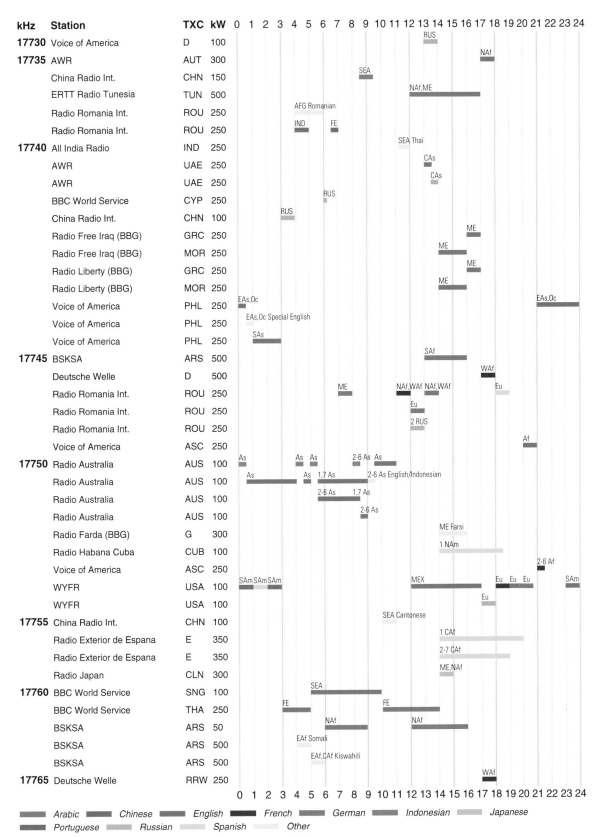

kHz	Station	TXC	kW	Schedule
17730	Voice of America	D	100	RUS (13-14)
17735	AWR	AUT	300	NAf (17-18)
	China Radio Int.	CHN	150	SEA (8-9)
	ERTT Radio Tunesia	TUN	500	NAf,ME (12-17)
	Radio Romania Int.	ROU	250	AFG Romanian (4-7)
	Radio Romania Int.	ROU	250	IND (4-5), FE (6-7)
17740	All India Radio	IND	250	SEA Thai (12-13)
	AWR	UAE	250	CAs (13-14)
	AWR	UAE	250	CAs (13-14)
	BBC World Service	CYP	250	RUS (6-7)
	China Radio Int.	CHN	100	RUS (3-4)
	Radio Free Iraq (BBG)	GRC	250	ME (15-16)
	Radio Free Iraq (BBG)	MOR	250	ME (13-15)
	Radio Liberty (BBG)	GRC	250	ME (16-17)
	Radio Liberty (BBG)	MOR	250	ME (13-15)
	Voice of America	PHL	250	EAs,Oc (0-1), EAs,Oc (21-24)
	Voice of America	PHL	250	EAs,Oc Special English (1-2)
	Voice of America	PHL	250	SAs (2-3)
17745	BSKSA	ARS	500	SAf (13-17)
	Deutsche Welle	D	500	WAf (17-18)
	Radio Romania Int.	ROU	250	ME (7-8), NAf,WAf (12-13), NAf,WAf (13-14), Eu (18-19)
	Radio Romania Int.	ROU	250	Eu (13-14)
	Radio Romania Int.	ROU	250	2 RUS (13-14)
	Voice of America	ASC	250	Af (20-21)
17750	Radio Australia	AUS	100	As (0-1), As (5-6), As (6-7), 2-6 As (8-9), As (9-10)
	Radio Australia	AUS	100	As (1-3), As (6-7), 1,7 As (7-8), 2-6 As English/Indonesian (9-12)
	Radio Australia	AUS	100	2-6 As (6-8), 1,7 As (8-9)
	Radio Australia	AUS	100	2-6 As (8-9)
	Radio Farda (BBG)	G	300	ME Farsi (13-16)
	Radio Habana Cuba	CUB	100	1 NAm (13-19)
	Voice of America	ASC	250	2-6 Af (21-22)
	WYFR	USA	100	SAm (0-1), SAm (1-2), SAm (2-3), MEX (13-15), Eu (17-18), Eu (18-19), Eu (19-20), SAm (22-24)
	WYFR	USA	100	Eu (18-19)
17755	China Radio Int.	CHN	100	SEA Cantonese (10-11)
	Radio Exterior de Espana	E	350	1 CAf (13-18)
	Radio Exterior de Espana	E	350	2-7 CAf (13-19)
	Radio Japan	CLN	300	ME,NAf (13-14)
17760	BBC World Service	SNG	100	SEA (5-9)
	BBC World Service	THA	250	FE (3-6), FE (10-13)
	BSKSA	ARS	50	NAf (6-9), NAf (13-16)
	BSKSA	ARS	500	EAf Somali (4-6)
	BSKSA	ARS	500	EAf,CAf Kiswahili (5-7)
17765	Deutsche Welle	RRW	250	WAf (17-18)

Arabic ▪ Chinese ▪ English ▪ French ▪ German ▪ Indonesian ▪ Japanese
Portuguese ▪ Russian ▪ Spanish ▪ Other

THE SHORTWAVE GUIDE

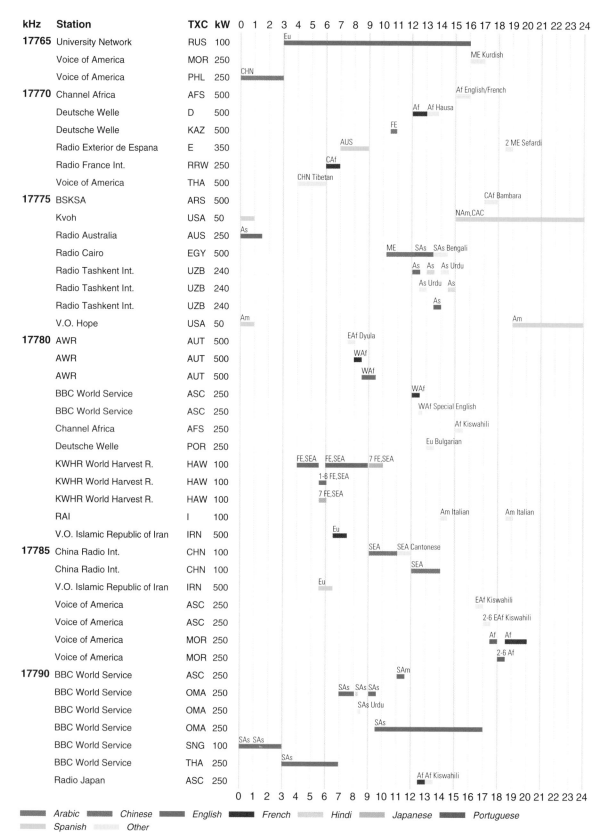

kHz	Station	TXC	kW
17765	University Network	RUS	100
	Voice of America	MOR	250
	Voice of America	PHL	250
17770	Channel Africa	AFS	500
	Deutsche Welle	D	500
	Deutsche Welle	KAZ	500
	Radio Exterior de Espana	E	350
	Radio France Int.	RRW	250
	Voice of America	THA	500
17775	BSKSA	ARS	500
	Kvoh	USA	50
	Radio Australia	AUS	250
	Radio Cairo	EGY	500
	Radio Tashkent Int.	UZB	240
	Radio Tashkent Int.	UZB	240
	Radio Tashkent Int.	UZB	240
	V.O. Hope	USA	50
17780	AWR	AUT	500
	AWR	AUT	500
	AWR	AUT	500
	BBC World Service	ASC	250
	BBC World Service	ASC	250
	Channel Africa	AFS	250
	Deutsche Welle	POR	250
	KWHR World Harvest R.	HAW	100
	KWHR World Harvest R.	HAW	100
	KWHR World Harvest R.	HAW	100
	RAI	I	100
	V.O. Islamic Republic of Iran	IRN	500
17785	China Radio Int.	CHN	100
	China Radio Int.	CHN	100
	V.O. Islamic Republic of Iran	IRN	500
	Voice of America	ASC	250
	Voice of America	ASC	250
	Voice of America	MOR	250
	Voice of America	MOR	250
17790	BBC World Service	ASC	250
	BBC World Service	OMA	250
	BBC World Service	OMA	250
	BBC World Service	OMA	250
	BBC World Service	SNG	100
	BBC World Service	THA	250
	Radio Japan	ASC	250

174

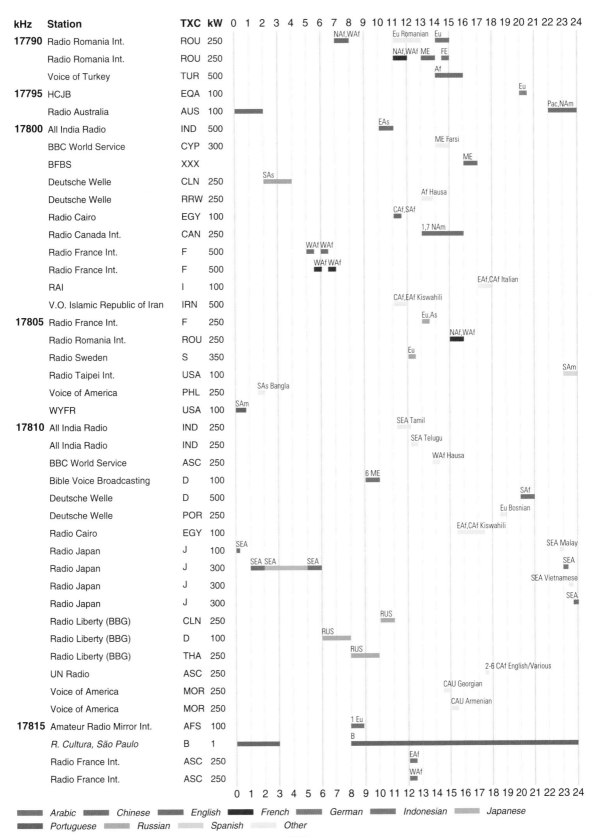

kHz	Station	TXC	kW
17790	Radio Romania Int.	ROU	250
	Radio Romania Int.	ROU	250
	Voice of Turkey	TUR	500
17795	HCJB	EQA	100
	Radio Australia	AUS	100
17800	All India Radio	IND	500
	BBC World Service	CYP	300
	BFBS	XXX	
	Deutsche Welle	CLN	250
	Deutsche Welle	RRW	250
	Radio Cairo	EGY	100
	Radio Canada Int.	CAN	250
	Radio France Int.	F	500
	Radio France Int.	F	500
	RAI	I	100
	V.O. Islamic Republic of Iran	IRN	500
17805	Radio France Int.	F	250
	Radio Romania Int.	ROU	250
	Radio Sweden	S	350
	Radio Taipei Int.	USA	100
	Voice of America	PHL	250
	WYFR	USA	100
17810	All India Radio	IND	250
	All India Radio	IND	250
	BBC World Service	ASC	250
	Bible Voice Broadcasting	D	100
	Deutsche Welle	D	500
	Deutsche Welle	POR	250
	Radio Cairo	EGY	100
	Radio Japan	J	100
	Radio Japan	J	300
	Radio Japan	J	300
	Radio Japan	J	300
	Radio Liberty (BBG)	CLN	250
	Radio Liberty (BBG)	D	100
	Radio Liberty (BBG)	THA	250
	UN Radio	ASC	250
	Voice of America	MOR	250
	Voice of America	MOR	250
17815	Amateur Radio Mirror Int.	AFS	100
	R. Cultura, São Paulo	B	1
	Radio France Int.	ASC	250
	Radio France Int.	ASC	250

Legend: Arabic — Chinese — English — French — German — Indonesian — Japanese — Portuguese — Russian — Spanish — Other

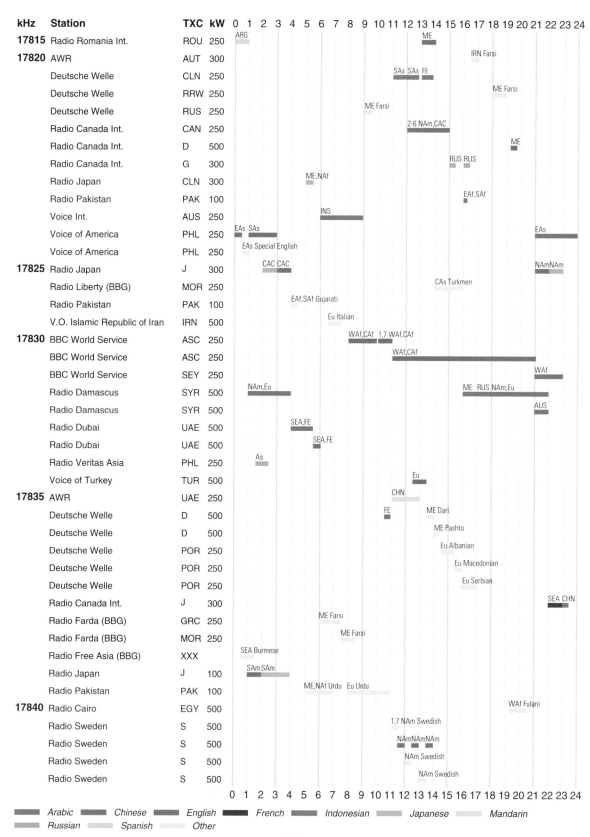

kHz	Station	TXC	kW	Schedule (0–24)
17815	Radio Romania Int.	ROU	250	ARG (0); ME (13–14)
17820	AWR	AUT	300	IRN Farsi (16–17)
	Deutsche Welle	CLN	250	SAs SAs FE (11–14)
	Deutsche Welle	RRW	250	ME Farsi (17–18)
	Deutsche Welle	RUS	250	ME Farsi (9–10)
	Radio Canada Int.	CAN	250	2-6 NAm,CAC (11–14)
	Radio Canada Int.	D	500	ME (19–20)
	Radio Canada Int.	G	300	RUS RUS (15–17)
	Radio Japan	CLN	300	ME,NAf (5)
	Radio Pakistan	PAK	100	EAf,SAf (16)
	Voice Int.	AUS	250	INS (6–9)
	Voice of America	PHL	250	EAs SAs (0–1); EAs (22–24)
	Voice of America	PHL	250	EAs Special English (1)
17825	Radio Japan	J	300	CAC CAC (2–4); NAmNAm (22–23)
	Radio Liberty (BBG)	MOR	250	CAs Turkmen (14–15)
	Radio Pakistan	PAK	100	EAf,SAf Gujarati (5–6)
	V.O. Islamic Republic of Iran	IRN	500	Eu Italian (7–8)
17830	BBC World Service	ASC	250	WAf,CAf (8–9); 1,7 WAf,CAf (10–11)
	BBC World Service	ASC	250	WAf,CAf (12–15)
	BBC World Service	SEY	250	WAf (21–22)
	Radio Damascus	SYR	500	NAm,Eu (1–4); ME RUS NAm,Eu (16–21)
	Radio Damascus	SYR	500	AUS (21–22)
	Radio Dubai	UAE	500	SEA,FE (5–6)
	Radio Dubai	UAE	500	SEA,FE (6–7)
	Radio Veritas Asia	PHL	250	As (2–3)
	Voice of Turkey	TUR	500	Eu (12–13)
17835	AWR	UAE	250	CHN (11–12)
	Deutsche Welle	D	500	FE (11)
	Deutsche Welle	D	500	ME Dari (13)
	Deutsche Welle	POR	250	ME Pashto (13–14)
	Deutsche Welle	POR	250	Eu Albanian (14)
	Deutsche Welle	POR	250	Eu Macedonian (15)
	Deutsche Welle	POR	250	Eu Serbian (15–16)
	Radio Canada Int.	J	300	SEA CHN (22–23)
	Radio Farda (BBG)	GRC	250	ME Farsi (6–7)
	Radio Farda (BBG)	MOR	250	ME Farsi (7–8)
	Radio Free Asia (BBG)	XXX		SEA Burmese (0–1)
	Radio Japan	J	100	SAm SAm (1–4)
	Radio Pakistan	PAK	100	ME,NAf Urdu (5–7); Eu Urdu (8–9)
17840	Radio Cairo	EGY	500	WAf Fulani (19–20)
	Radio Sweden	S	500	1,7 NAm Swedish (11–12)
	Radio Sweden	S	500	NAmNAmNAm (12–14)
	Radio Sweden	S	500	NAm Swedish (13–14)
	Radio Sweden	S	500	NAm Swedish (14–15)

Legend: *Arabic*   *Chinese*   *English*   *French*   *Indonesian*   *Japanese*   *Mandarin*   *Russian*   *Spanish*   *Other*

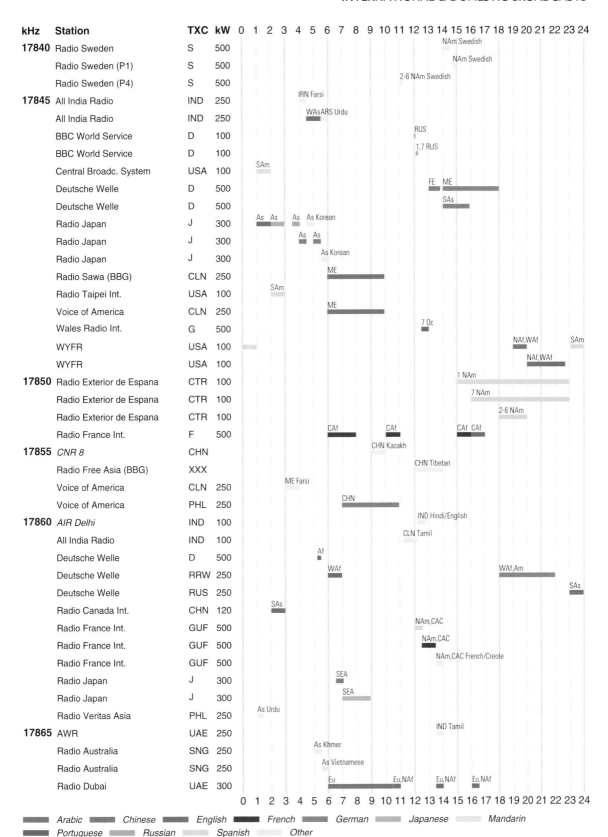

kHz	Station	TXC	kW
17840	Radio Sweden	S	500
	Radio Sweden (P1)	S	500
	Radio Sweden (P4)	S	500
17845	All India Radio	IND	250
	All India Radio	IND	250
	BBC World Service	D	100
	BBC World Service	D	100
	Central Broadc. System	USA	100
	Deutsche Welle	D	500
	Deutsche Welle	D	500
	Radio Japan	J	300
	Radio Japan	J	300
	Radio Japan	J	300
	Radio Sawa (BBG)	CLN	250
	Radio Taipei Int.	USA	100
	Voice of America	CLN	250
	Wales Radio Int.	G	500
	WYFR	USA	100
	WYFR	USA	100
17850	Radio Exterior de Espana	CTR	100
	Radio Exterior de Espana	CTR	100
	Radio Exterior de Espana	CTR	100
	Radio France Int.	F	500
17855	*CNR 8*	CHN	
	Radio Free Asia (BBG)	XXX	
	Voice of America	CLN	250
	Voice of America	PHL	250
17860	*AIR Delhi*	IND	100
	All India Radio	IND	100
	Deutsche Welle	D	500
	Deutsche Welle	RRW	250
	Deutsche Welle	RUS	250
	Radio Canada Int.	CHN	120
	Radio France Int.	GUF	500
	Radio France Int.	GUF	500
	Radio France Int.	GUF	500
	Radio Japan	J	300
	Radio Japan	J	300
	Radio Veritas Asia	PHL	250
17865	AWR	UAE	250
	Radio Australia	SNG	250
	Radio Australia	SNG	250
	Radio Dubai	UAE	300

Legend: *Arabic* · *Chinese* · *English* · *French* · *German* · *Japanese* · *Mandarin* · *Portuguese* · *Russian* · *Spanish* · *Other*

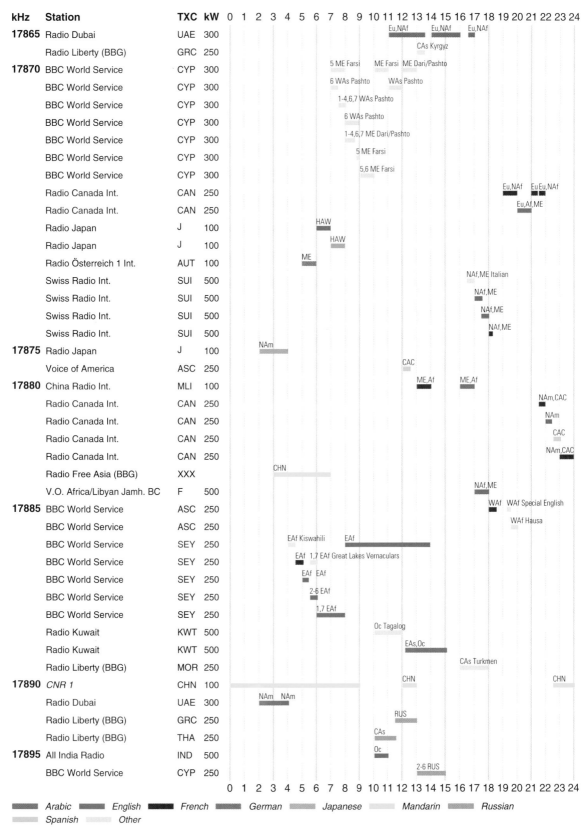

kHz	Station	TXC	kW	Schedule (0-24)
17865	Radio Dubai	UAE	300	Eu,NAf / Eu,NAf / Eu,NAf
	Radio Liberty (BBG)	GRC	250	CAs Kyrgyz
17870	BBC World Service	CYP	300	5 ME Farsi / ME Farsi / ME Dari/Pashto
	BBC World Service	CYP	300	6 WAs Pashto / WAs Pashto
	BBC World Service	CYP	300	1-4,6,7 WAs Pashto
	BBC World Service	CYP	300	6 WAs Pashto
	BBC World Service	CYP	300	1-4,6,7 ME Dari/Pashto
	BBC World Service	CYP	300	5 ME Farsi
	BBC World Service	CYP	300	5,6 ME Farsi
	Radio Canada Int.	CAN	250	Eu,NAf / Eu Eu,NAf
	Radio Canada Int.	CAN	250	Eu,Af,ME
	Radio Japan	J	100	HAW
	Radio Japan	J	100	HAW
	Radio Österreich 1 Int.	AUT	100	ME
	Swiss Radio Int.	SUI	500	NAf,ME Italian
	Swiss Radio Int.	SUI	500	NAf,ME
	Swiss Radio Int.	SUI	500	NAf,ME
	Swiss Radio Int.	SUI	500	NAf,ME
17875	Radio Japan	J	100	NAm
	Voice of America	ASC	250	CAC
17880	China Radio Int.	MLI	100	ME,Af / ME,Af
	Radio Canada Int.	CAN	250	NAm,CAC
	Radio Canada Int.	CAN	250	NAm
	Radio Canada Int.	CAN	250	CAC
	Radio Canada Int.	CAN	250	NAm,CAC
	Radio Free Asia (BBG)	XXX		CHN
	V.O. Africa/Libyan Jamh. BC	F	500	NAf,ME
17885	BBC World Service	ASC	250	WAf / WAf Special English
	BBC World Service	ASC	250	WAf Hausa
	BBC World Service	SEY	250	EAf Kiswahili / EAf
	BBC World Service	SEY	250	EAf 1,7 EAf Great Lakes Vernaculars
	BBC World Service	SEY	250	EAf EAf
	BBC World Service	SEY	250	2-6 EAf
	BBC World Service	SEY	250	1,7 EAf
	Radio Kuwait	KWT	500	Oc Tagalog
	Radio Kuwait	KWT	500	EAs,Oc
	Radio Liberty (BBG)	MOR	250	CAs Turkmen
17890	*CNR 1*	CHN	100	CHN / CHN / CHN
	Radio Dubai	UAE	300	NAm NAm
	Radio Liberty (BBG)	GRC	250	RUS
	Radio Liberty (BBG)	THA	250	CAs
17895	All India Radio	IND	500	Oc
	BBC World Service	CYP	250	2-6 RUS

Arabic　English　French　German　Japanese　Mandarin　Russian
Spanish　Other

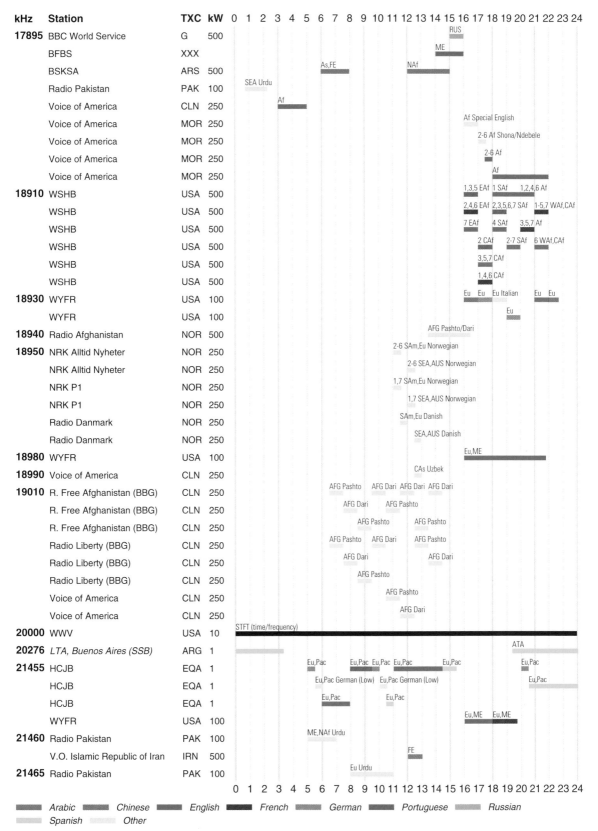

kHz	Station	TXC	kW
17895	BBC World Service	G	500
	BFBS	XXX	
	BSKSA	ARS	500
	Radio Pakistan	PAK	100
	Voice of America	CLN	250
	Voice of America	MOR	250
	Voice of America	MOR	250
	Voice of America	MOR	250
	Voice of America	MOR	250
18910	WSHB	USA	500
	WSHB	USA	500
	WSHB	USA	500
	WSHB	USA	500
	WSHB	USA	500
	WSHB	USA	500
18930	WYFR	USA	100
	WYFR	USA	100
18940	Radio Afghanistan	NOR	500
18950	NRK Alltid Nyheter	NOR	250
	NRK Alltid Nyheter	NOR	250
	NRK P1	NOR	250
	NRK P1	NOR	250
	Radio Danmark	NOR	250
	Radio Danmark	NOR	250
18980	WYFR	USA	100
18990	Voice of America	CLN	250
19010	R. Free Afghanistan (BBG)	CLN	250
	R. Free Afghanistan (BBG)	CLN	250
	R. Free Afghanistan (BBG)	CLN	250
	Radio Liberty (BBG)	CLN	250
	Radio Liberty (BBG)	CLN	250
	Radio Liberty (BBG)	CLN	250
	Voice of America	CLN	250
	Voice of America	CLN	250
20000	WWV	USA	10
20276	LTA, Buenos Aires (SSB)	ARG	1
21455	HCJB	EQA	1
	HCJB	EQA	1
	HCJB	EQA	1
	WYFR	USA	100
21460	Radio Pakistan	PAK	100
	V.O. Islamic Republic of Iran	IRN	500
21465	Radio Pakistan	PAK	100

Legend: Arabic, Chinese, English, French, German, Portuguese, Russian, Spanish, Other

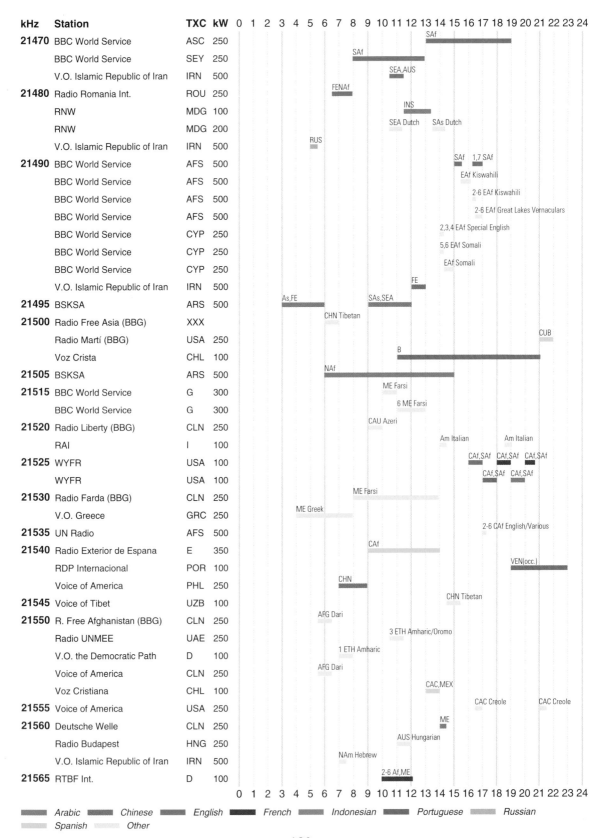

kHz	Station	TXC	kW
21470	BBC World Service	ASC	250
	BBC World Service	SEY	250
	V.O. Islamic Republic of Iran	IRN	500
21480	Radio Romania Int.	ROU	250
	RNW	MDG	100
	RNW	MDG	200
	V.O. Islamic Republic of Iran	IRN	500
21490	BBC World Service	AFS	500
	BBC World Service	AFS	500
	BBC World Service	AFS	500
	BBC World Service	AFS	500
	BBC World Service	CYP	250
	BBC World Service	CYP	250
	BBC World Service	CYP	250
	V.O. Islamic Republic of Iran	IRN	500
21495	BSKSA	ARS	500
21500	Radio Free Asia (BBG)	XXX	
	Radio Martí (BBG)	USA	250
	Voz Crista	CHL	100
21505	BSKSA	ARS	500
21515	BBC World Service	G	300
	BBC World Service	G	300
21520	Radio Liberty (BBG)	CLN	250
	RAI	I	100
21525	WYFR	USA	100
	WYFR	USA	100
21530	Radio Farda (BBG)	CLN	250
	V.O. Greece	GRC	250
21535	UN Radio	AFS	500
21540	Radio Exterior de Espana	E	350
	RDP Internacional	POR	100
	Voice of America	PHL	250
21545	Voice of Tibet	UZB	100
21550	R. Free Afghanistan (BBG)	CLN	250
	Radio UNMEE	UAE	250
	V.O. the Democratic Path	D	100
	Voice of America	CLN	250
	Voz Cristiana	CHL	100
21555	Voice of America	USA	250
21560	Deutsche Welle	CLN	250
	Radio Budapest	HNG	250
	V.O. Islamic Republic of Iran	IRN	500
21565	RTBF Int.	D	100

Chart labels (left to right by frequency row):

- 21470 BBC World Service ASC: SAf
- BBC World Service SEY: SAf
- V.O. Islamic Republic of Iran IRN: SEA,AUS
- 21480 Radio Romania Int. ROU: FENAf
- RNW MDG 100: INS
- RNW MDG 200: SEA Dutch / SAs Dutch
- V.O. Islamic Republic of Iran IRN: RUS
- 21490 BBC World Service AFS: SAf / 1,7 SAf
- BBC World Service AFS: EAf Kiswahili
- BBC World Service AFS: 2-6 EAf Kiswahili
- BBC World Service AFS: 2-6 EAf Great Lakes Vernaculars
- BBC World Service CYP: 2,3,4 EAf Special English
- BBC World Service CYP: 5,6 EAf Somali
- BBC World Service CYP: EAf Somali
- V.O. Islamic Republic of Iran IRN: FE
- 21495 BSKSA ARS: As,FE / SAs,SEA
- 21500 Radio Free Asia (BBG) XXX: CHN Tibetan
- Radio Martí (BBG) USA: CUB
- Voz Crista CHL: B
- 21505 BSKSA ARS: NAf
- 21515 BBC World Service G: ME Farsi
- BBC World Service G: 6 ME Farsi
- 21520 Radio Liberty (BBG) CLN: CAU Azeri
- RAI I: Am Italian / Am Italian
- 21525 WYFR USA: CAf,SAf / CAf,SAf / CAf,SAf
- WYFR USA: CAf,SAf / CAf,SAf
- 21530 Radio Farda (BBG) CLN: ME Farsi
- V.O. Greece GRC: ME Greek
- 21535 UN Radio AFS: 2-6 CAf English/Various
- 21540 Radio Exterior de Espana E: CAf
- RDP Internacional POR: VEN(occ.)
- Voice of America PHL: CHN
- 21545 Voice of Tibet UZB: CHN Tibetan
- 21550 R. Free Afghanistan (BBG) CLN: AFG Dari
- Radio UNMEE UAE: 3 ETH Amharic/Oromo
- V.O. the Democratic Path D: 1 ETH Amharic
- Voice of America CLN: AFG Dari
- Voz Cristiana CHL: CAC,MEX
- 21555 Voice of America USA: CAC Creole / CAC Creole
- 21560 Deutsche Welle CLN: ME
- Radio Budapest HNG: AUS Hungarian
- V.O. Islamic Republic of Iran IRN: NAm Hebrew
- 21565 RTBF Int. D: 2-6 Af,ME

Legend: Arabic — Chinese — English — French — Indonesian — Portuguese — Russian — Spanish — Other

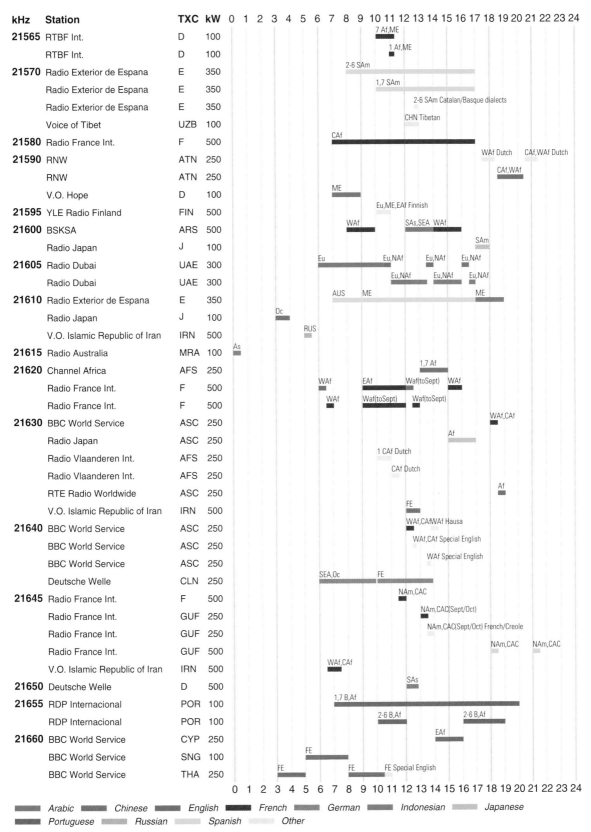

kHz	Station	TXC	kW	Details
21565	RTBF Int.	D	100	7 Af,ME
	RTBF Int.	D	100	1 Af,ME
21570	Radio Exterior de Espana	E	350	2-6 SAm
	Radio Exterior de Espana	E	350	1,7 SAm
	Radio Exterior de Espana	E	350	2-6 SAm Catalan/Basque dialects
	Voice of Tibet	UZB	100	CHN Tibetan
21580	Radio France Int.	F	500	CAf
21590	RNW	ATN	250	WAf Dutch / CAf,WAf Dutch
	RNW	ATN	250	CAf,WAf
	V.O. Hope	D	100	ME
21595	YLE Radio Finland	FIN	500	Eu,ME,EAf Finnish
21600	BSKSA	ARS	500	WAf / SAs,SEA WAf
	Radio Japan	J	100	SAm
21605	Radio Dubai	UAE	300	Eu / Eu,NAf Eu,NAf Eu,NAf
	Radio Dubai	UAE	300	Eu,NAf Eu,NAf Eu,NAf
21610	Radio Exterior de Espana	E	350	AUS ME ME
	Radio Japan	J	100	Oc
	V.O. Islamic Republic of Iran	IRN	500	RUS
21615	Radio Australia	MRA	100	As
21620	Channel Africa	AFS	250	1,7 Af
	Radio France Int.	F	500	WAf EAf Waf(toSept) WAf
	Radio France Int.	F	500	WAf Waf(toSept) Waf(toSept)
21630	BBC World Service	ASC	250	WAf,CAf
	Radio Japan	ASC	250	Af
	Radio Vlaanderen Int.	AFS	250	1 CAf Dutch
	Radio Vlaanderen Int.	AFS	250	CAf Dutch
	RTE Radio Worldwide	ASC	250	Af
	V.O. Islamic Republic of Iran	IRN	500	FE
21640	BBC World Service	ASC	250	WAf,CAfWAf Hausa
	BBC World Service	ASC	250	WAf,CAf Special English
	BBC World Service	ASC	250	WAf Special English
	Deutsche Welle	CLN	250	SEA,Oc FE
21645	Radio France Int.	F	500	NAm,CAC
	Radio France Int.	GUF	250	NAm,CAC(Sept/Oct)
	Radio France Int.	GUF	250	NAm,CAC(Sept/Oct) French/Creole
	Radio France Int.	GUF	500	NAm,CAC NAm,CAC
	V.O. Islamic Republic of Iran	IRN	500	WAf,CAf
21650	Deutsche Welle	D	500	SAs
21655	RDP Internacional	POR	100	1,7 B,Af
	RDP Internacional	POR	100	2-6 B,Af 2-6 B,Af
21660	BBC World Service	CYP	250	EAf
	BBC World Service	SNG	100	FE
	BBC World Service	THA	250	FE FE FE Special English

0 1 2 3 4 5 6 7 8 9 10 11 12 13 14 15 16 17 18 19 20 21 22 23 24

Legend: Arabic, Chinese, English, French, German, Indonesian, Japanese, Portuguese, Russian, Spanish, Other

THE SHORTWAVE GUIDE

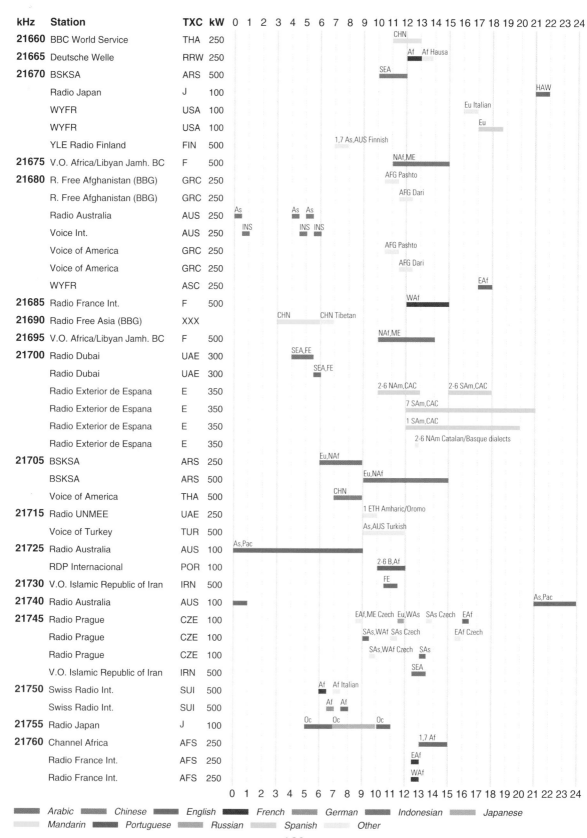

kHz	Station	TXC	kW	Schedule
21660	BBC World Service	THA	250	CHN
21665	Deutsche Welle	RRW	250	Af / Af Hausa
21670	BSKSA	ARS	500	SEA
	Radio Japan	J	100	HAW
	WYFR	USA	100	Eu Italian
	WYFR	USA	100	Eu
	YLE Radio Finland	FIN	500	1,7 As,AUS Finnish
21675	V.O. Africa/Libyan Jamh. BC	F	500	NAf,ME
21680	R. Free Afghanistan (BBG)	GRC	250	AFG Pashto
	R. Free Afghanistan (BBG)	GRC	250	AFG Dari
	Radio Australia	AUS	250	As / As As
	Voice Int.	AUS	250	INS / INS INS
	Voice of America	GRC	250	AFG Pashto
	Voice of America	GRC	250	AFG Dari
	WYFR	ASC	250	EAf
21685	Radio France Int.	F	500	WAf
21690	Radio Free Asia (BBG)	XXX		CHN / CHN Tibetan
21695	V.O. Africa/Libyan Jamh. BC	F	500	NAf,ME
21700	Radio Dubai	UAE	300	SEA,FE
	Radio Dubai	UAE	300	SEA,FE
	Radio Exterior de Espana	E	350	2-6 NAm,CAC / 2-6 SAm,CAC
	Radio Exterior de Espana	E	350	7 SAm,CAC
	Radio Exterior de Espana	E	350	1 SAm,CAC
	Radio Exterior de Espana	E	350	2-6 NAm Catalan/Basque dialects
21705	BSKSA	ARS	250	Eu,NAf
	BSKSA	ARS	500	Eu,NAf
	Voice of America	THA	500	CHN
21715	Radio UNMEE	UAE	250	1 ETH Amharic/Oromo
	Voice of Turkey	TUR	500	As,AUS Turkish
21725	Radio Australia	AUS	100	As,Pac
	RDP Internacional	POR	100	2-6 B,Af
21730	V.O. Islamic Republic of Iran	IRN	500	FE
21740	Radio Australia	AUS	100	As,Pac
21745	Radio Prague	CZE	100	EAf,ME Czech / Eu,WAs / SAs Czech / EAf
	Radio Prague	CZE	100	SAs,WAf SAs Czech / EAf Czech
	Radio Prague	CZE	100	SAs,WAf Czech / SAs
	V.O. Islamic Republic of Iran	IRN	500	SEA
21750	Swiss Radio Int.	SUI	500	Af / Af Italian
	Swiss Radio Int.	SUI	500	Af Af
21755	Radio Japan	J	100	Oc Oc Oc
21760	Channel Africa	AFS	250	1,7 Af
	Radio France Int.	AFS	250	EAf
	Radio France Int.	AFS	250	WAf

Legend (languages): Arabic, Chinese, English, French, German, Indonesian, Japanese, Mandarin, Portuguese, Russian, Spanish, Other

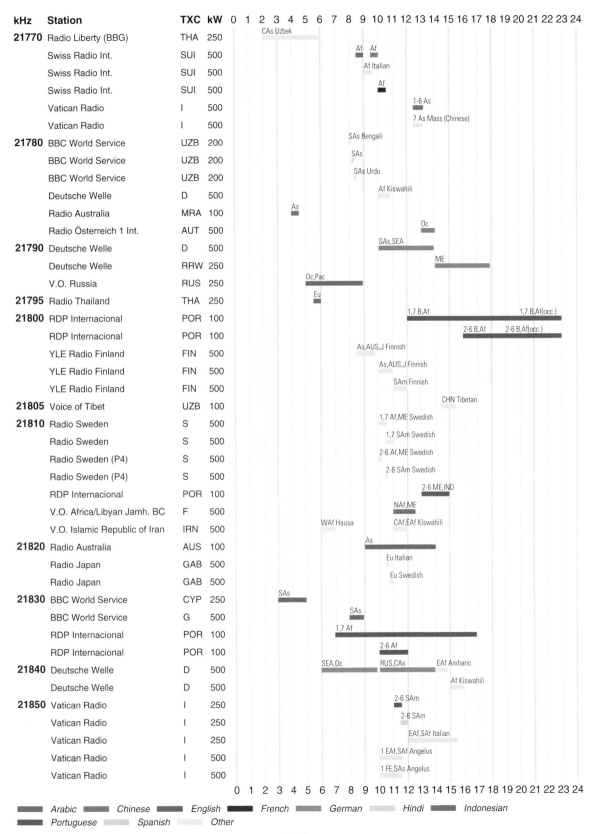

kHz	Station	TXC	kW	Details
21770	Radio Liberty (BBG)	THA	250	CAs Uzbek
	Swiss Radio Int.	SUI	500	Af Af
	Swiss Radio Int.	SUI	500	Af Italian
	Swiss Radio Int.	SUI	500	Af
	Vatican Radio	I	500	1-6 As
	Vatican Radio	I	500	7 As Mass (Chinese)
21780	BBC World Service	UZB	200	SAs Bengali
	BBC World Service	UZB	200	SAs
	BBC World Service	UZB	200	SAs Urdu
	Deutsche Welle	D	500	Af Kiswahili
	Radio Australia	MRA	100	As
	Radio Österreich 1 Int.	AUT	500	Oc
21790	Deutsche Welle	D	500	SAs,SEA
	Deutsche Welle	RRW	250	ME
	V.O. Russia	RUS	250	Oc,Pac
21795	Radio Thailand	THA	250	Eu
21800	RDP Internacional	POR	100	1,7 B,Af 1,7 B,Af(occ.)
	RDP Internacional	POR	100	2-6 B,Af 2-6 B,Af(occ.)
	YLE Radio Finland	FIN	500	As,AUS,J Finnish
	YLE Radio Finland	FIN	500	As,AUS,J Finnish
	YLE Radio Finland	FIN	500	SAm Finnish
21805	Voice of Tibet	UZB	100	CHN Tibetan
21810	Radio Sweden	S	500	1,7 Af,ME Swedish
	Radio Sweden	S	500	1,7 SAm Swedish
	Radio Sweden (P4)	S	500	2-6 Af,ME Swedish
	Radio Sweden (P4)	S	500	2-6 SAm Swedish
	RDP Internacional	POR	100	2-6 ME,IND
	V.O. Africa/Libyan Jamh. BC	F	500	NAf,ME
	V.O. Islamic Republic of Iran	IRN	500	WAf Hausa CAf,EAf Kiswahili
21820	Radio Australia	AUS	100	As
	Radio Japan	GAB	500	Eu Italian
	Radio Japan	GAB	500	Eu Swedish
21830	BBC World Service	CYP	250	SAs
	BBC World Service	G	500	SAs
	RDP Internacional	POR	100	1,7 Af
	RDP Internacional	POR	100	2-6 Af
21840	Deutsche Welle	D	500	SEA,Oc RUS,CAs EAf Amharic
	Deutsche Welle	D	500	Af Kiswahili
21850	Vatican Radio	I	250	2-6 SAm
	Vatican Radio	I	250	2-6 SAm
	Vatican Radio	I	250	EAf,SAf Italian
	Vatican Radio	I	500	1 EAf,SAf Angelus
	Vatican Radio	I	500	1 FE,SAs Angelus

Arabic   Chinese   English   French   German   Hindi   Indonesian
Portuguese   Spanish   Other

kHz	Station	TXC	kW	0	1	2	3	4	5	6	7	8	9	10	11	12	13	14	15	16	17	18	19	20	21	22	23	24
21850	Vatican Radio	I	500											1 Af,As Angelus														
25820	Radio France Int.	F	500										Waf(Sept/Oct)		Waf(Sept/Oct)													
	Radio France Int.	F	500													Waf(Sept/Oct)												
29810	*LTA, Buenos Aires (SSB)*	ARG	1													ATA												

English ▬▬▬ French ▬▬▬ Spanish ▭▭▭ Other ░░░

Reference Section

World Time Table

The differences marked + indicate the number of hours ahead of UTC. Differences marked – indicate the number of hours behind UTC. Variations from standard time during part of the year (in some countries referred to as Summer Time) are decided annually and may vary from year to year. N=Normal Time; S=Summer Time.

	N	S
Afghanistan	+4½	+4½
Alaska	-9	-8
	-10	-9
Albania	+1	+2
Algeria	+1	+1
Andorra	+1	+2
Angola	+1	+1
Anguilla	-4	-4
Antigua	-4	-4
Argentina (Ea.)	-3	-2
Argentina (rest)	-3	-3
Armenia	+4	+4
Aruba	-4	-4
Ascension Isl.	UTC	UTC
Australia		
Victoria & NSW	+10	+11
Queensland	+10	+10
Tasmania	+10	+11
N. Territory	+9½	+9½
S. Australia	+9½	+10½
(we. part)	+9	+10
W. Australia	+8	+8
Austria	+1	+2
Azerbaijan	+3	+4
Azores	-1	UTC
Bahamas	-5	-4
Bahrain	+3	+3
Bangladesh	+6	+6
Barbados	-4	-4
Belarus	+2	+3
Belgium	+1	+2
Belize	-6	-6
Benin	+1	+1
Bermuda	-4	-3
Bhutan	+6	+6
Bolivia	-4	-4
Bosnia/Hercegovina	+1	+2
Botswana	+2	+2
Brazil		
a) Oceanic Isl	-2	-2
b) Ea & Coastal	-3	-2
c) Manaos	-4	-3
d) Acre	-5	-4
Brunei	+8	+8
Bulgaria	+2	+3
Burkina Faso	UTC	UTC
Burundi	+2	+2
Cameroon	+1	+1
Canada		
a) NF, Labrador	-3½	-2½
(So. Ea.)		
b) Labrador (rest),	-4	-3
NS, NB, PEI		
c) ON, PQ	-5	-4
d) MB	-6	-5
e) AB, NWT	-7	-6
f) BC, YT	-8	-7
Cambodia	+7	+7
Canary Isl.	UTC	+1
Cape Verde Isl.	-1	-1
Cayman Isl.	-5	-4
Ce. African Rep.	+1	+1
Chad	+1	+1
Chile	-4	-3
China (P.R.)		
Beijing	+8	+9
Urumqi	+6	+7
Christmas Isl.	+7	+7
Cocos Isl.	+6½	+6½
Colombia	-5	-5
Comoro Rep.	+3	+3
Congo	+1	+1

	N	S
Congo (Dem. Rep)		
Kinshasa	+1	+1
Lubumbashi	+2	+2
Cook Isl.	-10	-9½
Costa Rica	-6	-5
Côte d'Ivoire	UTC	UTC
Croatia	+1	+2
Cuba	-5	-4
Cyprus	+2	+3
Czech Rep.	+1	+2
Denmark	+1	+2
Diego Garcia	+5	+5
Djibouti	+3	+3
Dominica	-4	-4
Dom. Rep.	-4	-4
Easter Isl.	-6	-5
East Timor	+8	+8
Ecuador	-5	-5
Egypt	+2	+2
El Salvador	-6	-6
Equatorial Guinea	+1	+1
Estonia	+2	+3
Ethiopia	+3	+3
Falkland Isl.	-4	-4
(Port Stanley)	-4	-3
Faroe Isl.	UTC	+1
Fiji	+12	+12
Finland	+2	+3
France	+1	+2
Gabon	+1	+1
Galapagos Isl.	-6	-6
Gambia	UTC	UTC
Georgia*	+4	+5
(* exc. Abkhasia	+3	+4)
Germany	+1	+2
Ghana	UTC	UTC
Gibraltar	+1	+2
Greece	+2	+3
Greenland		
Scoresbysund	-1	UTC
Thule area	-3	-3
Other areas	-3	-2
Grenada	-4	-4
Guadeloupe	-4	-4
Guam	+10	+10
Guatemala	-6	-5
Guiana (French)	-3	-3
Guinea (Rep.)	UTC	UTC
Guinea Bissau	UTC	UTC
Guyana (Rep.)	-3	-3
Haiti	-5	-4
Hawaii	-10	-10
Honduras (Rep.)	-6	-6
Hong Kong	+8	+8
Hungary	+1	+2
Iceland	UTC	UTC
India	+5½	+5½
Indonesia		
a) Java, Sumatra,		
W&C Kalimantan	+7	+7
b) Bali, S&E Kalimantan,		
Sulawesi, W. Timor	+8	+8
c) Moluccas,		
Irian Jaya	+9	+9
Iran	+3½	+4½
Iraq	+3	+4
Ireland	UTC	+1
Israel	+2	+3
Italy	+1	+2
Jamaica	-5	-4
Japan	+9	+9
Jordan	+2	+3

	N	S
Kenya	+3	+3
Kazakhstan	+6	+7
Kiribati	+12	+12
Korea (Rep.)	+9	+10
Korea (D.P.R.)	+9	+9
Kuwait	+3	+3
Kyrgyzstan	+5	+6
Laos	+7	+7
Latvia	+2	+3
Lebanon	+2	+3
Lesotho	+2	+2
Liberia	UTC	UTC
Libya	+1	+2
Lithuania	+2	+2
Lord Howe Isl.	+10½	+11
Luxembourg	+1	+2
Macau	+8	+8
Macedonia	+1	+2
Madagascar	+3	+3
Madeira	UTC	UTC
Malawi	+2	+2
Malaysia	+8	+8
Maldive Isl.	+5	+5
Mali	UTC	UTC
Malta	+1	+2
Marshall Isl.	+12	+12
Martinique	-4	-4
Mauritania	UTC	UTC
Mauritius	+4	+4
Mayotte	+3	+3
Mexico	-6	-6
(see country section		
for exceptions)		
Micronesia		
Truk, Yap	+10	+10
Pohnpei	+11	+11
Midway Isl.	-11	-11
Moldova	+2	+3
Monaco	+1	+2
Mongolia	+8	+9
Montserrat	-4	-4
Morocco	UTC	UTC
Mozambique	+2	+2
Myanmar	+6½	+6½
Namibia	+1	+2
Nauru	+11	+11
Nepal	+5¾	+5¾
Netherlands	+1	+2
Neth. Antilles	-4	-4
New Caledonia	+11	+11
New Zealand	+12	+13
Nicaragua	-6	-6
Niger	+1	+1
Nigeria	+1	+1
Niue	-11	-11
Norfolk Isl.	+11½	+11½
N. Marianas	+10	+10
Norway	+1	+2
Oman	+4	+4
Pakistan	+5	+5
Palau	+9	+9
Panama	-5	-5
Papua N. Guinea	+10	+10
Paraguay	-4	-3
Peru	-5	-4
Philippines	+8	+8
Poland	+1	+2
Polynesia (Fr.)	-10	-10
Portugal	UTC	+1
Puerto Rico	-4	-4
Qatar	+3	+3

	N	S
Reunion	+4	+4
Romania	+2	+3
Russia		
Moscow	+3	+4
Novosibirsk	+7	+8
Khabarovsk	+10	+11
Petropavlovsk	+12	+13
Rwanda	+2	+2
Samoa Isl.	-11	-11
S. Tomé	UTC	UTC
Saudi Arabia	+3	+3
Senegal	UTC	UTC
Serbia & Montenegro	+1	+2
Seychelles	+4	+4
Sierra Leone	UTC	UTC
Singapore	+8	+8
Slovakia	+1	+2
Slovenia	+1	+2
Solomon Isl.	+11	+11
Somalia	+3	+3
S. Africa	+2	+2
Spain	+1	+2
Sri Lanka	+6	+6
St. Helena	UTC	UTC
St. Kitts- Nevis	-4	-4
St. Lucia	-4	-4
St. Pierre	-3	-2
St. Vincent	-4	-4
Sudan	+2	+2
Suriname	-3	-3
Swaziland	+2	+2
Sweden	+1	+2
Switzerland	+1	+2
Syria	+2	+3
Tajikistan	+5	+5
Taiwan	+8	+8
Tanzania	+3	+3
Thailand	+7	+7
Togo	UTC	UTC
Tonga	+13	+13
Transkei	+2	+2
Trinidad	-4	-4
Tristan da Cunha	UTC	UTC
Tunisia	+1	+2
Turks & Caicos	-4	-4
Turkey	+2	+3
Turkmenistan	+5	+5
Tuvalu	+12	+12
Uganda	+3	+3
Ukraine	+2	+3
United Arab Em.	+4	+4
United Kingdom	UTC	+1
Uruguay	-3	-2
USA		
a) Eastern*	-5	-4
*) Indiana	-5	-5
b) Central	-6	-5
c) Mountain*	-7	-6
*) Arizona	-7	-7
d) Pacific	-8	-7
Uzbekistan	+5	+5
Vanuatu	+11	+12
Venezuela	-4	-4
Vietnam	+7	+7
Virgin Isl.	-4	-4
Wake Isl.	+12	+12
Wallis & Futuna	+12	+12
Yemen	+3	+3
Zambia	+2	+2
Zimbabwe	+2	+2

Directory of International Broadcasters

ADVENTIST WORLD RADIO (AWR AFRICA/ASIA) (Rlg)
SOUTH AFRICA
✉   AWR Africa Region, PO Box 75700, Garden View 2047, Gauteng, South Africa.
☎   +27 (11) 706 9576/9578
🖹   +27 (11) 706 8819
E-mail: africa@awr.org
Web: www.awr.org
LP: Region Director: Samuel Misiani; Engineer: Kurt Roberts.

ADVENTIST WORLD RADIO (AWR EUROPE) (Rlg)
UNITED KINGDOM
✉   AWR, "Whitegates", St Marks Road, Binfield, Berkshire. RG42 4AT, United Kingdom
☎   +44 (1344) 401401
🖹   +44 (1344) 401419
E-mail: europe@awr.org
Web: www.awr.org
LP: Region Director: Bert Smit.

ADVENTIST WORLD RADIO (KSDA) (Rlg)
GUAM
✉   P.O. Box 8990, Agat, Guam 96928
☎   +671 565 2000
🖹   +671 565 2983
E-mail: aproffice@awr.org
Web: www.awr.org
LP: Region Director: Akinori Kaibe; Site Manager and Chief Engineer: Brook Powers
V: QSL-card. Rp.
Ann: English: "From the Beautiful island of Guam in the West Pacific, this is Adventist World Radio, the Voice of Hope"

ADVENTIST WORLD RADIO, NORTH AMERICA (Rlg)
UNITED STATES
✉   AWR HQ: 12501 Old Columbia Pike, Silver Spring, MD 20904-6600, USA
E-mail: info@awr.org
Web: www.awr.org
Notes: AWR is the international radio service of the General Conference of Seventh Day Adventists.

AFRICA NO.1 (Comm)
GABON
✉   Africa No 1, BP 1, Libreville, Gabon
☎   +241 760001
🖹   +241 742133

E-mail: africa@africa1.com
Web: www.africa1.com
LP: President: Louis Barthélémy Mapangou; Dir. Délegué: Michel Koumbangoye, Pierre Devoluy; Dir. Tech: Gaston Ombolo Ki-Obo; Dir. Prgrs. & Adv: Augustin Letamba; Dir. Inf: Jean Valère Mbina Mandza.
V: QSL-letter (for own programmes only).
Notes: Main News: 0530, 0630, 0730, 1115 (W), 1215, 1700, 1830, 2200.

ALL INDIA RADIO (Gov)
INDIA
✉   External Services, P.O. Box 500, New Delhi-110001, India
☎   +91 (11) 3715411
🖹   +91 (11) 3710057
E-mail: airlive@air.org.in
Web: http://allindiaradio.org
LP: Director: P.M. Iyer
V: QSL-card. Reception reports on technical quality of the programmes are welcome and may be sent to the Director (Spectrum Management), All India Radio, Room No.204, Akashvani Bhavan, New Delhi-110001, India. Telefax: +91 (11) 23421062, 23421145.E-mail:faair@nda.vsnl.net.in
Ann: English: "This is the General Overseas Service of All India Radio"; Hindi: "Yeh Akashvani ki videsh prasaran sewa hai"; Tamil: "Idi Akashvani videsh sewai"; Sinhala: "Me All India Radio videshiya sevayai"; Nepali: "Yo All India Radio ho"; Dari: "Injaw Delhi"; Pashto: "Da All India Radio de"; Farsi: "Inja Delhi"; Indonesian: "Inilah All India Radio"

AMATEUR RADIO MIRROR INTERNATIONAL
SOUTH AFRICA
✉   P.O. Box 90438, Garsfontein 0042, South Africa
E-mail: armi@sarl.org.za
Web: www.sarl.org.za/public/ARMI/ARMI.asp
V: QSL-card.
Notes: AMRI is a weekly programme by the South African Radio League.

ARMED FORCES RADIO AND TELEVISION SERVICE (AFN) (Gov)
UNITED STATES
✉   AFRTS Broadcast Center, 1363 Z Street, Bldg. 2730, March Air Reserve Base, CA 92518-2017, USA.

☎ +1 (909) 413 2236
E-mail: QSL@mediacen.navy.mil
Web: www.myafn.net/radio/shortwave
V: QSL-card. Reception reports to:
QSL@mediacen.navy.mil
Ann: English: "You're listening to AFN"
Notes: AFRTS provides radio and television pro-
gramming to US service personnel and their
families serving outside the continental
United States.

AR RADIO INTERCONTINENTAL (Comm)
ARMENIA
✉ Varanants Street 28-34, 375070 Yerevan,
Armenia.
☎ +374 (1) 558010
E-mail: armen@arm.r.am
Web: www.mediacenter.am
LP: Director: Armen Amirian
V: QSL-card.
Notes: Rebroadcasts religious programmes from
Germany and Switzerland (in German).

BACK TO THE BIBLE (Rlg)
UNITED STATES
✉ PO Box 82808, Lincoln NE 68501, USA
☎ +1 (402) 464 7200
E-mail: info@backtothebible.org
Web: www.backtothebibble.org
LP: Dr Woodrow Kroll.
Notes: Operated by Good News Broadcasting
Association.

BANGLADESH BETAR (Gov)
BANGLADESH
✉ Bangladesh Betar, External Services, P.O. Box
2204, Dhaka 1000, Bangladesh
☎ +880 (2) 505113, 8625538
▤ +880 (2) 8612021
E-mail: dgbetar@bd.drik.net; rrc@aitlbd.net
LP: Mrs Dilruba Begum
V: QSL-letter. Re. To: Senior Engineer
(Research Wing)
IS: Local composition of violin and tanpura
Ann: English: "This is the External Service of
Bangladesh Betar"
Notes: DX Programme in English: 2nd & 4th Friday
in month at 1230-1300, 1815-1900

BBC WORLD SERVICE (Gov)
UNITED KINGDOM
✉ P.O. Box 76, Bush House, Strand, London
WC2B 4PH, United Kingdom
☎ +44 (20) 7240 3456
▤ +44 (20) 7557 1258
E-mail: worldservice.letters@bbc.co.uk
Web: www.bbc.co.uk/worldservice
LP: C.E.: Mark Byford; Deputy Dir., WS: Nigel
Chapman; Dir, News & Programme
Commissioning: Bob Jobbins; Dir., Regions:
Andrew Taussig; Head of English
Programmes Commissioning: Penny Tuerk;
Head of Resources Commissioning: Chris

Gill; Financial & Commercial Dir.: Andrew
Hind; Dir. Monitoring: Andrew Hills; Chief
Personnel Officer: Kate Poulton; Heads of
Region: Benny Ammar (Europe), Barry
Langridge (Africa & Middle East), Jerry
Timmis (Americas), Elizabeth Wright
(Asia/Pacific), David Morton (EurAsia);
Controller, Marketing & Communications:
Alan Booth.
V: Does not accept reception reports.
Ann: English: "This is London" or "You are listen-
ing to the World Service of the BBC"
Notes: BBC World Service programmes in English
and other languages are relayed by local
stations in many countries. Full details can be
found in BBC "On Air" magazine, available
on subscription (see end of entry). Available
on subscription - details from Focus on Africa
Magazine. Many other publications,
promotional items, videos and audio cassettes
available from BBC World Service
Information Centre & Shop, Bush House,
Strand, London WC2B 4PH

BFBS (BRITISH FORCES BROADCASTING
SERVICE) (Gov)
UNITED KINGDOM
✉ P.O. Box 903, Gerrards Cross, SL9 8TN,
United Kingdom
Web: www.bfbs.com

BIBLE VOICE BROADCASTING (Rlg)
UNITED KINGDOM
✉ Bible Voice Broadcasting Network, P.O.Box
200, Leeds, LS26 OWW, United Kingdom
E-mail :mail@biblevoice.org
Web: www.biblevoice.org
LP: Martin Thompson
Notes: Afilliated with High Adventure Ministries. On
July 1, 2002 High Adventure Gospel Com-
munication Ministries (Canada) and Bible
Voice (UK) partnered to provide shortwave
broadcasts into the Middle East, Europe,
Africa, China, India and South East Asia.
Currently, programming is aired for over 80
hours per week in 12 languages, under the
name of Bible Voice Broadcasting.

BROADCASTING BOARD OF GOVERNORS
(BBG) (Gov)
UNITED STATES
✉ 330 Independence Avenue SW, Washington,
DC 20237 USA
☎ +1 (202) 401 3736
▤ +1 (202) 401 6605
E-mail: jmower@ibb.gov
Web: www.bbg.gov
LP: Chairman: Kenneth Y. Tomlinson
Notes: On October 1, 1999, the Broadcasting Board
Governors (BBG) became the independent,
autonomous entity responsible for all U.S.
government and government sponsored, non-
military, international broadcasting. While the

"Broadcasting Board of Governors" is the legal name given to the Federal entity encompassing all U.S international broadcasting services, the day-to-day broadcasting activities are car-ried out by the individual BBG international broadcasters: the Voice of America (VOA), Radio Sawa, Radio Farda, Radio Free Afghanistan, Radio Free Iraq, Radio Free Europe/Radio Liberty (RFE/RL), Radio Free Asia (RFA), Radio and TV Marti, and WORLDNET Television, with the assistance of the International Broadcasting Bureau (IBB).

BBG / RADIO FARDA (Gov)
UNITED STATES
✉ 1201 Connecticut Avenue NW, Washington, DC 20036, USA. Studios: Vinohradska 1, 11000 Prague 1, Czech Republic.
E-mail: comment@radiofarda.com
Web: www.radiofarda.com
Ann: Farsi: "Radiyo Rupa Azad-Radiyo Azadi"
Notes: Radio Farda is a joint service of the Farsi lan-guage sections of RFE/RL and VOA.

BBG / RADIO FREE AFGHANISTAN (Gov)
UNITED STATES
✉ 1201 Connecticut Avenue NW, Washington, DC 20036, USA. Studios: Vinohradska 1, 11000 Prague 1, Czech Republic.
E-mail: afghan@rferl.org
Web: www.azadiradio.org
Ann: Pashto: "Da Azad Afghanistan Radyo"; Dari: "Radyo Afghanistani Azad"
Notes: Part of Radio Free Europe/Radio Liberty. The RFA service broadcasts in Pashto and Dari and has been active since January 2001.

BBG / RADIO FREE ASIA (Gov)
UNITED STATES
✉ 2025 M Street NW, Suite 300, Washington, DC 20036, USA
☎ +1 (202) 530 4900
🖨 +1 (202) 530 7794
E-mail: info@rfa.org
Web: www.rfa.org
LP: President: Richard Richter; V.P. Admin & Management: Craig Perry; V.P. Programming/Exec. Editor: Dan Southerland; Chief Financial Officer: Patrick Taylor; Dir. Tech. Operations: David Baden
V: QSL-letter, schedule and sticker. QSL-cards are being considered.
Notes: Active since September 1996. Aimed at listen-ers in E/SE Asia. Due to political sensitivities, RFA does not wish to disclose the location of it's transmitter sites.

BBG / RADIO FREE EUROPE-RADIO LIBERTY (Gov)
UNITED STATES
✉ 1201 Connecticut Avenue NW, Washington, DC 20036, USA. Studios: Vinohradska 1,

11000 Prague 1, Czech Republic.
E-mail: webmaster@rferl.org
Web: www.rferl.org
LP: President: Thomas Dine; Dir.: Luke Springer; Dir. News & Current Affairs: Robert McMahon; Public Relations Coordinator: Alena Fendrychova
V: QSL-card.
IS: RFE: Romanian: "Romanian Rhapsody" by G. Enescu. RL: Russian sce: "Hymn to Freedom" by Gretchaninoff
Ann: RFE, Sign on/Sign off: Romanian: "Aici e Radio Europa libera"; RL, Azeri: "Danyshyr Asadlyk Radiosu"; Armenian: "Chosume Azatutiun Rdiokayane"; Belarusian: "Havoryts Radyyo Svaboda"; Georgian: "Laparakobs Radio Thavisupleba"; Russian: "Govorit Radio Svoboda"; Tatar: "Monda Azatlyk Radiostansyese Suyili"; Ukrainian: "Hovorit Radio Svoboda"; Uzbeki: "Azadiq RadiosidanQapiramiz"; Turkish: "Qepleyaer Azatik Radiosi"; Tadjik: "Inja Radioi Aza: "; Kazakh: "Azzatiq Radioyosinan Sövlep Turmiz"; Kyrgyz: "Azattiq Radioyosinan Söylöbüz"

BBG / RADIO FREE IRAQ (Gov)
UNITED STATES
✉ 1201 Connecticut Avenue NW, Washington, DC 20036, USA
E-mail: iraq@rferl.org
Web: www.iraqhurr.org
Ann: Arabic: "Idha'at al-Iraq al-khar min Prag"

BBG / RADIO MARTÍ (Gov)
UNITED STATES
✉ Office of Cuba Broadcasting, 4201 NW 77th Ave, Miami, FL 33166, USA
☎ +1 (305) 9941720
🖨 +1 (305) 5974665
E-mail: martinoticias@ocb.ibb.gov
Web: www.martinoticias.com
LP: Dir. Office of Cuba Broadcasting: Herminio San Ramon; Dir.., Radio Marti: Roberto Rodriguez-Tejera; Dir.. Tech. Ops.: Michael Pallone; Dir. Programs: Martha Yedra ; Dir. News: William Valdez
Ann: Spanish: "Radio Marti"
Notes: Active since May 1985. Radio Martí is the Cuban radio broadcasting service of the US government's International Broadcasting Bureau (IBB).

BBG / RADIO SAWA (Gov)
UNITED STATES
✉ 330 Independence Avenue SW, Washington, DC 20547, USA. Studios: Vinohradska 1, 11000 Prague 1, Czech Republic.
☎ +1 (202) 619 2538
🖨 +1 (202) 619 1241
E-mail: comments@radiosawa.com
Web: www.radiosawa.com
Notes: Funded by the Broadcasting Board of

Governors (BBG). Radio Sawa is directed at a younger Arab audience, broadcasting 24/7 on SW, MW, FM and Satellite.

BROADCASTING SERVICE OF THE KINGDOM OF SAUDI ARABIA (Gov)
SAUDI ARABIA
✉	P.O. Box 570, Riyadh-1116, Saudi Arabia
☎	+966 (1) 4425170
▤	+966 (1) 4041692
V:	QSL-card (Rpt. to Frequency Mgr.).
IS:	'Ud' (Oriental Lute). Opens and closes with National Anthem
Ann:	Arabic: "Ithaa till Mamlakah till Aribiyah al-Saudiyah"; English: "This is the Broadcasting Service of the Kingdom of Saudi Arabia"

CBS RADIO TAIPEI INTERNATIONAL (Gov)
TAIWAN
✉	55 Pei'an Road, Tachih, Taipei 104, Taiwan
☎	+886 (2) 2885 6168
▤	+886 (2) 2885 2254
E-mail:	cbs@cbs.org.tw
Web:	www.cbs.org.tw
LP:	Chou Tien-jui
V:	QSL-card. Rec. acc.
Ann:	English: "This is Radio Taipei International"; Indonesian: "Inilah Radio Taipei Internasional"; Japanese: "Kochirawa Taipei Kokusai Hoso, CBS, Chukaminkoku Chuohosokyoku no nihongobangumi desu."; Korean: "Chigum yorobunkkesonum RTI angugo pangsong-ul tutko kyesimnida"; Mandarin: "Cheli shih CBS, Taipei Kuochi chih Sheng". For the Voice of Asia pro-gramme, English: "This is the Voice of Asia"; Mandarin: "Cheli shih CBS, Yachou chih Sheng".
Notes:	Some frequencies may change temporarily due to interference.

CHAN TROI MOI (Rlg)
VIETNAM
✉	P.O. Box 48 Nishi Yodogawa, Osaka 555-8691, Japan; P.O. Box 957, Cypress, CA 90630, USA
E-mail:	ctm@radioctm.com
Notes:	Operated by Japanese based Chan Troi Moi. Reported to have been active since 1993. Broadcasts in Vietnamese.

CHANNEL AFRICA (Gov)
SOUTH AFRICA
✉	P.O. Box 91313, Auckland Park 2006, South Africa
☎	+27 (11) 714 2255
▤	+27 (11) 714 2072
E-mail:	africancan@channelafrica.org
Web:	www.channelafrica.org
LP:	Exec. Editor: Hans-Dieter Winkens
V:	QSL-card. Rp. Re. to Data Section.
IS:	Birds chirping and native melody.

Ann:	English: "You are tuned to the English service of Channel Africa, broadcasting to Africa from Johannesburg, South Africa"

CHINA RADIO INTERNATIONAL (Gov)
CHINA
✉	P O Box 4216, Beijing 100040, China
☎	+86 (10) 688 91579
▤	+86 (10) 688 91582
E-mail:	crieng@cri.com.cn
Web:	www.cri.com.cn
LP:	General Director: Zhang Zhenhua; C.E.: Yu Jikai; Director, English Service: Xia Jixuan
V:	QSL-card.
Ann:	Arabic: "Idha'at as-Sin ad-Duwaliyah"; Chinese: "Zhongguo guoji guangbo diantai."; English: "This is China Radio International, broadcasting from Beijing"; Esperanto: "China Radio Internacia parolas en Pekino"; German: "Hier ist Radio China International"; Indonesian: "Inilah Radio CRI, China Radio Internasional"; Japanese: "Kochirawa Pekin Hoso, Chugoku Kokusai Hosokyoku desu"; Malay: "Inilah Radio Antarabangsa China, dalam bahasa Melayu"; Mongolian: "Hyatadyn Olon Ulsyn Radio"; Spanish: "Esta es Radio Internacional de China"; Swahili: "Hii ni Radio China kimataifa"; Vietnamese: "Day la dai phatthanh quoc te Trung quoc".

CHINA RADIO, TRUE LIGHT STATION (Clan)
CHINA
✉	53 Min Chuan West Road 9th Floor, Taipei, Taiwan 10418
E-mail:	readams@usa.net
LP:	Verification Signatory: Richard Adams
V:	QSL-letter.
Notes:	Also known as Zhen Guang Dian Tai. Schedule based on monitoring reports. Uses 9450kHz from March-September. Status unknown at time of publication.

CHRISTIAN VOICE INTERNATIONAL (Rlg)
ZAMBIA
✉	Vision International, Private Bag E606, Lusaka, Zambia
☎	+260 1 274251
▤	+260 1 274526
E-mail:	cvoice@zamnet.zm
Web:	www.christianvision.com
LP:	S.M.: Charles Maboshe; Tr. Engineer: John Kawele.
Ann:	English: "This is Radio Christian Voice"
Notes:	Vision International has been transmitting since December 1994 and also operates short-wave stations in Chile and Australia.

CYPRUS BROADCASTING CORPORATION (Comm)
CYPRUS
✉	P.O. Box 4824, 1397 Nicosia, Cyprus

☎ +357 (2) 422231
🖻 +357 (2) 314050
E-mail: rik@cybc.com.cy
Web: www.cybc.com.cy
LP: D.G.: vacant. P.D.: Panos Ioannides. Dir.
Tech. Sces.: Andreas Michaelides. Head of
Radio Prgrs.: Kyriacos Charalambides. Head
of Pub. & Int. Rel.: Nayia Roussou.
V: QSL-card. Rec. acc.
IS: "Avkoritssa" (guitar)
Ann: English: "Radiofonikon Idryma Kyprou"
Notes: Active since 1953. Station serves Cypriots liv-
ing in the UK.

DEJEN RADIO (Clan)
ETHIOPIA
E-mail: articles@ethiopiancommentator.com
Web: www.ethiopiancommentator.com/dejenradio
Ann: Tigrinyan: "Dejen radio".
Notes: Active since September 2001. Formerly
known as "Voice of the Tigrayans from North
America"

DEMOCRATIC VOICE OF BURMA (Clan)
MYANMAR
✉ P.O. Box 6720, N-0130 Oslo, Norway
☎ +47 22 86 84 86
🖻 +47 22 86 84 71
E-mail: acn@dvb.no
Web: www.dvb.no
Ann: Burmese: "Democratic Myanmar a-Than"
Notes: Active since July 1992. Operated by National
Coalition Government of the Union of
Burma and assisted by other organisations.
Broadcasts in Burmese with 15 mins per
transmission allocated to Mon, Arakan, Chin,
Shan, Karen, Karenni, Kayan and Kachin
languages.

DENGE MEZOPOTAMYA (VOICE OF
MESOPOTAMIA)(Clan)
IRAQ
✉ Based in Belgium.
☎ +32 (53) 648827
🖻 +32 (53) 680779
E-mail: info@denge-mezopotamya.com
Web: www.denge-mezopotamya-com
Ann: Kurdish: "Denge Mesopotamya"
Notes: Operated by Kurdistan Workers Party (PKK)
and backed by Syria. Active since May 2001.
Broadcasts in Kurdish.

DEUTSCHE WELLE
GERMANY
✉ D-50588 Cologne, Germany
☎ +49 (221) 3890
🖻 +49 (221) 3893000
E-mail: online@dw-world.de
Web: www.dw-world.de
V: QSL-card.
Ann: German: "Hier ist der Deutsche Welle"
Notes: Also webcasting and web television.

DIGITAL RADIO MONDIALE
CONSORTIUM (DRM)
SWITZERLAND
✉ DRM Project Office, P.O.Box 360, CH-1218
Grand-Saconnex, Geneva, Switzerland
☎ +41 (2) 27172718
🖻 +41 (2) 271727189513
E-mail: projectoffice@drm.org
Web: www.drm.org
LP: Project administrator: Anne Fechner
Notes: The DRM consortium, formed in 1998, con-
sists of broadcasters and manufacturers, with
the aim to develop and promote digital radio
broadcasting on the AM bands
(LW/MW/SW) below 30 MHz.

DTK T-SYSTEMS MEDIABROADCAST (Comm)
GERMANY
✉ Am Propsthof 51, D-53121 Bonn, Germany
☎ +49 (228) 7090
E-mail: mediabroadcast.marketing@t-systems.com
Web: www.t-systems.de
Notes: Conducts DRM tests on SW.

ECHO OF HOPE (Clan)
KOREA, NORTH (DPR)
Ann: Korean: "shimang-e meiari pangsong imnida"
Notes: Operational since June 1973 (known as the
Voice of Reunification prior to that date).
Operated by National Intelligence Service.
Broadcasts in Korean.

ERTT RADIO TUNISIA (Gov)
TUNISIA
✉ 71, ave. de la Liberté, Tunis 1002, Tunesia
☎ +216 (1) 287300
🖻 +216 (1) 785146
E-mail: info@radiotunis.com
Web: www.radiotunis.com
V: by letter (Re. in English, French to Dir. Gen.
des Telecommunications. Rec. acc.)
Ann: Arabic: "Idha'atu-l-gumhuriya at-tunisiyya
Notes: Relays of National programme on Shortwave.

EVANGELIUMSRADIO HAMBURG (Rlg)
GERMANY
✉ Postfach 920741, D-21137 Hamburg,
Germany
☎ +49 (40) 7027025
🖻 +49 (40) 85408861
E-mail: evangeliums-radio-hamburg@t-online.de
Web: www.evr-hamburg.de

EVANGELIUMS-RUNDFUNK (Rlg)
GERMANY
✉ Berliner Ring 62, D-35576 Wetzlar, Germany
☎ +49 (6441) 9570
🖻 +49 (6441) 957120
E-mail: info@erf.de
Web: www.erf.de

FANG GUANG MING RADIO (Rlg)
CHINA

✉ World Falun Dafa Radio, PO Box 93436, City of Industry, CA 91715, USA.
E-mail: editor@fgmtv.net
Web: www.falundafaradio.org; www.fgmtv.net (Audio)
Ann: Mandarin: "Shijie Falun Dafa Diantai" or "Zheli shi Fang Guang Ming Diantai"
Notes: Active since July 2000. Operated by Falun Gong members in USA. May also be known as "Fang Guang Ming" or "Great Brightness Radio".

FAR EAST BROADCASTING COMPANY-KFBS (SAIPAN) (Rlg)
NORTHERN MARIANAS

✉ Box 209, Saipan, MP 96950-0209
☎ +670 322 9088
📄 +670 322 3060
E-mail: saipan@febc.org
Web: www.febc.org
LP: Director: Chris Slabaugh. P.D.: Frank Gray. C.E.: Bob Springer. Head of Public Relations: Bob Stiles.
V: QSL-card. Rp. (2 IRC's). Rec. acc.
Notes: Daily prgr: in English: 2000-1200. News on the hour.

FEBA RADIO (Rlg)
UNITED KINGDOM

✉ Ivy Arch Road, Worthing BN14 8BX, United Kingdom
☎ +44 (1903) 237281
📄 +44 (1903) 205294
E-mail: info@feba.org.uk
Web: www.feba.org.uk
LP: C.E.: Mr John Bartlett
V: QSL-card. E-mail reports to: reception@feba.org.uk.

FEBC RADIO INTERNATIONAL (Rlg)
PHILIPPINES

✉ Box 1, Valenzuela City, Metro-Manila 0560, Philippines
☎ +63 (2) 292 5790
📄 +63 (2) 292 9430
E-mail: dev@febc.org.ph
Web: www.febc.ph
LP: Director: Efren M. Pallorina; Head of Prgrs: Peter McIntyre; C.E.: Romualdo M. Lintag; Head of P.R.: Priscilla R. Calica.
V: QSL-card. Rp. preferred (3 IRC's for Airmail).
Ann: English: "You are tuned to FEBC Radio International, we are your sound alternative"

GOSPEL FOR ASIA (Rlg)
UNITED STATES

✉ 1800 Golden Trail Court, Carrollton, TX 75010, USA
☎ +1 (800) 946 2742, +1 (972) 300 7777
E-mail: info@gfa.org
Web: www.gfaradio.org

HCJB WORLD RADIO AUSTRALIA (Rlg)
AUSTRALIA

✉ P.O. Box 6918, Melbourne, VIC 3001, Australia
E-mail: office@hcjb.org.au
Web: www.hcjb.org
LP: Ross Ramsay
Notes: Part of HCJB's expanding World Radio Network.

HCJB WORLD RADIO (Rlg)
UNITED STATES/ECUADOR

✉ USA Office: PO Box 39800, Colorado Springs, CO 80949-9800, USA. Ecuador Office: Casilla 17-17-69, Quito, Ecuador
☎ +1 (719) 590 9800
📄 +1 (719) 590 9801
E-mail: info@hcjb.org.ec
Web: www.hcjb.org
LP: S.M.: John E. Beck; P.D.: Alex Saks; Director of Local Radio: Mark Irwin; Eng. Director: Charles Jacobson; Satellite Prgr. Distribution Co-ord.: Remberto Fortich; F.M.: Douglas Weber; Marketing Director: (Contact Tom Narwold in Colorado Springs).
V: QSL-card. Rp (1 IRC), to assist with cost. Rec.acc. but tapes cannot be returned.
Notes: Transmissions in various Southern Asian languages are subject to change.

HIGH ADVENTURE MINISTRIES/VOICE OF HOPE (Rlg)
UNITED STATES

✉ P.O. Box 197569, Louisville, KY 40259, USA
☎ +1 (502) 968 7550
📄 +1 (502) 968 7580
E-mail: mail@highadventure.net
Web: www.highadventure.net
Notes: See also KVOH (USA). High Adventure Ministries has numerous offices covering the Americas, China, Europe, Russia, India, Middle East. From studios in Jerusalem, HAM covers most of the globe. Founded in 1979.

HMONG LAO RADIO (Clan)
LAO P.D.R.

✉ P.O. Box 2426, St. Paul, MN 55106, USA
Notes: First monitored in August 2002. Broadcasts in Hmong. Operated by US-based ULMD.

HRVATSKI RADIO/GLAS HRVATSKE (Gov)
CROATIA

✉ Prisavlje 3, 10000 Zagreb, Croatia
☎ +385 (1) 6342602
📄 +385 (1) 6343305
E-mail: shortwave@hrt.hr
Web: www.hrt.hr
LP: Editor in Chief: Ivana Jadresic.
V: QSL-letter. Re. in SINPO code to Zelimir Klasan, HRT, Prisavlje 3, 10000 Zagreb, Croatia.
Ann: Croatian: "Hrvatski Radio, Kratki Val"

IBC TAMIL RADIO (Comm)
UNITED KINGDOM
✉	Based in London, United Kingdom.
E-mail:	radio@ibctamil.co.uk
Web:	www.ibctamil.co.uk
LP:	Director of Programming: S.Shivaranjith
Ann:	Tamil: "Puligalin Kural"
Notes:	Originally formed by members of BBC Tamil service in 1997. IBC Tamil holds a UK broad-casting license.

IBRA RADIO (Rlg)
SWEDEN
✉	SE-14199 Stockholm, Sweden
☎	+46 (8) 608 9600
🖹	+46 (8) 608 9650
E-mail:	ibra@ibra.se
Web:	www.ibra.org
Notes:	Ibra (The Swedish Pentecostal Movements Radio Ministry) first broadcast on 29th July 1955 from Morocco.

INFORADIO (Comm)
NETHERLANDS
✉	P.O. Box 140, NL-5590 AC Heeze, The Netherlands
E-mail:	peter@inforadio.nl (management); mark@inforadio.nl (programs) or receptionreport@inforadio.nl
Web:	www.inforadio.nl (real audio available)
LP:	Peter Fris, Fokko van der Meijden and Mark Tasman.
Notes:	Broadcasts during summer months in Dutch for Dutch travellers/holidaymakers. Not active in 2003, next transmissions scheduled for summer 2004.

INFORMATION RADIO (US PSYOPS)
AFGHANISTAN
E-mail:	edward.shank@paharr.ang.af.mil
LP:	Verification Signatory: Lt. Edward Shank
V:	E-mail.
Ann:	Dari: "In radyio-i mau'lumati"; Pashto: "Da radyio mau'lumati."
Notes:	Operated by U.S. military.

JAKADA RADIO INTERNATIONAL (Clan)
NIGERIA
✉	Reported to be based in Spain
E-mail:	jakint2002@yahoo.com
Web:	www.jakadaradio.com
LP:	Yaro Yusuf Mamman.
Ann:	English: "This is Jakada Radio International".
Notes:	Operated by Alliance for Democracy. The station is owned by Oscar Marino Benjy Inc., "a legally registered media company in Europe with offices in London, Madrid and Frankfurt" and broadcasts in English. Active since May 2002.

KAIJ INTERNATIONAL (Rlg)
UNITED STATES
✉	Two If By Sea Broadcasting Corp, 22720 SE 410th Str, Enumclaw, WA 98022, USA
🖹	+1 (818) 606 1254
V:	QSL-card. Rp.
Ann:	English: "KAIJ, Dallas, Texas USA"
Notes:	Licensed since June 1995

KESTEDAMENA RADIO (RADIO RAINBOW) (Clan)
ETHIOPIA
✉	PO Box 140104, D-53056 Bonn, Germany
LP:	T.Assefa
Ann:	English: "Radio Rainbow, Voice of Peace and Brotherhood"; Amharic: "Ye Kestedamena Radio Ye Selamena Wendemamachenet Dimtse"
Notes:	Active since September 1997. Operated by GRAPECA - Groupe de Recherche et d' Action pour la Paix en Ethiopie et dans la Corne de l Afrique (RAGPEHA - Research and Action Group for Peace in Ethiopia / Horn of Africa).

KIMF (Rlg)
UNITED STATES
✉	9746 6th Street, Rancho Cucamonga, CA 91730, USA
🖹	+1 (909) 370 4862
LP:	Owner: James Planck; Frequency Consultant: George Jacobs.
Notes:	Expected to become operational in 2003/2004

KJES (Rlg)
UNITED STATES
✉	Our Lord's Ranch, 230 High Valley Rd., Vado, NM 88072, USA
☎	+1 (505) 233 2090
🖹	+1 (505) 233 3019
LP:	President: Fr Rick Thomas; G.M.: Michael Reuter
V:	QSL-card. Rp.
Ann:	English: "This is KJES Radio"
Notes:	Licensed since November 1992, orig. tests began in 1989.

KNLS INTERNATIONAL (Rlg)
ALASKA
✉	P.O. Box 473, Anchor Point, AK 99556, USA
☎	+1 (615) 371 8707
🖹	+1 (615) 371 8791
E-mail:	knls@aol.com
Web:	www.knls.org
LP:	President: Charles Caudill; S.M. & C.E.: Kevin K. Chambers; Exec. Prod.: Dale Ward; Freq. Coordinator: F.M. Perry; Dir..Prgr.: Dale Ward.
V:	QSL-card. Rec. acc. For E-mail reports, the subject should be "Reception Report" (no quotes). KLNS will provide one card for each report and will not verify transmissions from other operators who may air WCB programmes.
Ann:	English: "You are listening to station KNLS broadcasting from the top of the world,

Anchor Point, Alaska.

Notes: Owned and operated by World Christian Broadcasting Corporation. KNLS amends is schedule 4 times a year.

KOL ISRAEL (Gov)
ISRAEL

✉ P.O. Box 1082, Jerusalem 91010, Israel
☎ +972 (2) 248715
▤ +972 (2) 302327
E-mail: raphaelk@iba.org.il
Web: www.israelradio.org
LP: Director. & P.D.: Amnon Nadav; C.E.: David Cohen; Dir. Israel Radio International: Shmuel Ben-Zvi; Dir. of Liaison & Coordination: Raphael Kochanowski; Head of Western Broadcasting: Sarah Gabbai.
V: QSL-card.
Ann: Yiddish: "Hert zu der yisrael odizie im yiddish fyn yerushaalayion"; Ladino: "Emission de
yisrael en a langua Judeo-espagnol"; Spanish: "Esta es Kol Israel"; English: "Kol Israel - the voice of Israel, Jerusalem"; French: "Ici Kol Israel, radiodiffusion Israelienne"
Notes: Correspondence concerning technical matters should be sent to BEZEQ, The Israel Telecommunications Corp.Ltd, P.O. Box 29555, Tel Aviv 61290

KTBN SHORTWAVE RADIO (Rlg)
UNITED STATES

✉ PO Box 18147 Kearns, UT 84118, USA
☎ +1 (714) 731 1000
▤ +1 (714) 730 0661
E-mail: tbntalk@aol.com
Web: www.tbn.org
LP: S.M.: Johnny Mitchell; Dr. Paul F.Crouch; P.D.: Barry Phaeler; C.E.: Ben Miller; Head of P.R.: Rod Henke; Frequency Consultant: George Jacobs; QSL Manager: Laura Reyes.
V: QSL-card, rpt to QSL manager.
Ann: English: "This is the Trinity Broadcasting Network. This has been the superpower KTBN, Salt Lake City, Utah, USA"
Notes: Began operating in December 1990.

KVOH - VOICE OF HOPE (Rlg)
UNITED STATES

✉ High Adventure Ministries Inc, Box 100, Simi Valley, CA 93062, USA
☎ +1 (805) 520 9460
▤ +1 (805) 520 7823
E-mail: mail@highadventure.net
Web: www.highadventure.org
LP: C.E.: Jackie Mitchum Yockey; Manager: Paul Johnson (805) 520-9460; Chief Engineer: Paul W. Hunter (805) 520-9460; Programming Manager: Ralph McDevitt (805) 520-9460; Frequency Consultant George Jacobs.
V: QSL-card or letter.
Ann: English: "Voice of Hope" or "High Adventure

Ministries World Radio Network"
Notes: Active since 1979. KVOH is the Spanish language station on the High Adventure Ministry. See also "Voice of Peace", under UK.

KWHR - WORLD HARVEST RADIO (Rlg)
HAWAII

✉ P.O. Box 12, South Bend, IN 46624, USA
☎ +1 (574) 291 8200
▤ +1 (574) 291 9043
E-mail: whr@lesea.com
Web: www.whr.org (Real Audio available)
LP: G.M.: Peter Sumrall; C.E.: Douglas Garlinger; Freq Consultant: Douglas Garlinger; Sales/Program Coordinator: Joseph Brashier.
V: QSL-card.
Ann: English: "This is World Harvest Radio, the International service of LeSea Broadcasting Corporation. KWHR transmits from Naalehu, Hawaii in the United States of America."
Notes: Operated by LeSea Broadcasting, Inc. In addition to carrying it's own programs, airtime is hired to other religious and political organiza-tions. Programs may be carried in a variety of languages appropiate to the target area.

LAO NATIONAL RADIO (Gov)
LAO P.D.R.

✉ P.O. Box 310, Vientiane, Lao P.D.R..
☎ +856 (21) 212432
▤ +856 (21) 212430
E-mail: natradio@laonet.net
LP: D.G.: Bounthan Inthaxay; Deputy D.G.: Keungkham Vilayasith; M.D. (Tech.): D. Sisombath
V: QSL-card, no IRC's required.
IS: National Anthem
Notes: International service times variable on 7145kHz.

LASER RADIO UK (Comm)
UNITED KINGDOM

✉ BCM Aquarius, London WC1N 3XX, United Kingdom
E-mail: studio@laserradio.net
Web: www.laserradio.net
V: QSL-card.
Notes: Plans to move to new frequency (ex 9520 kHz).

MEZOPOTAMIAN RADIO AND TELEVISION (Clan)
IRAQ

✉ H.C. Andersens Boulevard 39, DK-1553 Copenhagen V, Denmark
☎ +45 7 026 0688
▤ +45 7 026 0788
E-mail: metv@metv.dk
Web: www.metv.dk
Notes: Active since September 2002.

NATIONAL VOICE OF CAMBODIA (Gov)
CAMBODIA

✉ 106 Preah Kossamak Street, Monivong Boulevard, Phnomh Penh, Cambodia
☎ +855 (23) 423369 or 422869
LP: Director, International Service: Mr In Chhay
V: QSL-letter.
Ann: Cambodian: "Vitthyu Cheat Kampuchea", English: "National Voice of Cambodia"

NEXUS IBA / IRRS-SHORTWAVE
ITALY

✉ P.O. Box 10980, I-20110 Milano, Italy
☎ +39 (2) 266 6971
🖷 +39 (2) 706 38151
E-mail: info@nexus.org
Web: www.nexus.org
LP: President: Alfredo Cotroneo; Vice-President: Flavia F. Cotroneo; P.R.: Ms Anna S. Boschetti
V: QSL-card. Rp.
IS: S/on: Triumphal Scene from Aida (Giuseppe Verdi); S/off: Prisoners' Chorus (Giuseppe Verdi).
Notes: Relays various radiostations and programmes. Nexus are now delivering Internet services from the USA, Germany, Italy and Russia, with future expansions in Africa and the Pacific planned in the next two years. For further information, see website. Alternate frequencies: 3980, 3985, 7120 and 7125kHz.

NHK WORLD RADIO JAPAN (Gov)
JAPAN

✉ NHK World Radio Japan, NHK 150-8001, Japan
☎ +81 (3) 3465 1111
🖷 +81 (3) 3481 1350
E-mail: info@intl.nhk.or.jp
Web: www.nhk.or.jp/nhkworld
LP: D.G.: K. Irisawa; Head of Programmes: K. Miyamoto
V: QSL-card.
IS: Opening music of Reg. Sce's "Sakura". Interval signal music "Kazoe Uta".
Ann: Chinese: "Zheli shi Riben Guoji Guangbo Diantai, NHK huanqiu guangbowang"; English: "This is Radio Japan, NHK World network, Tokyo"; German: "Hier ist Radio Japan, NHK WORLD"; Indonesian: "Inilah Radio Jepang, NHK World, siarang bahasa Indonesia"; Japanese: "Kochirawa NHK Warudo, Rajio Nippon, NHK no koku saihoso desu"; Korean: "Yeogineun Radio Ilbon, NHK Worldeumnida"

NRK - NORSK RIKSKRINGKASTING (Pub)
NORWAY

✉ N-0340 Oslo 3, Norway
☎ +47 2304 8444
🖷 +47 2304 7134
E-mail: alltid.nyheter@nrk.no
Web: www.nrk.no/radionorway

V: No verification of reports, but Radio Denmark sharing frequencies with NRK offer QSLs.
Notes: NRK no longer produce a dedicated foreign service, but feed their SW transmitter with domestic programmes. 0500-2100 UTC Monday-Friday is a relay of rolling news service NRK Alltid Nyheter. 2100-0500 UTC the main news from 1630 UTC is repeated every hour. During weekends SW relay NRK P1.

ORF - RADIO ÖSTERREICH INTERNATIONAL (Pub)
AUSTRIA

✉ Argentinierstrasse 30a, A-1040 Wien, Austria
☎ +43 (1) 50 101 16000 (Management)
🖷 +43 (1) 50 101 16006 (Management)
E-mail: roi.service@orf.at
Web: http://roi.orf.at (Live audio feed available)
LP: P.D.: Michael Kerbler; Heads of Prgr: Kerry Skyring (English); Lucien Giordani (French); Jan van der Brugge (Spanish); Oswald Klotz (Arabic and Internet); Katinka Fetes-Tosegi (Esperanto)
V: QSL-card. Rec. acc.
Ann: German: "Österreich 1"
Notes: Relays domestic programme "Österreich 1".

QUE HUONG RADIO (COUNTRY RADIO) (Clan)
VIETNAM

✉ 2670 S.White Road, Suite 165, San Jose, CA 95148, USA
☎ +1 (408) 223 3130
🖷 +1 (408) 223 3131
E-mail: qhradio@aol.com
Web: www.quehuongmedia.com
LP: S.M.: Nguyen Khoi
Ann: Vietnamese: "Que Huong Radio"
Notes: Operated by Vietnamese activists in USA. Active since November 1999. Broadcasts in Vietnamese.

RADIO AFGHANISTAN (Gov)
AFGHANISTAN

✉ P.O. Box 544, Kabul, Afghanistan

RADIO AFRICA INTERNATIONAL (UNITED METHODIST CHURCH) (Rlg)
UNITED STATES

✉ 475 Riverside Drive, New York, NY 10115, USA
☎ +1 (212) 870 3912
🖷 +1 (212) 870 3748
E-mail: radio@gbgm-umc.org
Web: www.gbgm-umc.org
Ann: English: "This is Radio Africa International - a production of the General Board of Global Ministries of the United Methodist Church."
Notes: Active since January 2001. Broadcasts in English and French. Operated by the General Board of the Global Ministries of the United Methodist Church.

RADIO AMHORO (ETHIOPIA) (Clan)
ETHIOPIA
✉ Postfach 510610, D-13366 Berlin, Germany
☎ +32 (2) 735 3915
▤ +32 (2) 735 3915
Ann: Amharic: "Kun Sagalee Bilisumma Omoroo"
Notes: Active since July 1988

RADIO AUSTRALIA (Gov)
AUSTRALIA
✉ P.O. Box 428G, Melbourne, VIC 3001,
 Australia
☎ +61 (3) 9626 1898 (management); Messages
 +61 (3) 9626 1825
▤ +61 (3) 9626 1899
E-mail: english@ra.abc.net.au
Web: www.abc.net.au/ra
LP: G.M.: Jean-Gabriel Manguy; P.D.: Tony
 Hastings; Transmission Manager: Nigel
 Holmes (Chief Engineer).
V: QSL-card.
Ann: English: "This is Radio Australia broadcasting
 from studios in Melbourne, Victoria"

RADIO AVAYE ASHENA (RADIO FAMILIAR
VOICE) (Clan)
IRAN (ISLAMIC REP. OF)
✉ P.O. Box 120228, D-10592 Berlin, Germany
E-mail: redaction@avayeashena.com
Web: www.avayeashena.com
Ann: Farsi: "Radyo Avaye Ashena"
Notes: Active since March 2002. Broadcasts in Farsi.

RADIO BAHRAIN (Gov)
BAHRAIN
✉ P.O. Box 702, Manama, Bahrain
☎ +973 781888
▤ +973 681544
E-mail: skhalid@bahrainradio.com
Web: www.moi.gov.bh
Notes: Relays domestic programme on SW.

RADIO BARABARI (RADIO EQUALITY) (Clan)
IRAN (ISLAMIC REP. OF)
E-mail: info@radiobarabari.net
Web: www.radiobarabari.net
Ann: Farsi: "Radio Barobari"
Notes: Active since May 2001. Broadcasts in Farsi

RADIO BAYRAK INTERNATIONAL (Gov)
CYPRUS
✉ BRT (Bayrak Radio Television Corporation),
 P.O. Box 417, Lefkosa, via Mersin 10, Turkey
E-mail: brt@cc.emu.edu.tr
Web: www.brt.gov.nc.tr
LP: Head: Ülfet Kortmaz
Notes: News at 1000 (English and Greek), 1200
 (Greek), 1215 (English), 1400 (English and
 Greek), 1600 (Arabic, Russian and German),
 1730 (English).

RADIO BUDAPEST (Gov)
HUNGARY
✉ Bródy Sándor u. 5-7, H-1800 Budapest,
 Hungary
☎ +36 (1) 328 7339
▤ +36 (1) 328 8838
E-mail: vanyolai@radio.hu
Web: www.kaf.radio.hu
LP: Dir., Foreign Broadcasting: Antal Réger. Vice
 Dir., Foreign Broadcasting: Zsuzsa Mesz ros;
 F.M.: Ferenc Horvath.
V: QSL-card. Rec. acc.
Ann: English: "This is Radio Budapest, Hungary";
 German: "Hier ist Radio Budapest".

RADIO BULGARIA (Gov)
BULGARIA
✉ 4, Dragan Tsankov Blvd., 1040 Sofia,
 Bulgaria
☎ +359 (2) 985241
▤ +359 (2) 650560
E-mail: rbul@nationalradio.bg
Web: www.nationalradio.bg
LP: D.G. (National Radio): Polya Stancheba;
 P.D.: Angel Nedialkov; Frequency Manager:
 Ivo Ivanov
V: by 6 new QSL-cards. Rp (1IRC) Rec. acc.
Ann: Albanian: "Ju flet Radio Bulgaria"; Bulgarian:
 "Tuk e Radio Bulgaria"; English: "This is
 Radio Bulgaria"; French: "Ici Radio
 Bulgarie"; German: "Hier spricht Radio
 Bulgarien"; Greek: "Akute to Radio Vulgaria";
 Russian: "V efire Radio Balgaria"; Serbian:
 "Radio Bugarska"; Spanish: "Esta es Radio
 Bulgaria"; Turkish: "Burasi Bulgaristan
 Radiosu"

RADIO CAIRO (Gov)
EGYPT
✉ P.O. Box 566, 11511 Cairo, Egypt
☎ +20 (2) 5789491
▤ +20 (2) 5789491
E-mail: englishprog@ertu.org
Web: www.ertu.org
LP: President, ERTU: Amin Bassiouni;
 Chairman. Eng. Sector: Ibrahim A. Ibrahim;
 Chairman. Sound Programme Sector: Farouk
 Susha; Chairman. of TV Programme Sector:
 Suhair El Attriby.
V: QSL-card.
Ann: General Prgr: "Idha'atu jumhuriya misr
 al'arabbiya min al-qahira"; English: "You are
 tuned to Radio Cairo"

RADIO CANADA INTERNATIONAL (Pub)
CANADA
✉ P.O. Box 6000, Montreal, Quebec, H3C 3A8,
 Canada
☎ +1 (514) 597 7500
▤ +1 (514) 597 7076
E-mail: rci@cbc.ca
Web: www.rcinet.ca
LP: Director: Jean Larin; Director of Eng.:
 Jacques Bouliane; News Editor: Derek Quinn
V: QSL-card. Re. handled by Canadian
 International DX Club.

196

IS: First bar of Canadian National Anthem.
Ann: English: "This is Radio Canada International"; French: "Ici Radio Canada International".
Notes: Canadian Broadcasting Corporation/Société Radio Canada. Publicly owned.

RADIO DAMASCUS (Gov)
SYRIA

✉ P.O. Box 4702, Damascus, Syria
☎ +963 (11) 720700
🖷 +963 (11) 2234336
LP: D.G.: Khudr Omran; Dir. Eng: M.Bara; Dir. Public Relations: Mrs. Awafet Haffar; Dir. Admin. & Finance: Zuheir Breidi.
V: QSL-card.
IS: Guitar
Ann: Arabic: "Idha'atu-l-gumhuriyati-l'arabiyya as-suriyya min dimashq". English: "This is the Syrian Arab Republic Broadcasting Service from Damascus"; French: "Ici Damas"; Hebrew: "Kol Damasek"

RADIO DANMARK (Pub)
DENMARK

✉ Radio Danmark, Radioavisen, Rosenorns Allé 22, DK-1999 Frederiksberg C, Denmark
☎ +45 3520 5784
🖷 +45 3520 5781
E-mail: Schedule and programme matters: rdk@dr.dk; Technical and reports: rdktek@dr.dk . By sending an empty email to schedule@dr.dk you'll get the current schedule emailed back.
Web: www.dr.dk/rdk or www.dr.dk/radiodanmark (real audio)
LP: Head of international service.: Jens Holme; Office: Jette Hapiach; Technical Adviser/Frequency Manager: Erik Køie.
V: QSL-card. RP preferred. Rec. (incl. RA and MP3 E-mail files) are accepted. Tapes are not returned.
Notes: All programmes are in Danish and are relays of the Home Service. News broadcasts at 11.00 and 17.30, plus sport, the first airings being at 12.30 and 18.30. The technical mailbag programme "Tune In" is heard every second Saturday from 12.48 to 17.48 UTC. Radio Denmark broadcasts via the facilities of Norkring (Norway).

RADIO DMR (Gov)
MOLDOVA

✉ ul. Rozy Lyuksemburg 10, MD-3300 Tiraspol, Moldova
☎ +373 (33) 35570
🖷 +373 (33) 32245
E-mail: radiopmr@inbox.ru
Ann: English: "Here is the Radio of the Dniester Moldavian Republic"
Notes: R. DMR is the state broadcasting service of the self-proclaimed Dniester Moldavian Republic (Transnistria) in Eastern Moldova.

RÁDIO ECCLÉSIA (Rlg)
ANGOLA

✉ C.P. 3579, Sao Paulo, Luanda, Angola
☎ +244 (2) 443041
🖷 +244 (2) 443093
E-mail: info@recclesia.org
Web: www.recclesia.org
V: QSL-E-mail.
Ann: Portuguese: "Rádio Ecclesia, uma Rádio para todo pais" or "Emmisora catolica de Angola".
Notes: Active since 1954 via various transmitters. Also relayed on FM. No longer authorised to broadcast from Angola.

RADIO ETHIOPIA (Gov)
ETHIOPIA

✉ Radio Ethiopia External Service, P.O. Box 654, Addis Ababa, Ethiopia
☎ +251 (1) 551011
🖷 +251 (1) 552263
LP: Head of International Service: Moges Taffese.
V: Letter. (IRC's acc.)
Ann: English: "This is R. Ethiopia broadcasting in English"
Notes: The Voice of Peace broadcast at 1100 and 1900 is a separate, UNICEF-funded, humanitarian service intended for Somalia

RADIO EXTERIOR DE ESPANA (Gov)
SPAIN

✉ Apartado 156.202, 28.080 Madrid, Spain
☎ +34 (91) 346 1149/1083
🖷 +34 (91) 346 1815
E-mail: ree.rne@rtve.es
Web: www.rtve.es/rne/ree
LP: Dir.: Javier Garrigos Fernandez; Asst. Dir.: Nuria Alonso Veiga; Head of Inf. Sce.: Eduardo Moyano Zamora; Head of Current Affairs Prgrs.: Aurora Sanchez Parez; Foreign Language Prgr.: Jose J. Amorena Zabalza; Tech. & Prgr. Secretary: Josefina E. Pera Marquez; Head of Freqs.: Jose MǏ Huerta Crisologo; Asst. Dir. Tech.: Enrique Martinez Torres.
V: QSL-card.
Ann: Spanish: "Radio Exterior de España; English: "This is Radio Exterior de Espana, broadcasting from Madrid"; French: "Radio Extérieure d'Espagne"; German: "Der Spanische Auslandssender Radio Exterior de España"; Arabic: "Idhaa'atu Isbania al-Jariyía".

RADIO FOR PEACE INTERNATIONAL (Rlg)
COSTA RICA

✉ Radio For Peace International, Apartado 88, 6150 Santa Ana, Costa Rica
☎ +506 (249) 1821
🖷 +506 (249) 1095
E-mail: info@rfpi.org
Web: www.rfpi.org.
LP: D.G.: Debra Latham; Station Manager: James Latham; Prgr. Co-ordinator (English): Joe Bernard

V: QSL-letter.
Ann: English: "This is Radio for Peace International, Costa Rica"
Notes: RFPI changes it's schedule 4 times a year. RFPI hosts programmes from numerous stations with topics such as disibility, womens rights and DXing.

RADIO FRANCE INTERNATIONALE (Gov)
FRANCE
✉ BP 9516, F-75016 Paris, France
☎ +33 (1) 42301212
🖷 +33 (1) 42303071
E-mail: english.service@rfi.fr
Web: www.rfi.fr
LP: President: Jean-Paul Cluzel; D.G.: Benoît Paumier; Dir. of Communication: Christine Berbudeau; Dir. of Inf.: Anne Toulouse; Editor-in-Chief (French Services): Henri Perilhou; Editor-in-Chief RFI Afrique: Jean-Karim Fall; Dir. Foreign Language Services.: Nicolas Levkov; Dir. of Prgrs.: Alex Taylor; Dir. Admin & Finance: Philippe Tarie; Dir. Human Resources: Catherine Dessein; Tech. Dir.: Patrice Berestetsky. Society, Nationale de Radiodiffusion sonore pour l',tranger (RFI)
V: QSL-card.
IS: "La Marseillaise"
Ann: French: "Radio France Internationale" or "Ici Paris, Radio France Internationale".

RADIO FREE VIETNAM (Clan)
VIETNAM
✉ P.O. Box 29245, New Orleans, LA 70189, USA
☎ +1 (504) 254 2303
🖷 +1 (504) 254 2305
E-mail: rfvla@aol.com
Web: www.radiofreevietnam.com
LP: Director in Chief: Vuong Ky-Son
Notes: Operated by Vietnamese activists in New Orleans, USA. Active since August 2001

RADIO FREE VIETNAM (Clan)
VIETNAM
✉ 12755 Brookhurst St., #104 Garden Grove, CA 92840, U.S.A
☎ +1 (714) 636 9977
🖷 +1 (714) 636 9513
E-mail: info@rfvn.com
Web: www.rfvn.com
Ann: Vietnamese: "Dai Viet Nam tui do".
Notes: Operated by the "Government of Free Vietnam", an opposition group based in California, USA. Active since August 1999. Broadcasts in Vietnamese. Can also be heard in the USA at various times on 1190kHz (CA), 1280kHz (AZ), 1570kHz (IL) and in Australia via 2VNR.

RADIO GEORGIA (Gov)
GEORGIA
✉ M Kostava Street 68, 380071 Tbilisi, Georgia

☎ +995 (32) 360063
🖷 +995 (32) 955137
E-mail: office@geotvr.ge
Web: www.geotvr.ge
LP: Editor-in-Chief: Shamil Cheabrishvili
V: QSL-card.
Ann: Georgian: "Laparakobs Tbilisi"; English: "This is Georgia"; Russian: "Govorit Radiostantsiya Gruziya"
Notes: Due to frequent problems with electricity supply, transmissions may to be irregular or at other times than scheduled.

RADIO HABANA CUBA (Gov)
CUBA
✉ Apartado 6240, La Habana, Cuba
☎ +53 (7) 791053
🖷 +53 (7) 795007
E-mail: rhc@radiohc.cu
Web: www.radiohc.cu
LP: M.D.: Milagro Hernandez Cuba; Head of Prgrs.: Ignacio Canel; C.E.: Luis Pruna
V: QSL-card.
Ann: English: "This is Radio Havana, Cuba"

RADIO IBRAHIM (Rlg)
CYPRUS
✉ P.O. Box 56500, 3312 Limassol, Cyprus
☎ +357 (99) 447768
🖷 +357 (25) 338394
E-mail: mail@radioibrahim.com
Web: www.radioibrahim.com
V: QSL-card. E-mail rpts to: dx@radioibrahim.com
Notes: Broadcasts in Classical Arabic and a number of Arabic dialects (which are shown in the schedule). Also operates on MW to serve the Gulf Area and via satellite for the entire Arab world.

RADIO INTERNATIONAL (Clan)
IRAN (ISLAMIC REP. OF)
✉ PO Box 1499, London WC1N 3XX, United Kingdom
☎ +44 (771) 461 1099
E-mail: radio7520@yahoo.com; azarmajedi@yahoo.com, ali_javadi@yahoo.com
Web: www.radio-international.org
LP: Ms Azar Majedi
Ann: Farsi: "Radio Anternacional-e" or "Im Radio International"
Notes: Active since December 1999. Operated by WCP (Worker Communist Party of Iran). Broadcasts in Farsi.

RADIO JORDAN (JRTV) (Gov)
JORDAN
✉ P O Box 1041 or 2333, Amman, Jordan
☎ +962 (6) 77311/9
🖷 +962 (6) 751503 (Admin) +962 (6) 788115 (Eng)
E-mail: rj@jrtv.gov.jo (prgrs); eng@jrtv.gov.jo

(technical); general@jrtv.gov.jo (general)
Web: www.jrtv.com
LP: D.G.: Ihsan Ramzi Shikim; Dir. Radio: Hashim Khuraysat. Dir; Eng: Fawzi Saleh; Prgr. Dir's: J. Hajjat (Arabic), Jawad Zada (English); Dir. Admin: A. Zabi; Dir. Int. Relations: Mrs. Fatima Masri.
V: QSL-card. Rec. acc. (Re. to P.O. Box 909, Amman).
Ann: Arabic: "Huna 'amman, Idha'atu-l-mamlaka al-urduniyya al-hashimiyya"; English: "This is Radio Jordan, broadcasting from Amman"

RADIO KOREA INTERNATIONAL (Gov)
KOREA, SOUTH (REP. OF)
✉ 18 Yeouido-dong, Yeongdeungpo-gu, Seoul 150-790, Republic of Korea
☎ +82 (2) 781 3650
🖺 +82 (2) 781 3694
E-mail: rki@kbs.co.kr
Web: http://rki.kbs.co.kr
LP: President: Park, Kwon Sang; D.G.: Kim, Min Gi; Exec. Dir.: Han, Hee Joo.
V: QSL-card.
IS: Korean children's song "Dar-a Dar-a Balgeun Dar-a (Oh, bright moon)" played by a glockenspiel. Original music "Dawn" composed by Kim Hee Jyo with KBS symphony orchestra.
Ann: Korean: "Yeogineun Daehanminguk Seoul-eseo Bonaedeurineun Hanguk Bangsong Gongsa, KBS-e Gukje Bangsong, Radio Hangug-imnida"; English: "This is Radio Korea International of the KBS"; Japanese: "Kochirawa Rajio Kankoku, KBS-no Kokusai Hoso desu"; Chinese: "Zheli shi Hanguo guoji guangbo diantai"; Indonesian: "Inilah siaran Radio Korea Internasional, KBS yang dipancarkan langsung dari ibu kota Republik Korea Seoul".

RADIO KUWAIT (Gov)
KUWAIT
✉ Dept of Frequency Management, PO Box 397, Safat 13004, Kuwait
☎ +965 2423774
🖺 +965 2456660
E-mail: radiokuwait@radiokuwait.org
Web: www.radiokuwait.org
LP: Asst. Under Secretary for Internal Media: Mrs. Amal Al-Hama; Asst. Under Secretary for TV Affairs: Saad Abdulaziz Ja'afer; Asst. Under Secretary for News Affairs and Political Prgrs: Mohamed Hamed A. Al-Qahtani; Technical Adviser: Yacoub Y. Dashty; Director of TV Engineering: Abdulazeez Al-Bagli; Director of Engineering Comms: Bader F. Al-Mazeedi; Director of Frequency Management: Nasser M. Al-Saffar; Director of Radio
V: QSL-folder. Rec. acc.
Ann: Arabic: "Huna al-Kuwait"; English: "This is Radio Kuwait"; Tagalog: "Radio Pinoy"

RADIO MARYJA (Rlg)
POLAND
✉ ul. Zwirki i Wigury 80, 87-100 Torun, Poland
☎ +48 (56) 6552361
🖺 +48 (56) 6552362
E-mail: radio@radiomaryja.pl
Web: www.radiomaryja.pl
V: QSL-card.

RADIO MEDITERRANEE INTERNATIONAL (MEDI 1) (Comm)
MOROCCO
✉ 3, rue Emsallah, 9000, Tangier, Morocco
☎ +212 (39) 936363
🖺 +212 (39) 935755
E-mail: medi1@medi1.com
Web: www.medi1.com
V: Letter.
Ann: French: "Ici Médi 1" or "Ici Radio Mediterranee Interantionale"
Notes: Daily programmes in Arabic and French. Can also be heard on a number of FM outlets in Morocco and France

RADIO MINSK (Gov)
BELARUS
✉ Cyrvonaja Street 4, 220807 Minsk, Belarus
☎ +375 (17) 239 5830
🖺 +375 (17) 284 8574
E-mail: radio-minsk@tvr.by
Web: www.tvr.by
LP: Director: Natalia Hlebus
V: QSL-card.
Ann: Belarusian: "Radiostancyja Belarus"; English: "Radio Belarus International" or "Radio Minsk".

RÁDIO NACIONAL DE ANGOLA (Gov)
ANGOLA
✉ C.P. 1329, Luanda, Angola
☎ +244 (2) 321638
🖺 +244 (2) 391234
E-mail: rna.dg@netangola.com
Web: www.rna.ao
Ann: English: "This is Luanda, the International service of the Angolan National Radio"

RADIO NEDERLAND WERELDOMROEP (INDEPENDENT FOUNDATION)
NETHERLANDS
✉ P.O. Box 222, NL-1200 JG Hilversum, Netherlands
☎ +31 (35) 672 4211
🖺 +31 (35) 672 4207 or +31 (35) 672 4239 (English Dept.)
E-mail: letters@rnw.nl
Web: www.rnw.nl
LP: D.G.: Lodewijk Bouwens; Creative Dir.: Jonathan Marks; Dir. of Finance & Logistics: Jan Hoek; Editor-in-Chief: Freek Eland; G.M., Training Centre: Mrs. Lem van Eupen; Head of Dutch Services: Peter Veenendaal; Head of English Service: Mike Shaw; Head of

Latin America Service: Jose Zepeda; Head of Indonesian Service: Indra Leihitu; Head of Television: Pieter Landman; Head of Music: Hans Quant; Head of Programme Distribution: Jan Willem Drexhage; Head of Marketing & Communications: Mrs. Nina Nomden-Stenekes

V: QSL-card. Reception reports appreciated if they follow the guidelines set down in "Writing Useful Reception Reports", available on the web at www.medianetwork.nl

Ann: English: "This is Radio Netherlands, the Dutch International Service"; Spanish: "Transmite R. Nederland desde la ciudad de Hilversum en Holanda";Indonesian: "Inilah Radio Nederland di Hilversum dengan siaran dalam bahasa Indonesia";

Notes: The station is also involved with music and television co-productions.

RADIO NEW ZEALAND INTERNATIONAL (Gov)
NEW ZEALAND

✉ P O Box 123, Wellington, New Zealand
☎ +64 (4) 474 1437
🖷 +64 (4) 474 1433
E-mail: info@rnzi.com
Web: www.rnzi.com
LP: C.E.: Dave Henderson; Tech. Manager: Adrian Sainsbury; Linden Clarke.
V: QSL-card. Rp (2 IRC's). Rec. not accepted.
Ann: Maori: "Te Reo Irirangi O te Moana-nui-a-kiwa"; English: "This is Radio New Zealand International, the voice of the Pacific"
Notes: Mainly in English with news in various Pacific languages. Usual Closedown is 1305 UTC. 6095kHz is used for occasional over-night broadcasts to the Pacific for sports commentaries or cyclone warnings. RNZI is closed for maintenence on the 3rd Thursday of every month.

RADIO PAKISTAN (Gov)
PAKISTAN

✉ Broadcasting House, Constitution Avenue, Islamabad 4400, Pakistan
☎ +92 (51) 921 0689
🖷 +92 (51) 920 1861
E-mail: cnoradio@isb.comsats.net.pk
Web: www.radio.gov.pk
LP: D.G: S Auwar Mehmood; D.Prgr: Inyatullah Baloch; Dir. N: Nazir Ahmed Bokhari; Dir.Tech:Muhammad Iqbel; Dir. Overseas Liaison: Muhammad Sharif
V: QSL-card. Int. Sce.: re to Controller: Freq. Management; Home Sces: Peshwar & Quetta by letter for direct rpts.
Ann: English: "This is Radio Pakistan"

RADIO PAYAM-E DOOST (RADIO MESSAGE FROM A FRIEND) (Rlg)
IRAN (ISLAMIC REP. OF)

✉ P.O.Box 765, Great Falls, Virginia 22066, USA

☎ +1 (703) 671 8888
🖷 +1 (301) 292 6947
E-mail: payam@bahairadio.org
Web: www.bahairadio.org
Ann: English: "Bahai Radio International" or "Radio Payam-e Doost"
Notes: Sponsored by members of the Baha'i Faith. Active since March 1994 and on SW since May 2001.

RADIO POLONIA (Pub)
POLAND

✉ P.O. Box 46, 00-977 Warszawa, Poland
☎ +48 (22) 645 9262
🖷 +48 (22) 645 5917
E-mail: radio.polonia@radio.com.pl
Web: www.radio.com.pl/polonia
LP: Head: Marek Traczyk
V: QSL-card.
Ann: Polish: "Tu Polskie Radio, Warszawa."; English: "This is Polish Radio, Warsaw" or "This is Radio

RADIO PRAGUE (Gov)
CZECH REPUBLIC

✉ Vinohradská 12, 12099 Praha 2, Czech Republic
☎ +420 (2) 242 18349
🖷 +420 (2) 2155 2903
E-mail: cr@radio.cz
Web: www.radio.cz
LP: Director: Miroslav Krupicka (mkrupick@cro.cz); Head of Prgrs: David Vaughan (dvaughan@cro.cz); C.E.: Oldrich Cip (cip@radio.cz).
V: QSL-card. Rec. acc.
Ann: English: "You are tuned to Radio Prague, the external service of Czech Radio"

RADIO ROMANIA INTERNATIONAL (Gov)
ROMANIA

✉ Radio Romania International, PO Box 111, General Berthelot Street 60-62 RO-70756, Bucharest
☎ +40 (1) 222 2556
🖷 +40 (1) 223 2613
E-mail: rri@radio.ror.ro
Web: www.rri.ro
LP: Deputy General Director of Radio Romania Society: Mr. Doru Vasile Ionescu. General Secretary of Radio Romania International Dpt: Mr. Dan
Ann: English: "This is Bucharest, Radio Romania International".

RADIO SANTEC (Rlg)
GERMANY

✉ Marienstrasse 1, D-97070 Würzburg, Germany
☎ +49 (931) 3903264
🖷 +49 (931) 3903195
E-mail: info@radio-santec.com
Web: www.radio-santec.com

Ann: English: "You are listening to Radio Santec"
Notes: Prgrammes may be aired under the following identities: The Cosmic Wave, The Word, The Universal Spirit.

RADIO SEDAYE IRAN (RADIO VOICE OF IRAN) (Clan)
IRAN (ISLAMIC REP. OF)

✉ KRSI, 9744 Wilshire Boulevard, Suite 207, Beverly Hills, CA 90212, USA
☎ +1 (310) 8882838
Web: www.krsi.net
Ann: Farsi: "Radyo Seda-ye Iran"
Notes: Operated by KRSI, California, USA. Active since December 1999

RADIO SINGAPORE INTERNATIONAL (Gov)
SINGAPORE

✉ P.O. Box 5300, Singapore 912899
☎ +65 6359 7662
▤ +65 6259 1357
E-mail: info@rsi.com.sg
Web: www.rsi.com.sg
Ann: Chinese: "Xinjiapo Guoji Guangbo Diantai"; Indonesian: "Radio Singapura Internasional"

RADIO SLOVAKIA INTERNATIONAL (Pub)
SLOVAK REPUBLIC

✉ P.O. Box 15, 81755 Bratislava
☎ +421 (7) 5727 3737
▤ +421 (7) 5249 6282
E-mail: englishsection@slovakradio.sk
Web: www.slovakradio.sk/rsi
Ann: English: "You are listening to Radio Slovakia International"
Notes: From October 1999, Radio Slovakia International started English language broadcasting on World Space AfriStar satellite to Africa. The "Various" languages shown in schedule are a combination of English, Slovakian, French and German.

RADIO SOLIDARITY (Clan)
ETHIOPIA

✉ Dade Desta, P.O. Box 60040, Washington, DC 20039, USA
E-mail: ethiopian@tisjd.net
Web: www.tisjd.net (Real Audio available)
Ann: Tigrinyan: "Radio Fathiriu"
Notes: Operated by the US based opposition group "Tigrayan International Solidarity for Justice and Democracy". Active since September 2001. Broadcasts in Tigrinyan.

RADIO SULTANATE OF OMAN (Gov)
OMAN

✉ Ministry of Information, P.O. Box 600, 113 Muscat
☎ +968 603888
▤ +968 604629
E-mail: tvradio@omantel.net.om
Web: www.oman-tv.gov.om
LP: D.G. Ibrahim Al Yahmadi; D.G.

(Engineering): Mohd Salim AL Morhouby; Dir. Freq.: Salim AL-Nomani; Dir. Foreign Sce.: Shakar Al-Araimi
V: QSL-folder.
Ann: Arabic: "Idha'atu Saltanat Oman min Muscat."; English: "This is Radio Sultanate of Oman"

RADIO SWEDEN (SR INTERNATIONAL) (Pub)
SWEDEN

✉ SE-10510 Stockholm, Sweden
☎ +46 (8) 784 7288
▤ +46 (8) 667 6283
E-mail: radiosweden@radiosweden.org
Web: www.radiosweden.org
LP: M.D.: Anne Sseruwagi.
V: QSL-card. Cassettes not accepted.
IS: "To the Wide, Wide World" (electronic music composed by Ralph Lundsten)
Ann: English: "This is Radio Sweden"; German: "Hier ist Radio Schweden, Stockholm"; Swedish: "Radio Sverige, Stockholm"; Russian: "Govorit Stokgolm"

RADIO TAJIKISTAN (Gov)
TAJIKISTAN

✉ Chapayev Street 31, 734025 Dushanbe, Tajikistan
☎ +992 (372) 277417
▤ +992 (372) 211198
Web: http://radio.tojikiston.com
LP: Editor: Nasrullo Ramazonov
V: Letter.
Ann: English: "This is Dushanbe, the capital of the Republic of Tajikistan"

RADIO TASHKENT INTERNATIONAL (Gov)
UZBEKISTAN

✉ Xorezm Street 49, 700047 Toshkent, Uzbekistan
☎ +998 (71) 1338920
▤ +998 (71) 1440021
E-mail: uztele@tkt.uz
Web: www.teleradio.uz
LP: Head: Sherzat Gulyamov
V: QSL-card.
Ann: Uzbek: "Toskentdan gapiramiz"; English: "Radio Tashkent calling" or "This is Radio Tashkent, broadcasting from the Republic of Uzbekistan"

RADIO THAILAND (Gov)
THAILAND

✉ Radio Thailand, World Service, Public Relations Department, Royal Thai Government, 236 Vibhavadi Rangsit Road, Din Daeng, Bangkok 10400,
☎ +66 (2) 277 1814
▤ +66 (2) 2741840
E-mail: amporns@mozart.inet.co.th
Web: www.prd.go.th
LP: D.G., Govt. Pub. Rel. Dept: Bangern Musikapong; Dir. Radio Thailand: Somphong

Visuttipat.
V: QSL-card.
IS: Gong
Ann: English: "This the World Service of Radio Thailand"

RADIO TIRANA (Gov)
ALBANIA
✉ Rruga "Ismail Qemali" 11, Tirana, Albania
☎ +355 (42) 22481
🖹 +355 (42) 22481
E-mail: dushiulp@yahoo.com
Web: http://rtsh.sil.at/foreign.htm
LP: Director: Martin Leka.
V: QSL-card.
Ann: Albanian: "Ju flet Tirana"; English: "This is Radio Tirana".

RADIO UKRAINE INTERNATIONAL (Gov)
UKRAINE
✉ Kreschatyk 26, 01001 Kyiv, Ukraine
☎ +380 (44) 2282534
🖹 +380 (44) 2287356
E-mail: vsru@nrcu.gov.ua
Web: www.nrcu.gov.ua
LP: Chief Editor: Viktor I. Nabrusko
V: QSL-card.
Ann: Ukrainian: "Hovorit Kyiv", "Vsesvitnya sluzhba Radio Ukrayiny" (World Service); English: "This is Radio Ukraine International"

RADIO UNMEE (Non-comm)
ETHIOPIA
✉ ECA Building, PO Box 3001, Addis Ababa, Ethiopia.
E-mail: bakari@un.org
Web: www.un.org/Depts/dpko/unmee/radio.htm
Notes: The Voice of the UN Peacekeeping Mission in Ethiopia and Eritrea. First began broadcasting in January 2001. Broadcasts in Amharic, Oromo, Tigrinya, Arabic, Tigrinyan and English.

RADIO VERITAS (Rlg)
SOUTH AFRICA
✉ Radio Veritas Productions, 36 Beelaerts St., 2139 Troyeville, South Africa
☎ +27 (11) 624 2516 or +27 (11) 624 2517
🖹 +27 (11) 614 7711
E-mail: info@radioveritas.co.za.
Web: www.radioveritas.co.za
LP: Director: Fr. Emil Blaser
Notes: Catholic station launched on 1st May 2002. The station has access to Vatican Radio programs and has rebroadcast rights for EWTN radio and TV

RADIO VERITAS ASIA (Rlg)
PHILIPPINES
✉ P.O. Box 2642, Quezon City 1166, Philippines
☎ +63 (2) 939 0012
🖹 +63 (2) 938 1940
E-mail: program-dept@rveritas-asia.org or technical@rveritas-asia.org
Web: www.rveritas-asia.org/technical
LP: Manager: Fr. Carlos S. Lariosa, SVD
V: QSL-card.
Ann: English: "This is Radio Veritas Asia, broadcasting from Quezon City, Philippines"

RADIO VILNIUS (Pub)
LITHUANIA
✉ Konarskio g. 49, LT-2600 Vilnius, Lithuania
☎ +370 (5) 2363021
🖹 +370 (5) 2363208
E-mail: radiovilnius@lrt.lt
Web: www.lrt.lt
LP: Head: Audrius Braukyla
V: QSL-card.
Ann: English: "This is R. Vilnius"; Lithuanian: "Vilniaus radijas uzsieniui"

RADIO VLAANDEREN INTERNATIONAAL (Pub)
BELGIUM
✉ P.O. Box 26, B-1043 Brussels, Belgium
☎ +32 (2) 741 3807
🖹 +32 (2) 741 6295/8336
E-mail: info@rvi.be
Web: www.rvi.be
LP: S.M.: Wim Jansen. Frequency Manager: Hector De Cuyper
V: QSL-card.
IS: "Tussen Maas en Schelde".
Ann: Flemish: "Dit is Radio Vlaanderen Internationaal"; English: "Brussels Calling"; French: "Ici Bruxelles"; German: "Hier ist Brüssel".

RADIO XORIYO (RADIO FREEDOM) (Clan)
ETHIOPIA
✉ 3401 Dufferin St P.O.Box 27618, Toronto, ON M3A 3B8, Canada
E-mail: radioxoriyo@ogaden.com
Web: www.ogaden.com/radio_Freedom.htm
Ann: Somali: "Ku soo dhawaada Radio Xoriyo Codkii Ummadda Ogadeniya"
Notes: Operated by ONLF. Active since May 2000. Broadcasts in Somali.

RADIO YARAN (Clan)
IRAN (ISLAMIC REP. OF)
✉ P.O. Box 1601, Simi Valley, CA 93062, USA
☎ +1 (818) 348 2766
🖹 +1 (818) 348 3627
E-mail: info@afnl.com
Web: www.afnl.com

RADIO YUGOSLAVIA (Gov)
SERBIA AND MONTENEGRO
✉ Radio Yugoslavia, Hilandarska 2/IV, 11000 Beograd, Serbia
☎ +381 (11) 3244455
🖹 +381 (11) 3232014

E-mail: radioyu@bitsyu.net
Web: www.radioyu.org
LP: Director: Milena Jokic; Head of Prgrs: Dr. Zivorad Djordjevic; C.E.: Rodoljub Medan/Predrag Graovac; Head of Public Relations: Aleksandar Popovic.
V: QSL-card. Rec. acc.
Ann: English: "This is Radio Yugoslavia, the International Radio of Serbia and Montenegro."

RADIODIFFUSION-TÉLÉVISION MAROCAINE (Gov)
MOROCCO
✉ 1, Rue El Brihi (or BP 1042), 10000 Rabat.
☎ +212 (3) 7766880
🖨 +212 (3) 7766888
E-mail: rc@maroc.net
Web: www.maroc.net/rc
LP: D.G.: Mohamed Tricha; Dir. Radio: Abderrahman Achour; Dir. TV: Mohamed Issari; Dir. Tech.: Jamal Eddine Tanane; Dir. Ext. Rel: Ali M'Barek; Dir. Finance & Administration:Mehdi Bouzekri.
V: QSL-card. Rec. acc. Rp.
Ann: Arabic: "Huna Ribat, idha'atu-l-mamlaka al Maghribiyya"
Notes: Relays domestic programme on SW.

RADIODIFUSIÓN ARGENTINA AL EXTERIOR (RAE) (Gov)
ARGENTINA
✉ Casilla de Correo 555, Correo Central, C1000WAF Buenos Aires, Argentina
☎ +54 (11) 43256368
🖨 +54 (11) 43259433
E-mail: rae@radionacional.gov.ar
Web: www.radionacional.gov.ar
LP: Directora Nacional: Patricia Barral; Directora RAE: Perla Damuri. Operated by Servicio Oficial de Radiodifusion (SOR)
V: QSL-card.
Ann: English: "This is R.A.E., the International Service of the Argentine Radio".
Notes: On Sat & Sun relays LRA1 R. Nacional

RADIOPHONIKOS STATHMOS MAKEDONIAS (ERA3) (Gov)
GREECE
✉ 2 odos Angelaki, GR-54621 Thessaloniki, Greece.
🖨 +30 31 236370
E-mail: info@ert3.gr
Web: www.ert3.gr

RADYO PILIPINAS OVERSEAS (Gov)
PHILIPPINES
✉ Philippine Broadcasting Sce, 4/F Bldg.,Visayas Ave., Q.C., Manila 1103, Philippines
☎ +63 (2) 924 2267
🖨 +63 (2) 924 2745
E-mail: pbs.pao@pbs.gov.ph

Web: www.pbs.gov.ph
LP: S.M.: Magtanggol Rodriguez
V: QSL-card. Rp. (2 IRC's). Rec. acc.

RAI INTERNATIONAL (Gov)
ITALY
✉ P.O. Box 320, I-00100 Roma, Italy
☎ +39 (06) 335 42526
🖨 +39 (06) 331 70767
E-mail: raiinternational@rai.it
Web: www.raiinternational.rai.it
LP: Director: Roberto Morrione.
V: QSL-card.
Ann: Italian: "RAI International programmi in lingua Italiana per.."; English: "This is the Italian Radio and Television Service Broadcasting from Rome"; French: "Ici la R. Italienne"; German: "Hier ist der Italienische Auslandssendedienst"
Notes: Sunday programmes may change time or be cancelled due to sports coverage.

RDP INTERNACIONAL (Pub)
PORTUGAL
✉ Avenue Duarte Pacheco 26, 1070-110 Lisbon, Portugal.
☎ +351 213 820000
🖨 +351 213 820165
E-mail: rdpinternacional@rdp.pt
Web: www.rdp.pt
LP: Director: Jaime Marques de Almeida
V: QSL-card. Rec. acc.
IS: Opens and closes with National Anthem, preceded by time gong.
Ann: Portuguese: "RDP Internacional-Rádio Portugal a emitir dos seus estúdios em Lisboa"
Notes: The programs ANTENA 1, ANTENA 3 RDP Africa and RDP Internacional may be listened in real audio at RDP home page (see website)

REMNANTS HOPE MINISTRY (Rlg)
UNITED STATES
✉ 4155 Strider Circle, Kannapolis, NC 28081, USA
☎ +1 (704) 932 4951
E-mail: pastortim@remnantshope.com
Web: www.remnantshope.com
LP: Pastor Tim Butler
Notes: Operated by a former Overcomer Ministry follower, to expose corruption within the Ministry.

RTBF INTERNATIONAL (Pub)
BELGIUM
✉ Local 3P09, 52 Bd Reyers, B-1044 Bruxelles, Belgium
☎ +32 (2) 737 4014
🖨 +32 (2) 737 3032
E-mail: rth@rtbf.be
Web: www.rtbf.be
LP: Director, Radio: Gerard Delacroix; RTBF: Ronald Theunen

Notes: Broadcasts in French to Europe and Africa. Some programmes to S. Africa may only be heard at weekends.

RTE RADIO WORLDWIDE (Gov)
IRELAND
✉ RTE, Dublin 4, Ireland.
☎ +353 (1) 208 2350
🖹 +353 (1) 208 3031
Web: www.rte.ie/radio/worldwide.html
LP: Chairman: Farrel Corcoran; D.G.: Bob Collins; M.D., Organisation & Development: Liam Miller; M.D., Commercial: Conor Sexton; Dir. of Finance: Gerry O'Brien; Dir. of Radio: Helen Shaw; Dir. of News: Ed Mulhall; Dir. of Public Affairs: Kevin Healy; Dir. of Corporate Affairs: Tom Quinn
Notes: RTE can be heard on the Internet with Real Audio and via telephone: +32 (2) 509 5050

SALAMA RADIO (Rlg)
UNITED KINGDOM
✉ Harvesttime Ministries, P.O Box 126, Chessington, Surrey, KT9 2WJ, United Kingdom
☎ +44 (208) 395 7425
E-mail: admin@salamaradio.org
Web: www.salamaradio.org
LP: Chairman: Dr Jacob Abdalla; Co-ordinator: Paul Gadzama
V: Letter (Rp.)
Notes: Operated by Harvest Time Ministries and broadcasts in ethnic languages to Africa. Active since August 2001.

SAWT AL MAHABBA (LA VOIX DE LA CHARITÉ) (Rlg)
LEBANON
✉ P.O. Box 850, Jounieh, Lebanon
☎ +961 (9) 918090
E-mail: email@radiocharity.org.lb
Web: www.radiocharity.org.lb
LP: Director: Father Fadi Tabet
V: Letter (Rp) or E-mail.
Notes: Founded in 1984 and active on SW since May 1996. The station promotes the Christian faith and broadcasts programmes of both theological and educational interest. As well as Arabic, the station broadcasts on FM daily in French and English, and once a week in various other languages including Armenian, Latin, Greek, African and Asian languages.

SAWT AL-ISLAH (VOICE OF REFORM) (Clan)
SAUDI ARABIA
✉ MIRA, BCM Box: MIRA, London WC1N 3XX, United Kingdom
☎ +44 (208) 452 0303
🖹 +44 (208) 452 0808
E-mail: info@islah.org
Web: www.miraserve.com
LP: Dr. Saad al-Faqih

Ann: Arabic: "Sawt al-Islah"
Notes: Broadcasts in Arabic. Operated by London based Movement for Islamic Reform in Saudi Arabia (MIRA). Active since December 2002. Schedule and frequencies may vary due to jamming.

SAWT LUBNAN AL-HURRIYA (VOICE OF FREE LEBANON) (Clan)
LEBANON
✉ Rassemblement Pour le Liban, Rue Sainte Anne, F-75002 Paris, France
E-mail: radio@tayyar.org
Notes: Active since November 22, 2002.

SCANDANAVIAN WEEKEND RADIO (Comm)
FINLAND
✉ P.O. Box 35, FIN-40321 Jyväskylä, Finland
☎ +358 400 995 559 Live while on air and SMS Service.
E-mail: info@swradio.net
Web: www.swradio.net
LP: Alpo Heinonen
V: QSL-card, Rp (2 IRC/USD)
Notes: SWR is the first private SW station in Scandanavia. SWR is on the air the first Saturday of every month for 24 hours, starting 0000 local time. (+3 UTC daylight saving time, +2 normal).

SEDAYE KOMALEH (VOICE OF KOMALAH) (Clan)
IRAN (ISLAMIC REP. OF)
✉ P.R., P.O.Box 3015, 145 03 Norsborg, Sweden
☎ +46 (70) 747 6542
🖹 +46 (18) 46 8493
E-mail: radiokom@radiokomaleh.com
Web: www.radiokomaleh.com
Notes: Operated by KOMALA (Kurdish Organisation of the Communist Party of Iran). Broadcasts in Farsi and Kurdish.

SLBC - RADIO SRI LANKA (Gov)
SRI LANKA
✉ P.O. Box 574, Independence Square, Colombo 7, Sri Lanka
☎ +94 (1) 696329
🖹 +94 (1) 695488
E-mail: slbcweb@sri.lanka.net
Web: www.infolanka.com/people/sisira.slbc.html
V: QSL-card. Rp.
IS: Melody on drums
Ann: English: "This is Radio Sri Lanka". During the news: "This is the Sri Lanka Broadcasting Corporation", "This is the All Asia Service of Radio Sri Lanka", "This is Radio Sri Lanka calling listeners in South East Asia, Japan the Far East and Australia".

SW RADIO AFRICA (Clan)
ZIMBABWE
✉ 105 Crystal Palace Rd, East Dulwich SE22

9ES, United Kingdom
☎ +44 (20) 8387 1441
E-mail: mail@swradioafrica.com
Web: www.swradioafrica.com
LP: Simon Surrey
Ann: English: "SW Radio Africa, Zimbabwe's Independent Voice" or "The Independant Voice of Zimbabwe"
Notes: Operated by a London based group of Zimbabwean exiles. Active since December 2001. Broadcasts in English.

SWISS RADIO INTERNATIONAL (Gov)
SWITZERLAND
✉ Giacomettistrasse 1, CH-3000 Bern 15, Switzerland
☎ +41 (31) 350 9551
🗎 +41 (31) 350 9544
E-mail: info@swissinfo.org
Web: www.sri.ch
LP: Dir.: Mr Nicolas D. Lombard; Vice-Dir. & Head of Multimedia: Mr Peter Hufschmid; Head of Marketing & Communication: Mrs Christine Dudle-Crevoisier; Editor in Chief: Mr Peter Salvisberg; Head of I.T. Department: Mr Pascal Dreer
V: QSL-card on request if reports include programme comments.
Ann: English: "Your tuned to Swiss Radio International in Bern"
Notes: SRI is cutting back its shortwave schedule.

T8BZ/KHBN - VOICE OF HOPE (Rlg)
PALAU
✉ P.O. Box 66, Koror, Republic of Palau 96940
☎ +680 488 2162
🗎 +680 488 2163
E-mail: v oh@broadcast.net
Web: www.highadventure.org/voh_china.html
LP: Rolland Lau +(680) 488-2162; Chief Engineer: Richard (Dick) Mann +(680) 544-6008 Voice/Fax; Frequency consultant: George Jacobs.
Ann: English: "Voice of Hope", "Holy Spirit Radio Station" or "High Aventure Ministries Global Broadcast Network"

THE ARABIC RADIO (SAWT AL WATAN) (Clan)
SYRIA
Web: www.arabicsyradio.org
Ann: Arabic: "Al-Idha'ah al-Arabiyyah"; "Sawt al Watan"
Notes: Broadcasts in Arabic. Appears to be anti Syrian Government. Also known as "Voice of Homeland (Sawt Al Watan)".

THE CARIBBEAN BEACON (Rlg)
ANGUILLA
✉ P.O. Box 690, Anguilla, BWI
☎ +1 (264) 497 4340
🗎 +1 (264) 497 4311
Web: www.drgenescott.org
LP: Owner: Dr. Gene Scott; C.E.: Kevin Mooney

V: QSL-card.
Notes: Carries programming from Dr Gene Scott's "University Network".

THE OVERCOMER MINISTRY (Rlg)
UNITED STATES
✉ P.O. Box 691, Walterboro, SC 29488-0007, USA
☎ +1 (843) 538 3892
🗎 +1 (843) 538 4202
E-mail: brotherstair@overcomerministry.com
Web: www.overcomerministry.com
LP: Brother R.G. Stair
Ann: English: "You have been listening to the International Broadcast - The Overcomer - This is the Voice of the Last Day Prophet of God"
Notes: Operated by Brother Ralph G. Stair. Can be heard on many shortwave and MW/FM stations at various times.

THE UNIVERSITY NETWORK (DR GENE SCOTT) (Rlg)
UNITED STATES
✉ PO Box 1, Los Angeles, CA 90053, USA
☎ +1 (818) 240 8151
E-mail: drgenescott@mail.drgenescott.org
Web: www.drgenescott.org/swave.htm (audio and video available)
LP: Dr Gene Scott
Notes: Dr Gene Scott can be heard over numerous radio stations, TV and satellite channels throughout the world.

THE VOICE (ASIA) (Rlg)
INDIA
✉ P.O. Box 1, Kangra-176001, Himachal, India
E-mail: mail@thevoiceasia.com
Web: www.thevoiceasia.com
Notes: Operated by Christian Vision, UK. Broadcasts in Hindi to the Indian Subcontinent. Began broadcasting in June 2002. Programs are produced in the UK and transmitted via Voice International's site in Darwin, Australia.

THE VOICE OF INDONESIA (Gov)
INDONESIA
✉ RRI Headquarters Jl.Medan Merdeka Barat 4-5, Jakarta 10110, Indonesia
☎ +62 (21) 3846817
🗎 +62 (21) 3457134
E-mail: rri@rrionline.com
Web: www.rrionline.com
V: QSL-card or letter. Listeners reports are welcomed.
Ann: English: "Radio Republic Indonesia" or "You are listening to Voice of Indonesia, the overseas service of Radio Republic Indonesia from Jakarta"

TOMORROWS NEWS TODAY (Rlg)
UNITED STATES
✉ P.O. Box 2100, Bowling Green,

KY 42102-2100, USA
☎ +1 (270) 843 354
E-mail: tntreporter@hotmail.com
Web: www.tnt777.com
Notes: Sponsored by Yahweh's Philadelphia Remnant Assembly.

TRANS WORLD RADIO (Rlg)
UNITED STATES
✉ P.O. Box 8700, Cary, NC 27512, USA
☎ +1 (919) 460 3700
▤ +1 (919) 460 3702
E-mail: info2@twr.org
Web: www.gospelcom.net/twr

TRANS WORLD RADIO - TWR SOUTH AFRICA (Rlg)
SOUTH AFRICA
✉ Private Bag 987, Pretoria 0001, South Africa
Web: www.twr.org.za

TRANS WORLD RADIO (GUAM) - KTWR (Rlg)
GUAM
✉ P.O. Box 8780, Agat, Guam 96928
☎ +1 (671) 477 9701
▤ +1 (671) 477 2838
E-mail: ktwrfreq@guam.twr.org
Web: http://gumwww.twr.org
V: QSL-card, 3 IRC's for airmail reply, one for surface mail. Rec. acc. Online report form: http://gumwww.twr.org/ktwrmonrep.htm
IS: We've a story to tell the Nations. played on an organ
Ann: English: "This is your Station for Inspiration, KTWR, Agana" or "This is the voice of Trans World Radio Pacific, Agana".

TRANS WORLD RADIO (TWR EUROPE) (Rlg)
MONACO
✉ B.P. 349, MC 98007, Monte Carlo, Monaco.
☎ +377 165600
▤ +377 165601
E-mail: info2@twr.org
Web: www.twr.org
Ann: English: "This is Monte Carlo. The following programme of Trans World Radio is in the.. language"

TRANS WORLD RADIO (TWR SWAZILAND) (Rlg)
SWAZILAND
✉ P.O. Box 64, Manzini, Swaziland (Letters to: P.O.Box 4232, Kempton Park, 1620 South Africa)
☎ +268 505 2781
▤ +268 505 5333
Web: www.twrafrica.org
LP: President: Thomas J. Lowell; Regional Dir (Africa): Rev. Stephen Boakye-Yiadom; Stn. Dir: Lee Lowell; Director of Prgrs.: Rev. Andrew Macdonald; C.E.: Steve Stavropolous; Freq. Manager: James Burnett.
V: QSL-folder. Rec. acc. preferably from target

areas. IRC's appreciated (3 IRC's for airmail reply).
IS: Last Bar bar of "We've a story to tell the Nations" on Hand Bells
Ann: English: "This is Trans World Radio - Swaziland" or "From the beautiful Kingdom of Swaziland, this is Trans World Radio"

UAE RADIO (DUBAI) (Gov)
UNITED ARAB EMIRATES
✉ P.O. Box 1695, Dubai, United Arab Emirates
☎ +971 (4) 370255
▤ +971 (4) 374111
E-mail: www.dubaitv.gov.ae/amara/FeedbackAr.asp (online form)
Web: www.dubaitv.gov.ae
LP: Director of Information: Sheikh Hasher al Maktoum. Ag.; D.G.: Ahmed Saeed al Gaoud; Controller of Radio: Hassan Ahmed; Controller of Eng: Abdul Rehman Al Ali.
V: QSL-card.
Ann: Arabic: "Idha'at al imarat al Arabiyyah al Mutahhida min Dubai"; English: "This is Emirates Radio from Dubai"
Notes: Times and frequencies are variable.

UNITED NATIONS RADIO (Gov)
UNITED STATES
✉ UN Radio, Secretariat Building, Room S-850A, New York, NY 10017, USA
☎ +1 (212) 963 5201
▤ +1 (212) 963 1307
E-mail: unradio@un.org
Web: www.un.org/av/radio
V: QSL-card. Rpt. to UNESCO Radio, 7 Place de Fontenoy, 75007, Paris, France.
Ann: English: "United Nations Radio in New York"
Notes: Active since 1946. Programmes are distributed to 185 countries in 15 languages. As well as shortwave radio, UN Radio is now using the internet to distribute its programmes.

VATICAN RADIO (Rlg)
VATICAN
✉ Vatican Radio, Piazza 3, I-00120 Vatican City
☎ +39 (6) 698 8305 or Int Rel: +39 (6) 6988 3945
▤ +39 (6) 698 83463
E-mail: promo@vatiradio.va; sedoc@vatiradio.va
Web: www.vaticanradio.org
LP: D.G.: Rev. Pasquale Borgomeo S.I; P.D.: Rev. Federico Lombardi S.I; T.D.: Rev. Lino Dan S.I; C.E.: Pier Vincenzo Giudici; Head of Int. Rel.: Mrs. Solange de Maillardoz
V: QSL-card.
Ann: Before all transmissions: Latin: "Laudetur Jesus Christus" (Praised be Jesus Christ), repeated in the language of the broadcast, then station identification. English: "This is the English programme of Vatican Radio".
Notes: Following concerns over electromagnetic pollution in the vicinity of the Vatican Radio transmitter site near Rome, some transmitters

may be operating on a reduced schedule and/or at reduced power. It has also been reported that Vatican Radio plans to transfer its high-power mediumwave transmissions to another location – possibly using relay facilities elsewhere in Europe. Other languages are available via internet. Programmes for overseas are relayed to Europe on MW. Vatican Radio broadcasts multilingual programmes within a given time slot. Some transmissions are on Sundays and Holy days.

VOICE INTERNATIONAL (Rlg)
AUSTRALIA
✉ P.O. Box 1104, Buderim, QLD 4556, Australia
☎ +61 (7) 5477 1555
▤ +61 (7) 5477 1727
E-mail: voice@voice.com.au
Web: www.voice.com.au
LP: S.M.: Raymond Moti; Operations Dir.: Terry Bennett; Christian Voice Australian Dir.: Mike Edmiston; Tech. Supervisor: Matthew Bodman; QSL signer: Mrs Lorna Manning
V: QSL-card, rp.
Notes: Christian Vision purchased the former R. Australia Darwin transmitter site in June 2000. Operated by Christian Vision, based in the UK. The new studio complex (Asia Pacific Broadcasting Centre) in Maroochydore, near Brisbane opened on 14th September 2002.

VOICE OF AFRICA/LIBYAN JAMAHIRIYA BROADCASTING (Gov)
LIBYA
✉ P.O. Box 9933, Soug al Jama, Tripoli, Libya
☎ +218 (21) 603191/5.
Web: www.ljbc.net
V: QSL-card.
Ann: Arabic: "Idha'at saout al-watan al'arabbiy al-Kabir"
Notes: Known as "Voice of the Arab Nation" prior to October 1998.

VOICE OF AMERICA (Gov)
UNITED STATES
✉ 330 Independence Ave SW, Washington, DC 20237, USA
☎ +1 (202) 260 4330
▤ +1 (202) 619 1241
E-mail: letters@voa.gov
Web: www.voa.gov
LP: VOA Director: David Jackson; Program Director: Mrs. Myrna Whitworth; Director of Broadcast Operations: Thomas R. Morgan
V: QSL-card.
Ann: At the start and end of the transmission period on each frequency: English: "This is the Voice of America, Washington, DC, signing on/off.". Before all foreign language programs: "This is the Voice of America. The following program is in.. (language)."

VOICE OF ARMENIA (Pub)
ARMENIA
✉ A.Manukyan Street 5, 375025 Yerevan, Armenia
☎ +374 (1) 558010
▤ +374 (1) 551513
E-mail: pr@armradio.am
LP: Director: Levon V. Ananikian.
V: QSL-card.
Ann: Armenian: "Yerevan e khosum" or "Yeterum Hayastani Dzaynn e"; English: "You are listening to the Voice of Armenia"; French: "Ici Erevan. Vous écoutez la Voix d'Armenie."

VOICE OF AZERBAIJAN (Gov)
AZERBAIJAN
✉ M.Hüseyn Street 1, 370011 Baki, Azerbaijan
☎ +994 (12) 398585
▤ +994 (12) 395452
E-mail: root@aztv.baku.az
Web: www.aztv.az
LP: Director.: Mrs Taman Bayatli-Oner.
V: QSL-card.
Ann: Azeri: "Danisir Baki"; English: "This is the Voice of Azerbaijan"

VOICE OF BIAFRA INTERNATIONAL (Clan)
NIGERIA
✉ 733, 15th Street NW, 3700 Washington DC 20005, USA
☎ +1 (202) 347 2983
E-mail: oguchi@mbay.net
Web: www.biafraland.com/vobi.htm
LP: M.D.: Oguchi Nkwocha
V: E-mail.
Ann: English: "This is Voice of Biafra International, coming to you from Washington D.C."
Notes: Operated by Washington D.C. based Biafra Foundation. Active since September 2001. Broadcasts in English and Igbo.

VOICE OF CHINA (Clan)
CHINA
✉ 2261 Morello Avenue, Suite A, Pleasant Hill, CA 94523, USA
☎ +1 (510) 687 2354
▤ +1 (510) 687 7396
E-mail: sdchina@aol.com
Ann: Mandarin: "Zhongguo zhi yin"
Notes: Operated by the "Foundation for China in the 21st Century". Active since April 1991. Broadcasts in Mandarin.

VOICE OF DEMOCRATIC ERITREA (Clan)
ERITREA
✉ Postfach 1946, D-65409 Rüsselsheim, Germany
Web: www.meskerem.net
Ann: Tigrinyan: "Demtsi Demcrasiyawit Writrea"; Arabic: "Sawt Eritrea al-Dimuqratuya - Sawtu jabhat al-Tahirir al-Eritrea".

VOICE OF ETHIOPIA (Clan)
ETHIOPIA

✉   P.O. Box 361, Fishers, IN 46038, USA
☎   +1 (317) 989 8326
E-mail: info@democracyfrontiers.org
Web: www.democracyfrontiers.org

VOICE OF ETHIOPIAN MEDHIN (Clan)
ETHIOPIA

✉   P.O. Box 111423, D-60049, Frankfurt,
      Germany
☎   +49 (693) 710 8253
Web: www.mehedin.com/radio
Ann: Amharic: "Yih Ye Ethiopia Medhin Dimts
      New"
Notes: Active since June 2000. Operated by
      Ethiopian Salvation Democratic Party, also
      known as Ethiopian Medhin Democratic
      Party. Broadcasts in Amharic.

VOICE OF GREECE (ERA5) (Gov)
GREECE

✉   Messogion 432, GR-153 42 Aghia Paraskevi
      Attikis Athens, Greece.
☎   +30 (10) 6066 308/ 6066 297
▤   +30 (10) 6066 309
E-mail: era5@ert.gr
Web: www.ert.gr (Live Audio)
LP: M.D.: Gina Vogiatzoglou; Head of Prgrs.:
      Angeliki Barka; Planning Engineer: D.
      Angelogiannis; Tech. Support: Char.
      Charalambopoulos.
V: QSL-card. Re in SINPO code.
Ann: Greek: "Era Pente, I Foni Tis Elladas"

VOICE OF HOPE (Clan)
SUDAN

✉   P.O. Box 33829, Kampala, Uganda
☎   +256 41 220334
E-mail: hope@africaonline.co.ug
Web: www.radiovoiceofhope.net
Ann: English: "This is Radio Voice of Hope, the
      voice for the voiceless in Southern Sudan"
Notes: Active since December 2000. The programme
      is produced at Catholic Church studios in
      Kampala, sponsored by New Sudan Council
      of Churches, Uganda and in Hilversum,
      Holland.

VOICE OF ISLAMIC REVOLUTION IN IRAQ
(Clan)
IRAQ

✉   27a Old Gloucester St., London WC1N 3XX,
      United Kingdom
☎   +44 (171) 371 6815
▤   +44 (171) 371 2886
E-mail: sciri@btinternet.com
Web: www.sciri.org
LP: Muhammad Baqir al-Hakim or Dr. Hamid Al
      Bayati (SCIRI UK)
Ann: Arabic: "Sawt al-Thawrah al-Islamiyah fi al-
      Iraq"
Notes: Operated by: Shii Supreme Council of the

Islamic Revolution of Iraq (SCIRI). Active
since 1983. Broadcasts in Arabic.

VOICE OF JAMMU KASHMIR (Clan)
INDIA

✉   P.O.Box 102, Muzaffarabad (Azad Kashmir),
      Pakistan
E-mail: harakat@muslimsonline.com
Web: www.ummah.org.uk/kashmir
LP: Islam-ud Din Butt
Ann: Kashmiri: "Sada-i Hurriyat-i Jammu
      Kashmir"; English: "This is Radio Hurriyat-i
      Jammu Kashmir"
Notes: Active since 1999. Operated by Harkat ul-
      Mujahideen and backed by Pakistan and Al
      Qaeda. Broadcasts in Kashmiri, English and
      Urdu. Previously known as "Voice of
      Independant Kashmir"

VOICE OF KAMPUCHEA KROM (KHMER
KROM) (Clan)
VIETNAM

✉   Khmer Kampuchea Krom Federation, PO
      Box 6239, Lakewood, CA 90714, USA
☎   +1 (562) 598 5431
▤   +1 (562) 598 8089
E-mail: vokk@khmerkrom.org
Web: www.khmerkrom.org/radio
Ann: Vietnamese: "Thini Vitthayu Samleng Khmer
      Kampuchea Krom"
Notes: Operated by the Khmer Kampuchea Krom
      Federation. Broadcasts in Khmer. Active since
      1st June 2001.

VOICE OF KOREA (Gov)
KOREA, NORTH (DPR)

✉   Voice of Korea, Pyongyang, Democratic
      Republic of Korea
☎   +850 (2) 381 6035
▤   +850 (2) 381 4416
V: QSL-card.
IS: Song of General Kim Il Sung. Opening &
      closing music for Korean sce: Nat. Anthem
Ann: Korean: "Joson Jung-ang Pangsong-imnida"
      or "Pyongyang Pangsong-imnida"; Arabic:
      "Sawt Kuriya"; Chinese: "Chaoxian zhi
      Sheng Guangbo Diantai"; English: "This is
      Voice of Korea"; French: "La Voix de la
      Coree"; German: "Stimme Koreas"; Japanese:
      "Choson nokoe hoso desu"; Russian: "Golos
      Korei"; Spanish: "Aqui la Voz de Corea"

VOICE OF MALAYSIA (Gov)
MALAYSIA

✉   P.O. Box 11272, 50740 Kuala Lumpur,
      Malaysia
☎   +60 (3) 22887824
▤   +60 (3) 22847594
E-mail: vom@rtm.net.my
Web: www.rtm.net.my/vom
LP: Controller: Stephen Sipaun.
V: QSL-card. Rec. acc.
IS: First bar of Nat. Anthem "Negara Ku" (Chimes)

Ann: Malay: "Inilah Suara Malaysia"; English: "This is the Voice of Malaysia"

Notes: Tagalog programmes are produced at RTM Kota Kinabalu, Sabah. Chinese programmes are in Mandarin except for news in Cantonese at 1100.

VOICE OF MONGOLIA (Gov)
MONGOLIA

✉ P.O. Box 365, Ulaanbaatar 13, Mongolia

☎ +976 (1) 327900

🖷 +976 (1) 323096

E-mail: mr@mongol.net

Web: www.mongol.net/vom

LP: Director, Foreign Service: Mrs B. Narantuya; Head of International Relations Section: Ms Bolor Purevdorj; Protocol Officer: Mrs Jamyanjamts

V: QSL-card (9 different cards available). Re. to relevant language section. Rec. acc. (non returnable), Rp (2 IRC's or $1) appreciated.

Ann: Mongolian: "Ullaanbaataraas yaridz baina"; English: "This is the Voice of Mongolia"

VOICE OF NATIONAL SALVATION (Clan)
KOREA, SOUTH (REP. OF)

✉ Kankoku Minzoku Minshu Sensen, Grenier Osawa 107, 40 Nando-cho, Shinjyuku-ku, Tokyo, 162-0837, Japan

E-mail: ndfsk@campus.ne.jp

Web: www.ndfsk.dyn.to

Ann: Korean: "Kuguge sori pangsong-imnida."; English: "This is the English language service of the Voice of National Salvation, the National Democratic Front of South Korea."

Notes: Active since 1985 (Active since 1967 as Radio Station of the South Korean Democratic National League for Liberation, and from June 1970 as the Voice of the Revolutionary Party for Reunification). Operated by National Democratic Front of South Korea. Broadcasts in Korean and English.

VOICE OF NIGERIA (Gov)
NIGERIA

✉ Broadcast House, Ikoyi, PMB 40003, Falomo, Lagos, Nigeria

☎ +234 (1) 269 3078

🖷 +234 (1) 269 1944

E-mail: vonlagos@fiberia.com

Web: www.voiceofnigeria.org

LP: D.G.: Taiwo Allimi; Pgr. Dir.: Alhaji Ayodele Sulaiman; News Dir.: Ben Egbuna.

V: QSL-card.

IS: As Home Sce. Also bells playing the first bars of the National Anthem 15 mins. before the commencement of each block period.

Ann: English: "This is the Voice of Nigeria".

Notes: Founded in 1961. Has studios in Lagos and Abuja (the Federal capital).

VOICE OF OROMIYAA (Clan)
ETHIOPIA

E-mail: a-jelil.abdella@pca.state.mn.us

Web: www.voiceoforomiyaa.com

Ann: Amharic: "Sagalee Oromiyaa"

Notes: Operated by a US based group. Active since December 2001. Broadcasts in Afaan Oromo.

VOICE OF OROMO LIBERATION (Clan)
ETHIOPIA

✉ P.O. Box 510620, D-13366 Berlin, Germany

☎ +49 (30) 4941036

🖷 +49 (30) 4943372

E-mail: sbo13366@aol.com

Web: www.oromoliberationfront.org/sbo.html

LP: Secretary, SBO Committee: Taye Teferra

Ann: Amharic: "Kun Sagalee Bilisummaa Oromoo"

Notes: Active since July 1988. Operated by Oromo Liberation Front. Broadcasts in Amharic and Afaan Oromo. Has broadcast from USA (WWCR and WHRI).

VOICE OF RUSSIA (Gov)
RUSSIA

✉ Pyatnitskaya 25, 115326 Moscow, Russia

☎ +7 (095) 950 6278

🖷 +7 (095) 230 2828

E-mail: etters@vor.ru

Web: www.vor.ru

LP: M.D.: Armen Oganesyan

V: QSL-card.

Ann: English: "This is Moscow, You are tuned to the World Service of the Voice of Russia"

VOICE OF SOUTHERN AZERBAIJAN (Clan)
IRAN (ISLAMIC REP. OF)

✉ Vosa Ltd., P.O. Box 108, A-1193 Wien, Austria

☎ +31 (307) 192189

Notes: The Voice of Southern Azerbaijan is believed to be operated by the National Revival Movement of Southern Azerbaijan. Active since January 2003. Broadcasts in Farsi.

VOICE OF THE DEMOCRATIC PATH TO ETHIOPIAN UNITY (Clan)
ETHIOPIA

✉ Finote Democracy, PO Box 88675, Los Angeles, CA 90009, USA.

E-mail: efdpu@finote.org

Web: www.finote.org (on demand audio)

Ann: Amharic: "Yih Finote Demokrasi Ye-Ethiopia Andinet Dimts"

Notes: Active since January 2000. Operated by Finote Democracy, a group of human rights motivated Ethiopians living in the Netherlands, USA and Germany. Broadcasts in Amharic.

VOICE OF THE ISLAMIC REPUBLIC OF IRAN (Gov)
IRAN (ISLAMIC REP. OF)

✉ P.O. Box 19395-6767, Teheran, Iran

☎ +98 (21) 2042808

🖷 +98 (21) 2051635

E-mail: webmaster@irib.com

Web: www.irib.com
LP: President : Dr Ali Larijani; Director, Int. Relations / Broadcasting Union: M.Safdari.
V: QSL-card. Rec. acc.
Ann: English: "This is Tehran, the Voice of the Islamic Republic of Iran"; Persian: "Inja Tehran ast, seda-ye jomhuri-ye eslami-ye Iran"; French: "Ici Tehran la Voix de la République Islamique de l'Iran"; Russian: "Govorit Tegeran, Golos Islamskoy Respubliki Iran"; Turkish: "Burasi Tehran, Iran Islam"; Arabic: "Huna Tehran saut al-gumhuriyati-l-islamiyya fi l-Iran".
Notes: Extra broadcasts in Azeri, Tajik and Turkish for Ramadan (6th November to 6th December) are included in the schedule.

VOICE OF THE MEDITERRANEAN (Gov)
MALTA
✉ Chircop Building Floor 2, Valley Road, Birikirkara, BKR14, Malta
☎ +356 22797000
📄 +356 22797111
E-mail: info@vomradio.com
Web: www.vomradio.com
LP: M.D.: Richard Muscat
V: QSL-card (online form available on website)
Ann: English: "Voice of the Mediterranean", "Radio Melita" or "Valetta Calling"

VOICE OF THE PEOPLE (Clan)
KOREA, NORTH (DPR)
Ann: Korean: "Yoginun Pyongyang-eso ponaedurinun inminuisoribang-imnida".
Notes: Active since June 1985. Operated by the Korean Workers Union (Korean Armed Forces). Broadcasts in Korean.

VOICE OF TIBET (Clan)
CHINA
✉ Wellhavensgate 1, N-0166 Oslo, Norway
☎ +47 (22) 11 49 80
📄 +47 (22) 11 49 88
E-mail: votibet@online.no
Web: www.vot.org
Notes: Broadcasts in Tibetan dialects including u-Tsang, Amdo, Kham. Active since July 1996.

VOICE OF TURKEY (Gov)
TURKEY
✉ P.K. 333, 06443 Yenisehir-Ankara, Turkey
☎ +90 (312) 490 9842
📄 +90 (312) 490 9846
E-mail: genel.sekreterlik@trt.net.tr
Web: www.trt.net.tr
LP: D.G.: Yucel Yener; Dep. D.G. (Eng): Haluk Buran; Dep. D.G. (Prgr): Bulent Varol; Dep. D.G. (Admin): Yalcin Yeniaras; Sec. Gen.: Ayhan Karapars; International Technical Relations: Elif Soyata Arslan.
V: QSL-card.
Ann: English: "This is the Voice of Turkey"; German: "Hier ist die Stimme der Turkei"; Turkish: "Burasi Turkiyenin Sesi Radyosu"

VOICE OF VIETNAM (Gov)
VIETNAM
✉ 58 Quan Su Street, Hanoi, Vietnam
☎ +84 (4) 8240044
📄 +84 (4) 8261122
E-mail: qhqt.vov@hn.vnn.vn
Web: www.vov.org.vn
LP: Director: Dao Dinh Tuan.
Ann: English: "This is the Voice of Vietnam, broadcasting from Hanoi".
Notes: Some Voice of Vietnam frequencies are subject to variation and usage can be erratic.

VOIX DE L'ORTHODOXIE (GOLOS PRAVOSLAVIYA) (Rlg)
FRANCE
✉ BP 416-08, F-75366 Paris CEDEX 08, France
☎ +7 (812) 3232867 (Russia)
📄 +7 (812) 3232867 (Russia)
Web: www.russie.net/orthodoxie/vo
V: QSL-card.

VOZ CRISTIANA/VOZ CRISTA (Rlg)
CHILE
✉ Castilla 490, Santiago 3, Chile.
☎ +56 (2) 855 7046
📄 +56 (2) 855 7053
E-mail: comentarios@vozcristiana.com
Web: www.vozcristiana.com (Spanish) or www.vozcrista.com (Portuguese)
LP: World Director: Terry Bennett; Reg. Americas Director: Juan Mark Gallardo; General Operations Director (Chile): Samuel Svensson; Programming Director, David Gozalez.
Ann: Spanish: "Radio Voz Cristiana"
Notes: Active since July 1998. Operated by Christian Vision, which is based in the UK and also operates stations in Zambia, Australia and India.

VT MERLIN COMMUNICATIONS LTD.
UNITED KINGDOM
✉ 20 Lincoln's Inn Fields, London WC2A 3ED, United Kingdom
☎ +44 (20) 7969 0000
📄 +44 (20) 7936 6223
E-mail: marketing@merlincommunications.com
Web: www.vtplc.com/merlin
LP: C.E.: Fiona Lowry; Marketing Manager: Laura Jelf

WALES RADIO INTERNATIONAL (Gov)
UNITED KINGDOM
✉ Wales Radio International / Radio Rhyngwladol Cymru, Pros Kairon, Crymych, Pembrokeshire, SA41 3QE, Wales, United Kingdom
☎ +44 (1437) 563361
📄 +44 (1239) 831390
E-mail: jenny@wri.cymru.net
Web: http://wri.cymru.net
Ann: English: "Wales Radio International"

WBCQ - THE PLANET (Rlg)
UNITED STATES
- ✉   97 High Street, Kennebunk, MA 04043, USA
- ☎   +1 (207) 985 7547
- **E-mail:** wbcq@gwi.net
- **Web:** www.wbcq.us
- **LP:** Owner Allan Weiner; Frequency Consultant: George Jacobs
- **V:** by QSL-card rp (1 IRC) to WBCQ, 274 East Road, Monticello, Maine 04760 USA.
- **Ann:** English: "This is WBCQ, in Monticello, Maine broadcasting from the USA"
- **Notes:** Has separated into WBCQ 1, 2 and 3 programmes. Active since September 1998. When not broadcasting it's own programmes, WBCQ hires airtime on it's transmitters to other broadcasters.

WBOH - WORLDWIDE BEACON OF HOPE (Rlg)
UNITED STATES
- ✉   Fundamental Broadcasting Network, 520 Roberts Road, Newport, NC 28570, USA
- **E-mail:** fbn@clis.com
- **LP:** Pastor Clyde E.Eborn
- **Web:** www.fbnradio.com (WinAmp audio available)
- **Notes:** Operated by Fundamental Broadcast Network, who also own and operate WTJC (see separate entry). Future plans include services in Spanish and Brazillian

WEWN - GLOBAL CATHOLIC RADIO (Rlg)
UNITED STATES
- ✉   5817 Old Leeds Rd., Irondale, AL 35210-2164, USA
- ☎   +1 (205) 271 2900
- 🖷   +1 (205) 271 2926
- **E-mail:** gtapley@ewtn.com
- **LP:** S.M.: Richard Jones; Frequency Manager: Joseph A. Dentici
- **V:** QSL-card. Rp (3 IRC's).
- **Ann:** English: "This is WEWN, Global Catholic Radio, Birmingham Alabama, USA"
- **Notes:** Operated by the Eternal Word TV Network. Began broadcasting in December 1992

WHRA - WORLD HARVEST RADIO (Rlg)
UNITED STATES
- ✉   P.O. Box 12, South Bend, IN 46624, USA.
- ☎   +1 (574) 291 8200
- 🖷   +1 (574) 291 9043
- **E-mail:** whr@lesea.com
- **Web:** www.whr.org (Real Audio available)
- **LP:** G.M.: Peter Sumrall; C.E.: Douglas Garlinger; Frequency Consultant: Douglas Garlinger; Sales/Program Coordinator: Joseph Brashier.
- **V:** QSL-card.
- **Ann:** English: "This is World Harvest Radio, the International Service of LeSea Broadcasting Corporation. WHRA transmits from Greenbush, Maine in the United States of America"
- **Notes:** Operated by LeSea Broadcasting, Inc. In addition to carrying its own programmes, airtime is hired to other religious and political organizations. Programmes may be carried in a variety of languages appropiate to the target area.

WHRI - WORLD HARVEST RADIO (Rlg)
UNITED STATES
- ✉   P.O. Box 12, South Bend, IN 46624, USA.
- ☎   +1 (574) 291 8200
- 🖷   +1 (574) 291 9043
- **E-mail:** whr@lesea.com
- **Web:** www.whr.org (Real Audio available)
- **LP:** G.M.: Peter Sumrall; C.E.: Douglas Garlinger; Frequency Consultant: Douglas Garliner; Sales/Program Coordinator: Joseph Brashier.
- **V:** QSL-card
- **Ann:** "This is World Harvest Radio, the International Service of LeSea Broadcasting Corporation. WHRI transmits from Noblesville, Indiana in the United States of America."
- **Notes:** Operated by LeSea Broadcasting Inc. In addition to carrying its own programmes, airtime is hired to other religious and political organizations. Programmes may be carried in a variety of languages appropriate to the target area.

WINB - WORLD INTERNATIONAL BROADCASTERS (Rlg)
UNITED STATES
- ✉   P.O. Box 88, 2900 Windsor Road, Red Lion, PA 17366, USA
- ☎   +1 (717) 244 5360
- 🖷   +1 (717) 246 0363
- **E-mail:** info@winb.com
- **Web:** www.winb.com
- **V:** QSL-card and schedule
- **Ann:** English: "This is WINB, Red Lion, Pennsylvania in the United States of America"

WJIE SHORTWAVE (Rlg)
UNITED STATES
- ✉   P.O. Box 197309, Louisville, KY 40259, USA
- ☎   +1 (502) 968 1220
- 🖷   +1 (502) 964 4228
- **E-mail:** wjiesw@hotmail.com
- **Web:** www.wjiesw.com
- **Notes:** Station was previously known as WJCR. Has moved from Upton to Louisville. Relays programmes from the FM Gospel station (WJIE) in Louisville. An international outreach of Evangel World Prayer Center, Louisville, KY USA

WMLK (Rlg)
UNITED STATES
- ✉   P.O. Box C, Bethel, PA 19507, USA
- ☎   +1 (717) 933 4518
- **E-mail:** aoy@wmlkradio.net
- **Web:** http://wmlkradio.net

LP: Director: Elder Jacob O. Meyer; C.E.: Gary McAvin.
V: QSL-card. Rp.
Ann: English: "This is Radio Station WMLK"

WMR - WORLD MUSIC RADIO (Comm)
DENMARK
✉ P.O. Box 112, DK-8900 Randers, Denmark
☎ +45 (70) 222 222
🖷 +45 (70) 222 888
E-mail: wmr@wmr.dk
Web: www.wmr.dk
LP: Owner and General Manager: Stig Hartvig Nielsen
V: QSL-card. Rp preferred. Rec. acc.
Ann: English: "International music radio - this is World Music Radio"

WRMI - RADIO MIAMI (Comm)
UNITED STATES
✉ P.O. Box 526852, Miami, FL 33152, USA
☎ +1 (305) 559 9764
🖷 +1 (305) 559 8186
E-mail: info@wrmi.net
Web: www.wrmi.net
LP: G.M.: Jeff White; C.E.: Indalecio Espinosa.
V: QSL-card. Reception reports greatly appreciated.
Ann: English: "This is WRMI, Radio Miami International"

WRNO WORLDWIDE (Comm)
UNITED STATES
✉ P.O. Box 895, Fort Worth, TX 76101, USA
☎ +1 (504) 889 2424
🖷 +1 (504) 889 0602
E-mail: hope@goodnewsworld.org
Web: www.goodnewsworld.org
LP: C.E.: Jack Bruce; Head of P.R.: David Schneider; Frequency Consultant: George Jacobs.
V: QSL-card. Rp. (2 IRC's).
IS: "When the Saints go marching in"
Ann: English: "From New Orleans, Louisiana, your listening to WRNO Worldwide broadcasting from the United States of America"

WSHB (HERALD BROADCASTING SYNDICATE) (Rlg)
UNITED STATES
✉ Shortwave Broadcasts, PO Box 1524, Boston MA 02117-1524, USA
☎ +1 (617) 450 2929
E-mail: Letters and reception reports: letterbox@csps.com
Web: www.tfccs.com/gv/CSPs/Herald/bdcst/bdcst.jhtml
LP: Broadcast Director: Catherine Aitken-Smith; WSHB Station Manager: C. Ed Evans (Email:evansc@wshb.com); Frequency and Production Coordinator: Tina Hammers; Chief Engineer: Damian Centgraf; Transmission Engineer: Tony Kobatake; QSL Coordinator: Cindy Reihm.
Ann: English: "This is WSHB, Cypress Creek, South Carolina in the United States of America"
Notes: Toll Free tel. no. in US: (800) 288 7090 (extension 2060 to hear recorded frequency information, or 2929 for Shortwave Helpline to request printed schedules and information). Mix of religious programs 7 days a week in English, Spanish, German, French, Russian and Portuguese, including the Sunday Church Service [English only].

WTJC (WORKING TILL JESUS COMES) (Rlg)
UNITED STATES
✉ Grace Missionary Baptist Church, 520 Roberts Road, Newport, N.C. 28570, USA.
☎ +1 (252) 223 6088
E-mail: fbn@clis.com
Web: www.fbnradio.com (WinAmp audio available)
LP: Pastor Clyde E.Eborn; Engineer: David Robinson
Ann: English: "WTJC Newport, North Carolina USA"
Notes: Active since September 1999. Operated by Fundamental Broadcasting Network (FBN)

WWBS (Rlg)
UNITED STATES
✉ Radio Station WWBS, 965 Hickory Ridge Drive, Macon, GA 31204-1018, USA
☎ +1 (912) 745 1485
E-mail: wwbs@bmi.net
Web: www.wwbs.org
LP: Owner: Charles C. Josey (deceased)
Ann: English: "WWBS Macon, Georgia, United States of America"
Notes: Active since December 1998.

WWCR - WORLDWIDE CHRISTIAN (Rlg)
UNITED STATES
✉ 1300 WWCR Avenue, Nashville, TN 37218, USA
☎ +1 (615) 255 1300
🖷 +1 (615) 255 1311
E-mail: wwcr@aol.com or askwwcr@wwcr.com
Web: www.wwcr.com
LP: President: Fred P. Westenberger; G.M.: George McClintock; Operations Manager: Adam W. Lock, Sr.
V: QSL-card. Rp. preferred (SAE or one IRC). Rec. acc.
Ann: English: "This is World Wide Christian Radio-WWCR, Nashville, Tennessee USA"
Notes: 2 x 100kW transmitters are used exclusively by the Overcomer Ministry and the University Network. Permit holder WNQM Inc.

WWRB (Rlg)
UNITED STATES
✉ P.O. Box 7, Manchester, TN 37349-0007, USA

☎ +1 (931) 841 0492
E-mail: dfrantz@tennessee.com
Web: www.wwrb.org
LP: Owner & C.E.: Dave Frantz
V: QSL-card.
Notes: Active since July 1995 as WGTG. Callsign changed to WWFV in November 2000. Station relocated to Manchester, Tennessee and redesignated as WWRB. WWFV in McCayesville is kept as a backup station.

WYFR - FAMILY RADIO (Rlg)
UNITED STATES
✉ Family Radio, 290 Hegenberger Rd, Oakland, CA 94621, USA
☎ +1 (510) 568 6200
🖹 +1 (510) 633 7983
E-mail: shortwave@familyradio.com
Web: www.familyradio.com
LP: President: Harold Camping; Dir. Tech.: Wesley D. Becker; Prgr. Mgr.: Thomas A. Schaff; Dir. Int. Rel.: Richard H.Homeres; International Manager: David Hoff.
V: QSL-card. No tapes.
Ann: English: "This is your Family Radio, International Broadcast Station WYFR, Okeechobee, Florida, theUnited States of America"; German: "Dies ist Ihr Familien Radio, die Internationale Radio-Station WYFR, in Okeechobee, Florida, der Vereinigten Staatan von Amerika"; Spanish: "Esta es Family Radio WYFR con el Sonido de la Nueva Vida transmitiendo desde Okeechobee, Florida, y con estudios de produccion ubicados en Oakland, California, Estados Unidos de Norte America".

XERMX - RADIO MÉXICO INTERNACIONAL (Gov)
MEXICO
✉ Apartado Postal 21-300, Mexico D.F. 04021, Mexico
☎ +52 56281700
🖹 +52 560 46753
E-mail: rmi@imer.com.mx
Web: www.imer.gob.mx
LP: Martin Rizo Gavira
Ann: English: "On air, from Mexico to the World, Radio Mexico International"

YLE RADIO FINLAND (Pub)
FINLAND
✉ FIN-00024 Yleisradio, Finland
☎ +358 (9) 14801
🖹 +358 (9) 1480 1169
E-mail: rfinland@yle.fi
Web: www.yle.fi/rfinland
LP: Head of International Radio: Mr Juhani Niinisto; Managing Editor for foreign language broadcasting: Mrs Heidi Zidan; Managing Editor for external broadcasting in Finnish and Swedish: Mr Pertti Seppä; Systems Manager: Mr Janne Nieminen.

V: All correspondence concerning technical verifications should be addressed to Digita Ltd. YLE does not verify technical reports.
Notes: YLE international services are intended for Finnish nationals living abroad. Broadcasts in English, German and French were stopped in late 2002. YLE Capital FM, (YLE Capital FM is a registered subname of YLE, the Finnish Broadcasting Co.), Helsinki 97.5 MHz, broadcasts relays of many internationally available radio stations in several languages, as well as YLE programming in English and Russian.

Clubs for DXers & Listeners

This section lists non-commercial hobby clubs serving international radio enthusiasts. Some are oriented to programme listening, and others to DXing, the reception of low power or distant stations. In most cases, bulletins are produced on a regular basis. Sample copies are generally available for return postage (3 or 4 IRCs). Unless otherwise stated, all clubs are international, publish bulletins in English (E.; F = French, G = German, I = Italian, S = Spanish) and cover all aspects of the hobby. This list does not include clubs run by commercial publications or by individual broadcasters.

Europe

European DX Council, P.O. Box 18120, 50129, Firenze, Italy. Umbrella organization of DX Clubs in Europe. Secretary General Luigi Cobisi, Assistant Secretary General: Paolo Morandotti E-mail: sg@edxc.org Web: www.edxc.org

AUSTRIA: **Assoziation Junger DXer in Österreich,** ADXB-OE, Postfach 1000, A-1082 Wien. Tel. +43 2287 5162. (G) E-mail: adxbsuess@aon.at Web: www.adxb-oe.org Publication: *ADXB-OE Rundschreiben* quartely. Annual DX camp.

BELGIUM: **DX-Antwerp**, P.O. Box 16, B-2660 Hoboken (Flemish). E-mail: dxa@dxa.be Web: www.dxa.be – **Radio Contact**, Avenue des Croix de Guerre, 94, 1120 Brussels E-mail: fred.quentin@infonie.be

BULGARIA: **Association of Balkan Cross-band DXers**, ABCDX, c/o Rumen Pankov, P.O.Box 199, 1000 Sofia-C – **Bulgarian DX Club plus Satellite**, c/o Ivan Penev, P.O. Box 47, Sofia 11, Bulgaria 1111. E-mail: penev@internet-bg.net

CZECH REPUBLIC/SLOVAK REPUBLIC: **Czechoslovak DX Club**, c/o Václav Dosoudil, Horní 9, 768 21 Kvasice. (Czech/Slovak/E). E-mail: mail@csdxc.cz Web: www.csdxc.cz Publication: *DX Revue* (Czech)

DENMARK: **Danish Shortwave Club International**, Tavleager 31, DK-2670 Greve. E-mail: kaj.bredahl@post.tele.dk Web: www.dswci.org – **Dansk DX Lytter Klub**, P.O.Box 392, DK-8100 Aarhus C (Danish). E-mail: ddxlk@ddxlk.net Web: www.ddxlk.net

Publication: *DX-Focus (alt om radio)* (Danish)

FINLAND: **Suomen DX-liitto**, P.O.Box 454, 00101 Helsinki (Umbrella organization of Finnish language DX Clubs) E-mail: toimisto@sdxl.org (Finnish/E) Web: www.sdxl.org Publication: *Radiomaailma* (Finish/E) – **Finlands Svenska DX-Förbund r.f**, P.O. Box 9, 68601 Jakobstad (Umbrella organization of various Swedish-language DX clubs in Finland). E-mail: cristere@alcom.aland.fi Web: www.saunalahti.fi/~bkl/fsdxf.htm (Swedish).

FRANCE: **Amitié Radio**, B.P.56, 94002 Creteil Cédéx (F). E-mail: roland.paget@wanadoo.fr Web: http://amitie.multimania.com Publication: *A l'Ecoute du Monde* (F) – **Association Union des Ecouteurs Français**, B.P. 31, 92242 Malakoff Cédéx (F, also publishes magazine on disk in F/E) E-mail: tsfinfo@magic.fr Web: www.radiocom.org – **Monde & Radiodiffusion**, 65, Montée des Princes, F-84100 Orange (F/E) – **Radio Club de la Poste**, Marcel Lecerf, 13 avenue St Michel, 54220 Malzéville – **Radio Club du Perche**, 12 rue du Grand Thuret, 72320 Greez sur Roc. – **Radio Club International Ondes Courtes**, 19 lot Saturne, 26120 Malissard – **Radio DX Club d'Auvergne**, Centre Municipal P. et M. Curie, 2 bis, Rue du Clos Perret, 63100 Clermont-Ferrand. Tel/Fax: +04 73 37 08 46.

GERMANY: **ADDX** (der Assoziation deutschsprachiger Kurzwellen-hörer) (G). Jointly issues *Radio-Kurier* magazine with AGDX. E-mail: dfp@addx.de. Web: www.addx.de – **AGDX** (Arbeitsgemeinschaft DX e.V.) is the umbrella organisation of German-speaking shortwave listeners clubs. The member clubs from Germany and Austria are:
- **ADXB-DL,** Assoziation Junger DXer e.V.

(Germany)
- **ADXB-OE,** Assoziation Junger DXer in Oesterreich (Austria)
- **KWFS,** Kurzwellenfreunde Sachsen (Germany)
- **WWDXC,** Worldwide DX Club (Germany; has non-German-speaking members world-wide)

Members of the AGDX clubs receive the German-language magazine *Radio-Kurier – weltweit horen*, an international magazine for long-distance broadcast reception. An English-language DX magazine is available from WWDXC. The postal address of AGDX e.V. is: P.O. Box 1214, 61282 Bad Homburg, Germany. Tel: +49 6102 800 999.

National Clubs:
ADDX e.V., Postfach 130124,D 40551 Düsseldorf .Tel. +49 211 790636. Fax: +49 211 793272. (G) E-mail: kurier@addx.de Web: www.addx.de – **Assoziation Junger DXer, adxb-DL e.V.**, c/o Thomas Schubaur, Am Hansenhohl 9, 86470 Thannhausen. Tel. +49 8281 798230 (18-20 MEZ). Fax: +49 8281 798231. (G) E-mail: DL1TS@t-online.de Web: home.t-online.de/home/dl1ts/adxb-dl. – **Arbeitsgemeinschaft DX, AGDX e.V.**, Postfach 1214, 61282 Bad Homburg, Die Dachorganisation deutschsprachiger DX-Klubs. Tel. +49 89 31221852 (abends). (G) E-mail: agdx@swl.net Web: www.swl.net/agdx – **Power from Heaven, Christlicher DX Club PfH**, c/o Marcel Goerke, Friedrich-Ebert-Str.11, 52249 Eschweiler (G). E-mail: pfh_club@web.de Web: www.evr-hamburg.de – **UKW/TV AK, UKW/TV Arbeitskreis der AGDX**, c/o H.-J. Kuhlo, Wilhelm Leuschner Str. 293B. 64347 Griesheim. Tel. +49 6155 66300. Büro (Mo-Fr 8-15h): Tel. +49 6151 836486 Fax: +49 6151 835090. (G) E-mail: sekretariat@ ukwtv.de Web: www.ukwtv.de **WWDXC**, Postfach 1214, 61282 Bad Homburg, Tel. +49 6172 390918 (abends). Fax: +49 6102 800999. E-mail: mail@wwdxc.de Web: www.wwdxc.de

Regional Clubs (all G):
Berliner Empfangsamateure e.V., Verein für Rundfunk-Fernempfang, Postfach 200113, 13511 Berlin. Tel. +49 30 3323953. Fax: +49 30 3323953. E-mail: W.G.Lehmann@t-online.de – **Deutscher Welt Radio Club, DWRC e.V.**, c/o Bernd Schilling, Hüling 11, D 53332 Bornheim. Tel. +49 2222 62517 – **East and West Radio Club, EAWRC**, c/o Adolf Schwegeler, Bahnhofstr. 56, 50374 Erftstadt , Tel. +49 2235 45046. Fax: +49 2235 45046. E-mail: info@eawrc.de Web: www.eawrc.de – **Eastside**

DX, EDX, c/o Jens Adolph, Postfach 100137, 04001 Leipzig E-mail: ctu33@gmx.de – **Hamburger Freunde des Rundfunkfernempfangs**, c/o Dieter Schäfer, Am Sportplatz 18, 24629 Kisdorf. Tel. +49 4193 93407 (to 1900 hr). E-mail: DL1LAD@darc.de – **Kurzwellenfreunde Brand**, c/o Hans-Jürgen Schmelzer, Mitterteicher Str. 15, 95643 Tirschenreuth. E-mail: hugotir@t-online.de – **Kurzwellenclub Schwalmtal, A KWCS e.V.**, c/o Helmut Reitzer Jr., Willy Rösler Str. 41, 41366 Schwalmtal. Tel. +49 2163 30052. Fax: +49 2163 30052. E-mail: dk0kws@qsl.net Web: www.qsl.net/dk0kws – **Kurzwellen-freunde Rhein-Ruhr, KWFR e.V.**, Postfach 101555, 45815, elsenkirchen E-mail: kwfr.ge@t-online.de Web: www.kwfr.de – **Kurzwellenfreunde Sachsen, KWFS**, c/o AGDX e.V., Postfach 1214, D 61242 Bad Homburg, Anfragen via AGDX e.V. (s.u.) bzw. via. E-mail: DK5TL@qsl.net Web: www.swl.net/agdx/KWFS.html – **Kurzwellen-freunde Wuppertal, KWFW**, Postfach 220342, D 42373 Wuppertal. Tel.: +49 202 602472 (abends). Fax: +49 202 606742 – **Neandertal DX-Club**, c/o Veit Pelinski, Morper Allee 34, D 40699 Erkrath Oldenburger – Kurzwellenfreunde, c/o Olaf C. Hänßler, Sandweg 98, 26135 Oldenburg, Tel. +49 441 12632. E-mail: olaf.haenssler@informatik. uni-oldenburg.de – **Radio Hörer Club International, RHCI e.V.**, c/o Hans-Joachim Brustmann, Straße am Park 16, 04209 Leipzig. Tel. +49 341 4211160. E-mail: info@ww-rundfunk.de Web: www.rhci.de – **Rhein-Main-Radio-Club, RMRC e.V.**, Postfach 1168, D 65432 Flörsheim am Main. E-mail: eamrmrc@t-online.de Web: www.rmrc.de – **Kurzwellenhörerklub Saar, SWLCS**, Postfach 1230, D 66585 Merchweiler. Tel. +49 6825 8380. Fax: +49 6825 8380 E-mail: p_hell@freenet.de Web: www.swlcs.de

HUNGARY: **Hungarian DX Club**, Beke ut 85, 2519 Piliscsev (Hungarian) E-mail: tibor.szilagyi@skf.com

IRELAND: **Irish DX Radio Club**, c/o Edward Dunne, 17 Anville Drive, Kilmacud, Stillorgan, Co. Dublin (E) E-mail: irishdxclub@hotmail.com Publication: *MediaWatch* (E) by e-mail.

ITALY: **Associazione Italiana Radioascolto (A.I.R.)**, C.P. 1338, IT-10100 Torino A.D.(TO) (I) Web: www.arpnet.it/~air E-mail: air@ arnet.it – **BCL Sicilia Club**, c/o Roberto Scaglione, C.P. 119, Succ. 34, IT-90144 Palermo (PA) (I) Web: www.bclnews.it E-mail: info@ bclnews.it – **Coordinamento del Radioascolto, (CO.RAD)**, c/o Dario Monferini (Umbrella organisation of various Italian DX Clubs operating only on the WEB) (I) Web: www.corad.net/E-mail: co.rad@ tiscalinet.it – **Gruppo Ascolto MondoRadio DX club**, c/o Salvo Miccichè, Via Alighieri 27, IT-97018 Scicli (RG) (I) Web: listen.to/mondoradio E-mail: mondoradio.dx@tiscali.it – Gruppo d'Ascolto della Marca Trevigiana, C.P. 3, Succ. 10, IT-31100 Treviso (TV) (I) Web: www.geocities.com/CapeCanaveral/Hall/2875/GA MT E-mail: gamt@ntt.it – **Gruppo d'Ascolto due Mari**, C.P. 1099, IT-74100 Taranto Centro (TA) (I/E) Web: fly.to/gadm E-mail: tarantodx@ hotmail.com – **Gruppo d'Ascolto Radio dello Stretto**, c/o Giovanni Sergi, Via Crotone 33, IT-98149 Messina (MS) (I) – **Gruppo Radioascolto Liguria**, c/o Riccardo Storti, Via Sapri 34/51, IT-16134 Genova (GE) (I). Web: www.gral.0catch. com/ E-mail: ristort@tin.it – **Mediterraneo Radio Club (MRC)**, c/o Alberto Lo Passo, Casella Postale 172, IT-96100 Siracusa (SR) (I) (only local activity) – **Play-DX**, c/o Dario Monferini, Via Davanzati 8, IT-20158 Milano (MI) (I/E/S) (specialises in difficult DX) Web: listen.to/playdx E-mail: playdx@hotmail.com – **Quelli del Faiallo**, (Independent Group operating only on the Web) c/o:Enrico Oliva, Web: www.faiallo.org/faiallo.html E-mail: info@faiallo.org

NETHERLANDS: **Benelux DX Club**, Postbus 583, 3700 AN ZEIST, The Netherlands. E-mail: h.louwsma@chello.nl Web: www.bdxc.nl Publication: *BDXC-Bulletin* (Dutch/E)

NORWAY: **DX Listeners' Club**, c/o Bernt Erfjord, Hellemyrasen 1, N-4628 Kristiansand (Norwegian). E-mail: dxn@dxlc.com Web: www.dxlc.com

RUSSIA: **Club of DX-ers**, c/o Vadim Alexeew, P.O.Box 65, 125581 Moscow (Russian/E/G) E-mail: gusev@itep.ru. Web: www.radio.hobby.ru. -- **Irkutsk DX Club**: c/o Feodor Brazhnikov, P.O. Box 3036, 664059 Irkutsk. E-mail: feodor@ pp.irkutsk.ru Web: www.irkutsk.com/radio -- **Russian DX League**, c/o Anatoly Klepov, ul. Tvardovskogo 23-365, 123458 Moscow (Russian/E) E-mail: dx-league@mtu-net.ru Web: rusdx.narod.ru -- **Sankt-Peterburg DX Club**, c/o Alexey Osipov, P.O.Box 46, 195213 Sankt-Peterburg. E-mail: dxspb@softhome.net – **Tomsk DX Club**, c/o Vladimir Kovalenko, Tomsk (Russian/ English). E-mail: tomskdx@mail.ru – **Novosibirsk DX Club**, c/o Igor Yaremenko, Novosibirsk (Russian/English). E-mail: dxer@yandex.ru Web: www.dxing.hotbox.ru

SPAIN: **Asociación DX Barcelona**, P.O. Box 335, 08080 Barcelona (S). E-mail: adxb@ redestb.es Web: www.redestb.es/adxb/ Publication: *Mundo DX* (S) – **Asociación Española de Radioescucha (AER)**, P.O. Box 4031, 28080 Madrid (S). E-mail: sedano@lander.es Web: www.aer-dx.org – **Mediterranean DX Group**, P.O. Box 4212. 41080 Sevilla E-mail: medidx@svq.servicom.es Web: www.geocities.com/ SiliconValley/4847/

SWEDEN: **Sveriges DX Förbund**, Box 3108, SE-103 62 Stockholm. (Swedish, umbrella organization of over 30 clubs). E-mail: ordf@sdxf. org Web: www.sdxf.org – **Arctic Radio Club**, c/o Tore Larsson, Frejagatan 14A, SE-521 43 Falköping (MW, Swedish/E) E-mail: tore.larsson@beta.telenordia.se

SWITZERLAND: **Radio-und Fernseh-Club Basel und Umgebung, RFCB**, Postfach 67, CH 4027 Basel, Switzerland, E-mail: info@rfcb.ch Web: www.rfcb.ch

UNITED KINGDOM: **British DX Club**, 126 Bargery Rd, Catford, London SE6 2LR E-mail: secretary@bdxc.org.uk Web: www.bdxc.org.uk Publication: *Communication* (E) – **International Shortwave League**, c/o John Raynes, 267 Pelham Rd., Immingham N.E. Lincs DN40 1JU (Also for amateur radio enthusiasts). E-mail: www.geocitiesiswl@g0bwg.freeserve.co.uk Web: www. freespace.virgin.net/nigel.dyche/ – **Medium Wave Circle**, 59 Moat Lane , Luton LU3 1UU (LW/MW only) E-mail: contact@mwcircle.org Web: www.mwcircle.org – **World DX Club**, 17 Motspur Drive, Northampton NN2 6LY. E-mail: mark@dxradio. demon.co.uk. North American secretary: Mr. R. A. D'Angelo, 2216 Burkey Drive, Wyomissing, PA 19610, USA. Publication: *Contact* (E).

Africa

IVORY COAST: **DX-Ivoire**, c/o Jibirila Liasu, 20 B.P. 197 Abidjan 20 (F).

NIGERIA: **Africa DX Association**, c/o Mr. Friday I. Okoloise, NITEL, P.M.B. 23, Lafia, Plateau State.

TUNISIA: **Club des Auditeurs et de l'Amitié**, c/o De Riadh Sakka, Route de Gremda Merkez Sahnoun, 3012 Sfax (F).

SOUTH AFRICA: **South African DX Club**, P.O. Box 18008, Hillbrow 2038.

TOGO: **Club Inter Amitié Radio**, CCF B.P.

2090, Lomé (F) – Groupe Endoc, B.P. 2667, Lomé (F)

UGANDA: The International DX Club of East Africa, c/o Ouma Samuel, PB 565, Iganga. E-mail: amivivek@africaonline.co.ug, or 077444201@ mtnconnect.co.ug

Asia

BANGLADESH: International Radio Listeners Club, Konabari, P.O.Nilnagor, Gazipur, Dhaka (E/Bengali) – **'Rose' DW Listeners Club**, c/o Ashik Eqbal "Tokon", Luximpur Greater Rd, GPO Box 56, Rajshahi 6000. E-mail: rosedwlc@librad.net Web: www.geocities. com/rosedwlc – **Aurora Listeners' Club**, c/o Miss Kakali Rani, Harida Khalsi-6403, Madhnagar-Natore-6400, Bangladesh – **Harida Khalsi-A.L. Club**, c/o Mr. PK. Biral Kumar, Madhnagar-6403, Natore-6400, Bangladesh – **Online DX Forum**, c/o MD Azizul Alam Al-Amin, Gourhanga, Ghoramara, Rajshahi-6100. E-mail: mtech@ rajbd.com.

PAKISTAN: Pakistan Aafaqie Lehrain Society (PALS), c/o Asrar Chaudhary, Dusehra Ground, Sheikhupura, 39350, Punjab, Pakistan. Newsletter *Aafaqie Lehrain*, Urdu with an English-language section.

INDIA: Youth International Radio Listeners' Club (Y.I.R.L.C.), c/o Mr. Pranab Kumar Roy, Vill +PO, Shyamnagar, Via Palashipara 741155, Nadia, West Bengal. E-mail: yirlc@india.com Publication: *Radio Monitors' Guide* – **Paribar Bandhu SWL Club**, c/o Mr. Anand Mohan Bain, Indira Colony, Baloda Bazal, Raipur-M.P. PIN-493332 India – **Universal DX League**, c/o Kanwarjit Sandhu, P.O. Box: 1128, Chandigarh-160 015. E-mail: udxl@hotmail.com – **World DX Club & Library**, c/o Baidyanath Upadhyaya, At Khairabarigaon, P.O. Khawrang, Udalguri 784509, Darrang, Assam – **Apollo DX International**, c/o Deepak Kumar Das, Dholi Sakra-843105, Distt-Muzaffarpur (Bihar), India. E-mail: deepakdx@ rediffmail.com – **Chaudhary Srota Sangh**, c/o Santosh Kumar (President), Kharauna Jairam, Kharauna Dih, 843113 – Distt-Muzaffarpur (Bihar), India – **El Nino Electronics DX Club**, c/o Partha Sarathi Goswami, Kishalay, College Road, Siliguri 734 401, Darjeeling, West Bengal. E-mail: info@elnino.gq.nu Web: www.elnino.gq.nu or http://dear.to/dx – **World DXing Club**, c/o Mr Madhab Ch. Sagour, 60 Jan Md Ghat Road, P.O.

Box Naihati 743 165, 24 Parganas (North), West Bengal. E-mail: madhabdxing@yahoo.co.in – **Globe Radio DX Club, GRDXC**, c/o Harjot Singh Brar, P.O Box 158, Chandigarh 160 017. E-mail: drdxc@hotmail.com Web: www.geocities. com/grdxc/index/html – **Ardic DX Club**, Room 6, Sherif Mansion, 10 Nallathambi Street, Triplicane, Chennai-600 005, Tamil Nadu. E-mail: ardicdxclub@yahoo.co.in Web: www.geocities.com/ ardicdxclub/vanoli Publications: *Dxers Guide* (E), *Sarvadesa Vanoli* (Tamil).

INDONESIA: Indonesian DX Club, P.O. Box 2001 DPPS, Depok 16432, Indonesia (E/Indonesian). E-mail: idxc@hotmail.com Web: www.idxc.org.

JAPAN: Japanese Association of DXers, P.O. Box 1766, Tokyo 100-91 (J/E) – **Asian Broadcasting Institute**, C.P.O. Box 1334, Tokyo 100-8693 (J). E-mail: abi@246.ne.jp Web: www.246.ne. jp/~abi/ – **Japan BCL Federation**, 1-9-23 Tadao, Machida City, Tokyo 194-0035. E-mail: mywave@m2.ocv.ne.jp (J) – **Japan Short Wave Club**, P.O. Box 29, Sendai Central 980-8691 (J/E) Fax: +81 22 227 4194 E-mail: jswchq@ hotmail.com – **Indonesian DX Circle Japan**, c/o Atusnori Ishida, 1-16-201 Teranishi, Saichi-cho, Iwakura-shi, Aichi 42-0036 (J) – **Nagoya DXers Circle**, c/o Shigenori Aoki, 2-51 Kasumori-cho, Nakamura-ku, Nagoya 453-0855 Web: http://newswire.ndxc.org – **Radio Nuevo Mundo**, c/o Tetsuyu Hirahara, 5-6-6 Nukuikita, Koganei-shi, Tokyo 184-0015.

NEPAL: Friendship Radio Club, Tanki Sinwari 5, District Morang, Biratnagar – **Listeners' Club of Nepal**, (Reg. No.144), P.O Box 126, Biratnagar-4.

PAKISTAN: National Society of Pakistani DXers, E-161/1, Iqbal Park, opposite Adil Hospital, Defence Housing Society Rd, Lahore Cantt. – **Pakistani Shortwave Listeners Association**, 38/2-Habib Colony, Bahawalpur 63108. Tel. 00 92 621 887590. Fax 00 92 621 874060. E-mail: imranmehr-pswla@hotmail.com. – **Pakistan: Wonderful World of Shortwave (WWSW)**, c/o Baber Shehzad, 43 Habib Colony, Bahawalpur-63108 E-mail: baber73@yahoo.com, Publication: *News Letter of Pakistani DX-ers* by e-mail.

SRI LANKA: Union of Asian DXers, c/o Victor Goonetilleke, "Shangri-La" 298 Kolamunne, Piliyandala.

TAJIKISTAN: DX Listeners' Club, c/o Ibrahim Rustamov, Navobod Str. 39-39, 735920 Isafara, Tajikistan. (E/G/Russian/Tajik/Farsi)

Pacific

South Pacific Association of Radio Clubs
E-mail: clarkb@sparc. org.nz. (Umbrella organization of most Australian and NZ DX Clubs).

AUSTRALIA: Australian Radio DX Club Incorporated, c/o John Wright, 15 Olive Crescent, Peakhurst, 2210 NSW. E-mail: dxer@ fl.net.au Web: www.ardxc.fl.net.au – **The Electronic DX Press** (EDXP), E-mail: edxp-subscribe@topica.com, info@edxp.org Web: http://edxp.org – **The Australian DXing Association**, c/o Bob Padula, 404 Mont Albert Road, Mont Albert, Victoria 3127. E-mail: ausdx@edxp.org Web: http://ausdx.edxp.org

NEW ZEALAND: New Zealand Radio DX League, P.O. Box 3011, Auckland. Web: http://radiodx.com/spdxr/subscription – **New Zealand DX Radio Association**, c/o R.Dickson, 88 Cockerell Str, Brockville, Dunedin.

North America

Association of North American Radio Clubs (ANARC), 2216 Burkey Drive, Wyomissing, PA 19610-1553. Umbrella organization for most North American DX Clubs. Send 2 IRC's for more information about the association and its member clubs. Web: www.anarc.org

CANADA: Canadian International DX Club, P. O. Box 67063-Lemoyne, St. Lambert, PQ J4R 2T8. E-mail: CIDXclub@yahoo.com Web: www.anarc.org/cidx/ – **Club Ondes Cortes du Quebec**, 5120 35ème rue, Grand-Mère, PQ G9T 3N6 (F). E-mail: dduplessis@ infoteck.qc.ca – **Ontario DX Association**, P.O. Box 161, Willowdale Postal Station 'A', ON M2N 5S8. E-mail: odxa@compuserve.com Web: www.odxa.on.ca

MEXICO: Audio Pico DX Club, c/o César Granillo, Apartado 309, 94301 Orizaba, Ver – **Club DX Miguel Auza**, c/o Luis Antero Aguilar, Apartado Postal 38, 98330 Miguel Auza, Zacatecas – **Consultorio DX**, c/o Miguel Angel Rocha Gámez, Ap. Postal 31, 31820 Ascensión, Chih – **Nayarit DX Club**, P.O. Box 62, 63001 Tepic. E-mail: naydx@tepic.megared.net.mx – **Sociedad de Ingenieros Radioescuchas**, c/o Cesar Fernandez de Lara Garcia, Ap. Postal 203, Admon. No.1, C.P. 91701 Veracruz, Ver.

E-mail: fedela@hotmail.com or fedela@veracruz.podernet.com.mx

USA: All Ohio Scanner Club, 20 Philip Drive, New Carlisle, OH 45344-9108 (non-broadcast Public Service Bands only). E-mail: n8oay@sprintmail.com Web: www.aosc.rpmdp.com/ – **American Shortwave Listeners Club**, 16182 Ballad Lane, Huntington Beach, CA 92649-2204. E-mail: wdx6aa@earthlink.net Web: http://communitylink.ocnow.com/groups/aswlc – **Association of Clandestine Enthusiasts**, P.O. Box 11201, Shawnee Mission, KS 66207-0201. Web: www.frn.net/ace/ – **Chicago Area DX Club**, c/o Edward G.Stroh, 53 Arrowhead Drive, Thornton, IL 60476 – **DecaloMania**, 9705 Mary NW, Seattle. WA 98117 (for collectors of station promotional items and airchecks). E-mail: bytheway@atk.com. Web: www.anarc.org/ decal/ – **DX Audio Service**, P.O. Box 164, Mannsville, NY 13661-0164 (for sight-impaired listeners). E-mail: gnbc@wcoil.com Web: www.wcoil.com/~gnbc/ – **DX South Florida**, 1673 Palace Drive, Clearwater, FL 34616-1833 – **Fine Tuning**, 779 Galiliea Ct., Blue Springs, MO 64014 – **International Radio Club of America**, P.O Box 1831, Perris, CA 92572-1831 (mediumwave only) E-mail: phil_tekno@yahoo.com Web: www. geocities.com/Heartland/5792/info.htm – **Lone Star DX Association**, c/o Jim Bass, 2709 Monarch Drive, Arlington, TX 76006 E-mail: jwbass@autoeloan.com Web: www.dxer.org/lsdxa – **Longwave Club of America**, 45 Wildflower Road, Levittown, PA 19057 Web: www. users.aol.com/lwcanews/ – **Memphis Area Shortwave Hobbyists**, P.O. Box 3888, Memphis, TN 38173 – **Miami Valley DX Club (MVDXC)**, Box 292132, Columbus, OH 43229 E-mail: dhammer@freenet.columbus.oh.us. Web: www.anarc.org/ mvdxc/ – **Michigan Area Radio Enthusiasts**, P.O. Box 530933, Livonia, MI 48153-0933 E-mail: xx024@detroit.freenet.org Web: www.detroit. freenet.org /sigs/l-radio/ – **Minnesota DX Club (MDXC)**, 16330 Germane Ct W Rosemount, MN 55068. Web: www.anarc.org/mdxc/ – **National Radio Club**, P.O. Box 164, Mannsville, NY 13661-0164 (mediumwave only.) E-mail: plsbcbdxer@aol.com Web: www. nrcdxas.org – **North American Shortwave Association**, 45 Wildflower Road,

Levittown, PA 19057 (SW only) E-mail: NASWA1@aol.com Web: www.anarc.org/ naswa – **Pacific Northwest, British Columbia DX Club**, c/o Bruce Portzer, 6546 19th Ave NE, Seattle WA 98115. E-mail: bytheway@atk.com Web: www.anarc.org/pnbcdxc – **Radio Communications Monitoring Association**, P.O. Box 542, Silverado, CA 92676 (utility stations) – **Southern California Area DXers**, c/o Bill Fisher Sr., 6398 Pheasant Drive, Buena Park, CA 90620-1356. E-mail: billfishernow@netzero. net Web: http://communitylink.ocnow.com/groups/scads – **Washington Area DX Association**, 606 Forest Glen Rd, Silver Spring, MD 20901 – **Worldwide TV-FM DX Association**, PO Box 501, Somersville, CT 06072 Web: www.anarc.org/wtfda.

South America

ARGENTINA: **Grupo Radioescucha Argentino**, Casilla de Correos 1159, 1000 Buenos Aires (S) E-mail: dxline@arnet.com.ar – **Grupo DX Suquia**, c/o Carolina J.G. Vandenberghe, Estafeta Rivera Indarte, C.C. No.26, 5149 Córdoba – **Asociación DX del Littoral**, Casilla 406, Rosario (Santa Fé)

BRAZIL: **Clube DX-ista da Amazônia**, a/c Djaci Franklin Soares da Silva, Tv. Angustura 1961, apto. 1205 – Pedreira, 66087-710 Belem, PA. E-mail: cdxa@ig.com.br Web: http://www.geocities.com/ cdxpara – **DX Clube do Brasil**, C.P. 384, 09701-970 São Bernardo do Campo (P) E-mail: dxcp@yahoogroups.com Web: www.ondascurtas.com – **Santa Rita DX Clube**, Caixa Postal 4, 58300-970 Santa Rita, Paraiba. E-mail: srdxc@bol.com.br Web: www.srdxclube.hpg .com.br.

CHILE: **Amigos Radioescuchas de Santiago**, Casilla 183, La Cisterna, Santiago 14 (S). E-mail: dxars@yahoo.com / hlopez@interaccess.cl – **Departamento de Radioescucha**, c/o Héctor Frías Jofre, Casilla 9571, correo 21, Santiago. E-mail: ce3fzl@latinmail.com (S) – **Club Diexista de Chile**, Calle 3 Ponienta 55, Talca. E-mail: chiledxclub@mixmail.com Web: www.lanzadera.com/ chiledxclub

SURINAME: **Suriname DX Club International**, Bechaniestraat 58, Paramaribo.

URUGUAY: **Asociación Diexman Uruguay**, P.O. Box 6008, C.P. 11000 Montevideo (S). E-mail: cx4ban@adinet.com.uy or rialv@network.bbs.com.uy – **DX Club**

Montevideo, Calle Batovi 2068, 11800 Montevideo. E-mail: dxclubmontevideo@yahoo.com

VENEZUELA: **Asociación Diexista de Venezuela**, P.O. Box 65657, Caracas 1066-A (S) E-mail: marl1@hotmail.com – **Club Diexistas de la Amistad**, c/o Ing. Santiago San Gil Gonzáles, P.O. Box 202, Barinas 5201-A, Estado Barinas. (S). E-mail: americaenantena@hotmail.com Web: www.lanzadera.com/radioficionado – **Venezuelan QSL Help**, c/o Winter Monges, P.O. Box 1.116, Barquisimeto 3001-A, Lara. E-mail: wintermonges @yahoo.com Web: www.members.tripod.com/~wintermonges/ index.html

Standard Time & Frequency Transmissions

What are STFTs?

Standard Time and Frequency Transmissions (STFTs) are continuous transmissions of 'beeps' or 'pips' every second, with the time in UTC announced on the minute. They are often referred to as 'Time Signals'. Most stations broadcast for 24 hours a day while some other stations run for a few hours each day or, in some cases, only on certain days of the week.

Using STFTs

STFTs are invaluable aids for the radio user. Not only do they allow listeners to synchronise station clocks to UTC but they are also a handy tool for checking propagation and reception paths. Their most useful role for serious shortwave listeners, however, is for the periodical performance checks needed to ensure that equipment is performing as it should and to test for receiver error.

Checking Performance

As you will be able to see from the black bars in the lists, it is possible to carry out tests on a variety of frequencies ranging from 2500 to 20000kHz. First select an appropriate set of STFT's, perhaps by saving them into a set of memory channels if your radio has the facility. A quick check can be made for the characteristic ticks and pulses to ensure that there is a good reception path and that the STFT is currently active, before moving on to the tests themselves. Don't forget that it is essential to allow the radio to warm up for an hour or so before starting these tests.

The object of the exercise is to mix the incoming STFT signal with an internally generated signal from the radio's Beat Frequency Oscillator (BFO) and then tune the radio until the resulting whistle drops down to zero. This process of tuning for 'zero beat' then ensures that the radio is on exactly the same frequency as the transmission – any error shown on the dial will be the receiver error and any drift in tone will be receiver drift.

For most radios it will suffice to select either upper or lower sideband mode with a wide filter setting and use a loudspeaker or pair of headphones having a good low frequency audio performance. Note that many SW receivers have internal speakers which are not very good when it comes to reproducing low notes so are useless for this procedure.

As you carefully tune down and hear the note drop, you should find that the S-meter needle starts to fluctuate. This means that you are very close to 'zero beat'. The lower the rate of S-meter needle movement, the nearer you are to the end point. At this stage you will probably not be able to hear much in the headphones apart from a near silent carrier, and the meter will be your best guide as it will be indicating the 'phase error' between the STFT and your BFO. When your needle moves at its slowest rate, you have reached zero beat. Make a note of the dial reading as this will show the receiver error.

For those radios with particularly good filters, which may prevent a 'zero beat' approach, a similar technique can be used by switching to CW mode.

This method requires the use of an audio digital frequency meter – ask around at your radio club and you will probably find one. First of all, refer to the manual and find the 'CW offset' frequency. This is the audio frequency which the receiver will produce when it is exactly tuned to the carrier of the STFT. Common values are 600Hz to 800Hz. Some radios allow the user to programme the CW offset, so make sure that it has not been changed before you start the tests.

The procedure is essentially the same as has just been described, except that you will be tuning the radio until the DFM reads 600Hz exactly, then the receiver dial will show you any error.

By repeating this test on a number of frequencies, you can be confident that your radio is accurate, or at least be aware of any errors or developing problems, and of course these are handy techniques for testing a radio you are considering buying

Selected Internet Resources

General Pages

1000 Lakes DX Page:
www.geocities.com/Colosseum/Park/3232/dx.htm

DXing.com: www.dxing.com

DXing.info: www.dxing.info

The Electronic DX Press (EDXP): http://edxp.org
(also Australian regional interest)

**Funkenhauser's Whamlog & Medium Wave DX
Radio Links**: home.inforamp.net/~funk

Hard-Core-DX: www.hard-core-dx.com

Links for DX-listeners:
www.geocities.com/Colosseum/Park/3232/dxlinks/index.htm

Radio-portal: www.radio-portal.org

Radio Homepage of Martin Schöch: www.schoechi.de

Russkiy SWL DX Sait (Russian): www.radio.hobby.ru

World of Radio: www.worldofradio.com

Regional

Broadcasting in South & South East Asia:
asiaradio.crosswinds.net

Central American & Caribbean Broadcasting:
www.stevenwiseblood.com

Patepluma Radio: donmoore.tripod.com

Databases

ADDX Non-English Frequency List:
dxworld.com/cgi-bin/hfp.sh

DX & Other Interesting Programs Guide:
dxworld.com/cgi-bin/speedx.sh

Eike Bierwirth's DX page: www.eibi.de.vu

Eldorado for LA Dxers:
members01.chello.se/mwm/eldorado/index.html

European Medium Wave Guide:
users.pandora.be/hermanb/Emwg

International Listening Guide:
www.ilgradio.com/ilgradio.htm

NA on the air: www.pp.clinet.fi/~ejh/radio/koje.htm

Online DX Logbook: www.odxl.org.uk

Pacific Asian Log: www.qsl.net/n7ecj

Prime Time Shortwave:
www.triwest.net/~dsampson/shortwave

Shortwave Radio Resource Center:
shortwave.hfradio.org

WWW Shortwave Listening Guide:
www.anarc.org/naswa/swlguide

News Sources and Bulletins

BC-DX Top News: www.wwdxc.de/topnews.htm

Clandestine Radio Watch:
groups.yahoo.com/group/crwatch

DX-listening Digest:
www.worldofradio.com/dxldmid.html

Electronic DX Press:
members.tripod.com/~bpadula/edxp.html

Eurolog:
www.geocities.com/Colosseum/Park/3232/eurolog.htm

Moskovskiy Informatsionniy DX-Bulletin (Russian):
www.internews.ru/rts

Shortwave Bulletin: www.hard-core-dx.com/swb

Signal: dxsignal.by.ru/indexen.htm

Ydun's MW Page: www.ydunritz.com

Receiver Software

Computer Supported SW Listening:
home.wxs.nl/~jarkest

Software Geared to the Shortwave Hobbyist:
www.fineware-swl.com

Identifying

Interval Signals Archive: www.intervalsignals.net

National Anthems: www.thenationalanthems.com

Propagation

Solar Terrestrial Activity Report: www.dxlc.com/solar

Space Weather Alerts, Warnings and Forecasts:
sec.noaa.gov

Mailing Lists

A-DX Mailingliste (German): www.ratzer.at/A-
DX/faq.html

AM, FMTV lists: www.nrcdxas.org/email

Conexion Digital (Spanish):
www.topica.com/lists/conexion.digital

Cumbre DX List:
cs2.ralabs.com/mailman/listinfo/cumbredx (regular
contributions)

dx_bistro (Russian): www.topica.com/lists/dx_bistro

FM & TV DX Lists: www.fmdx.com

Groups.yahoo.com: active_dx (Russian) – AM-SW-
DXing – atividadedx (Portuguese) – condiglist (Spanish)
– dexismo (Portuguese) – dxradio – globe-radio-dx –
mwdx – playdx2001 (Italian) – RealDX – shortwave
dxing – shortwavelistening – shortwaves – swedx
(Swedish)

Hard-Core-DX Email List: www2.hard-core-
dx.com/mailman/listinfo/hard-core-dx

NoticiasDX (Spanish):
es.groups.yahoo.com/group/NoticiasDX

SW Programs: www.topica.com/lists/swprograms

Search Engine

Radio-Portal: www.radio-portal.org

Selected Broadcasts for DXers & Listeners

Station	Lang.	Prog.	Frequencies
AWR, Abu Dhabi	E	Wavescan	9720, 9810, 15160, 17740, 17870
AWR, Meyerton	E	Wavescan	3215, 3345, 9520, 15105
AWR, Moosbrunn	E	Wavescan	9820, 9775, 17780
AWR, Slovakia	E	Wavescan	7130
Channel Africa, Meyerton	E	Amateur Radio Mirror	3215, 7082usb, 9750, 21560
Deutsche Welle, Cologne	G/E	Radio World DX Meeting	7285, 9470, 9615, 9765, 9815, 11645, 11690, 11890, 11965, 13605, 15275, 15410, 17765, 21790
KSDA, Guam	E	AWR Wavescan	9385, 11560, 11850, 11930, 11975, 11980, 12105, 15215, 15235, 15275, 15320, 17630
KWHR, Hawaii	E	Dxing with Cumbre	11565, 17510, 17780
Polish Radio, Warsaw	E	Multimedia	5995, 7285, 9525, 11820
Radio Australia	E	Feedback	7240, 9500, 9580, 9660, 11880, 15240, 15415, 15515, 17715, 17750, 21740
Radio Australia	E	Media Report	5995, 9475, 9580, 9660, 11650, 11660, 11695, 11880, 12080, 13620, 15230, 15240, 15415, 17580, 17715, 17750, 17795, 21725, 21740, 21820
RAI, Vienna	F	Flash des Ondes	5945, 6155, 13730
RAI, Vienna	G	DX Telegram	6155, 13730
RAI, Vienna	G	Hotline	5945, 6155, 13730
RAI, Vienna	G	InterMedia	5945, 6155, 13730
RAI, Vienna	G	Matrix	5945, 6155, 13730
Radio Belarus, Minsk	G		6070, 7105, 7210
Radio Budapest	E/G	DX Corner	3975, 6025, 7135, 9570, 9590, 9835, 11720, 11890
Radio Bulgaria, Sofia	E/G	Calling Dxers	5800, 7400, 7500, 9400, 11600, 11700, 11900, 12000, 13600, 15170, 15700, 17860
RCI, Montreal	E	Maple Leaf Mailbag	5850, 5995, 9755, 11690, 11965, 12015, 15305, 15325, 15510 9805, 11725, 13650, 15150, 17860
RCI, Montreal	E	CIDX Report	5850, 5995, 11695, 11965, 12015, 15325, 15510, 17860
REE, Madrid	E	Radio Waves	9570, 9840, 15385
RFPI, Costa Rica	E	Continent of Media	7445usb, 15039
RFPI, Costa Rica	E	World of Radio	7445usb, 15039
Radio Havana Cuba, Havana	E	Dxers Unlimited	6000, 6195, 9820, 11670
Radio Korea Int., Seoul	E/G	Worldwide Friendship	3955, 5975, 7275, 9515, 9535, 9570, 9650, 9870, 11810, 13670, 15575
Radio New Zealand Int.	E	Mailbox/S. Pacific DX Rp.	6095, 9885, 11980, 15160, 17675
Radio Prague	G		5930, 5990, 6055, 7345, 9880
Radio Tashkent	E	(Listener's Letterbox)	5025, 5955, 5975, 6025, 7105, 7215, 7285, 9715, 11905
Radio Tashkent	G		5025, 5035, 5060, 7105, 9540, 11905
Radio Ukraine Int., Kiev	E	World on the Radio Dial	5905, 12040, 15415
Radio Ukraine Int., Kiev	G		5905, 7240
RVI, Brussels	E	Radio World	5985, 9865, 9925, 13690, 13710, 15565
RAE, Buenos Aires	E/G	Dxers Special	9690, 11710, 15345

Station	Lang.	Prog.	Frequencies
RRI Bucharest	G		5960, 7130, 7195, 9570, 11940, 15245, 17745
Vatican Radio	E	(Listener's Letterbox)	7305, 7335, 9600, 9605, 9865, 11830, 13765, 15235
Voice of America	E	Main Street	4960, 5745, 6035, 6080, 6080, 7115, 7195, 7290, 7370, 9530, 9575, 9590, 9635, 9760, 9770, 11705, 11725, 11805, 11820, 11835, 11965, 11995, 12080, 13620, 13650, 15205, 15240, 15425, 17740, 17820, 17895
Voice of Mediterranean Malta	E	(Listener's Letterbox)	7440, 9850
Voice of Turkey	E	DX Corner	6020, 7240, 9525, 9655, 9890, 17690, 17815
Voice of Turkey	G		9745, 21530
WBCQ	E	Off the Hook	7415
WBCQ	E	The Real Amateur R. Show	7415
WBCQ	E	Tom and Darryl	7415
WBCQ, Maine	E	World of Radio	7415, 17495
WHRA	E	Dxing with Cumbre	7580, 17650
WHRI, Indiana	E	Dxing with Cumbre	5745, 7315, 9495, 13760
WJIE, Kentucky	E	World of Radio	7490
WRMI Miami	E/Sp	Viva Miami/AWR Wavescan	9955, 15725
WWCR, Nashville	E	Spectrum	3210, 5070
WWCR, Nashville	E	World of Radio	3210, 5070, 9475, 15825
WWCR, Nashville	E	Ask WWCR	5070, 9475, 12160, 15825
WWCR, Nashville	Sp	Mundo Radial	9475
XERMX Mexico	E	DXperience	9705 11770

DETAILS

It is not possible here to give the full details for all the broadcasts. For details of times and days of broadcasts to your region of the world, see amongst others:

SWL/Media programs in English and German, Worldwide DX Club: www.wwdxc.de
DX Programmes, British DX Club: www.bdxc.org.uk
Selected English language DX/SWL/Media Programs on SW: www.worldofradio.com/dxpgms.html

and then, for quick reference write the details of your favourite programmes here.

Station	kHz	Day	Hours

Country Codes used in this guide

Code	Country
ABW	Aruba
AFG	Afghanistan
AFS	South Africa
AGL	Angola
AIA	Anguilla
ALA	Alaska
ALB	Albania
ALG	Algeria
AND	Andorra
ARG	Argentina
ARM	Armenia
ARS	Saudi Arabia
ASC	Ascension
ATA	Antarctica
ATG	Antigua and Barbuda
ATN	Netherlands Antilles
AUS	Australia
AUT	Austria
AZE	Azerbaijan
AZR	Azores
B	Brazil
BAH	Bahamas
BDI	Burundi
BEL	Belgium
BEN	Benin
BER	Bermuda
BFA	Burkina Faso
BGD	Bangladesh
BHR	Bahrain
BIH	Bosnia and Herzegovina
BIO	Chagos Islands
BLR	Belarus
BLZ	Belize
BOL	Bolivia
BOT	Botswana
BRB	Barbados
BRM	Myanmar
BRU	Brunei Darussalam
BTN	Bhutan
BUL	Bulgaria
CAF	Central African Rep.
CAN	Canada
CBG	Cambodia
CHL	Chile
CHN	China
CHR	Christmas Island
CKH	Cook Island
CLM	Colombia
CLN	Sri Lanka
CME	Cameroon
COD	Congo (Dem. Rep.)
COG	Congo
COM	Comoros
CPV	Cape Verde
CTI	Cote d'Ivoire
CTR	Costa Rica
CUB	Cuba
CVA	Vatican
CYM	Cayman Islands
CYP	Cyprus
CZE	Czech Republic
D	Germany
DJI	Djibouti
DMA	Dominica
DNK	Denmark
DOM	Dominican Rep.
E	Spain
EGY	Egypt
EQA	Ecuador
ERI	Eritrea
EST	Estonia
ETH	Ethiopia
F	France
FIN	Finland
FJI	Fiji
FLK	Falkland Islands
FRO	Faroe Islands
FSM	Micronesia
G	United Kingdom
GAB	Gabon
GDL	Guadeloupe
GEO	Georgia
GHA	Ghana
GMB	Gambia
GNB	Guinea-Bissau
GNE	Equatorial Guinea
GRC	Greece
GRD	Grenada
GRL	Greenland
GTM	Guatemala
GUF	French Guiana
GUI	Guinea
GUM	Guam
GUY	Guyana
HAW	Hawaii
HKG	Hong Kong
HND	Honduras
HNG	Hungary
HOL	Netherlands
HRV	Croatia
HTI	Haiti
I	Italy
ICO	Cocos-Keeling Islands
IND	India
INS	Indonesia
IRL	Ireland
IRN	Iran (Islamic Rep. of)
IRQ	Iraq
ISL	Iceland
ISR	Israel
J	Japan
JMC	Jamaica
JOR	Jordan
KAZ	Kazakhstan
KEN	Kenya
KGZ	Kyrgyzstan
KIR	Kiribati
KOR	Korea, South (Rep. of)
KRE	Korea, North (DPR)
KWT	Kuwait
LAO	Lao P.D.R.
LBN	Lebanon
LBR	Liberia
LBY	Libya
LCA	Saint Lucia
LHW	Lord Howe Island
LIE	Liechtenstein
LSO	Lesotho
LTU	Lithuania
LUX	Luxembourg
LVA	Latvia
MAC	Macao
MAU	Mauritius
MCO	Monaco
MDA	Moldova
MDG	Madagascar
MDR	Madeira
MDW	Midway Island
MEX	Mexico
MHL	Marshall Islands
MKD	Macedonia
MLA	Malaysia
MLD	Maldives
MLI	Mali
MLT	Malta
MNG	Mongolia
MOZ	Mozambique
MRA	Northern Marianas
MRC	Morocco
MRT	Martinique
MTN	Mauritania
MWI	Malawi
NCG	Nicaragua
NCL	New Caledonia
NFK	Norfolk Island
NGR	Niger
NIG	Nigeria
NIU	Niue
NMB	Namibia
NOR	Norway
NPL	Nepal
NRU	Nauru
NZL	New Zealand
OCE	French Polynesia
OMA	Oman
PAK	Pakistan
PAL	Palestine
PAQ	Easter Island
PHL	Philippines
PLW	Palau
PNG	Papua New Guinea
PNR	Panama
POL	Poland
POR	Portugal
PRG	Paraguay
PRU	Peru
PTR	Puerto Rico
QAT	Qatar
REU	Reunion
ROU	Romania
RRW	Rwanda
RUS	Russia
S	Sweden
SCN	St. Kitts and Nevis
SDN	Sudan
SEN	Senegal
SEY	Seychelles
SHN	St. Helena
SLM	Solomon Islands
SLV	El Salvador
SMA	American Samoa
SMO	Western Samoa
SMR	San Marino
SNG	Singapore
SOM	Somalia
SPM	St. Pierre and Miquelon
SRL	Sierra Leone
STP	Sao Tome and Principe
SUI	Switzerland
SUR	Suriname
SVK	Slovak Republic
SVN	Slovenia
SWZ	Swaziland
SYR	Syria
TCA	Turks and Caicos Islands
TCD	Chad
TGO	Togo
THA	Thailand
TJK	Tajikistan
TKM	Turkmenistan
TMP	East Timor
TON	Tonga
TRC	Tristan da Cunha
TRD	Trinidad and Tobago
TUN	Tunisia
TUR	Turkey
TUV	Tuvalu
TWN	Taiwan
TZA	Tanzania
UAE	United Arab Emirates
UGA	Uganda
UKR	Ukraine
URG	Uruguay
USA	United States
UZB	Uzbekistan
VCT	St Vincent & Grenadines
VEN	Venezuela
VIR	U.S. Virgin Islands
VRG	British Virgin Islands
VTN	Vietnam
VUT	Vanuatu
WAL	Wallis and Futuna Isl.
WES	Western Sahara
XXX	Unknown or not specified
YEM	Yemen
YUG	Serbia and Montenegro
ZMB	Zambia
ZWE	Zimbabwe

Target Area Codes

Code	Target Area
Af	Africa
Am	Americas
As	Asia
CAC	Central America & the Caribbean
CAf	Central Africa
CAs	Central Asia
EAf	Eastern Africa
Eu	Europe
FE	Far East
GEN	General or no specific target area
ME	Middle East
Med	Mediterranean
NAf	Northern Africa
NAm	North America
Oc	Oceania/Australasia
Pac	Pacific
SAf	Southern Africa
SAs	South Asia
SAm	South America
SEA	South East Asia
UNK	Unknown or not specified
WAf	Western Africa